A Development of Quantum Mechanics

Fundamental Theories of Physics

A New International Series of Monographs on The Fundamental Theories of Physics: Their Clarification, Development and Application

A Development of Quantum Mechanics

Based on Symmetry Considerations

by

George H. Duffey

Department of Physics,
South Dakota State University

D. Reidel Publishing Company

A MEMBER OF THE KLUWER ACADEMIC PUBLISHERS GROUP

Dordrecht / Boston / Lancaster

Library of Congress Cataloging in Publication Data

CIP

Duffey, George H.
A development of quantum mechanics based on symmetry considerations.

(Fundamental theories of physics)
Bibliography: p.
Includes index.
1. Quantum theory. I. Title. II. Series.
QC174.12.D83 1983 530.1'2 83-6208
ISBN 90-277-1587-4

Published by D. Reidel Publishing Company,
P.O. Box 17, 3300 AA Dordrecht, Holland.

Sold and distributed in the U.S.A. and Canada
by Kluwer Academic Publishers,
190 Old Derby Street, Hingham, MA 02043, U.S.A.

In all other countries, sold and distributed
by Kluwer Academic Publishers Group,
P.O. Box 322, 3300 AH Dordrecht, Holland.

Printed in The Netherlands

Table of Contents

Preface

The theory of quantum mechanics continues to appear arbitrary and abstruse to new students; and to many veterans, it has become acceptable and useable only because it is familiar. Yet, this theory is at the basis of all modern physics, chemistry, and engineering, describing, as it does, the behavior of the submicroscopic particles making up all matter. So it needs to be presented more effectively to a diverse audience.

The primary question is, I believe, 'What can be considered self-evident?' Indeed, what do certain key experiments reveal about the workings of nature? How can we consider that some probabilities are not a result of our ignorance, but instead, fundamental properties?

We must pay particular attention to the subject of what we can do, what we cannot do, and what we can and cannot observe. We can prepare a homogeneous beam of almost independent particles by boiling electrons out of a metal and accelerating them by a given potential drop. We cannot follow an electron individually in the beam without introducing conditions that destroy the beam's homogeneity, but we can determine when electrons arrive at a given position.

Such arrivals are found to be governed by probability. In the homogeneous beam, the resulting probability density ρ is constant. There is thus symmetry over displacement along the beam, and similarly, over time. But to describe propagation along the beam, we need an independent function. A simple choice is to consider ρ, the square of the absolute value of a complex function Ψ. The phase of the function is involved in describing propagation.

To be consistent, we make each infinitesimal change $d\Psi$ meet the symmetry requirements over space and time. We also suppose that, for pure motion in one direction, each part of Ψ exerts the same effect on $d\Psi$ as each other part. Furthermore, the influences on $d\Psi$ vary directly with Ψ. We thus have a simple symmetry over Ψ.

To determine whether the constructed form is suitable, we again turn to experiment. Investigators find that homogeneous beams are diffracted exactly as this Ψ would be. Furthermore, one of the parameters in the theory satisfies de Broglie's equation, which photons in light are known to obey.

From photoelectric determinations, we induce a relationship for the other parameter. In Chapter 3 we consider how the arguments need to be modified to allow for a varying potential. In Chapter 4 this discussion is expanded to cover motion over more than one variable.

Translatory motion is considered first because it is the simplest. Rotatory motion is in many respects similar, so it is considered next. Symmetry arguments are invoked in constructing the variation of Ψ over θ from that over φ. The simplest kind of varying field is that for the harmonic oscillator; so its treatment appears next. Then we go to the hydrogen-like atom, the model in terms of which other atoms and nuclei are understood.

Operator theory is developed only after we have gained some confidence in treating the simpler systems.

Because particles of the same species are indistinguishable, a corresponding symmetry exists in multiparticle systems. This has profound effects on the properties of such systems in atoms, molecules, and thermodynamic arrays.

The material is all presented at a level suitable for junior and senior students in physics and chemistry. Engineers who work with molecular, atomic, and electronic processes would also benefit from the course.

There are nearly 20 problems at the end of each chapter. These have been developed through actual use in undergraduate classes.

General Introduction

In our early years, we develop a commonsense view of the world based on inductions from the experiences we have in common with all other people. Each individual conceives a reality external to himself or herself in which objects occupy definite positions in space at any given time. These move, and accelerate or decelerate, in response to the action of forces.

The replacement of qualitative observations with quantitative ones, made with the help of various instruments, has led to more and more profound inductions: the science of geometry developed from the measurement of agricultural fields and other areas; Newtonian mechanics developed from astronomical observations and measurements; Maxwellian electromagnetism developed from electrical studies and measurements; and the Law of Definite Proportions, an indirect support for the atomic theory of matter developed from chemical measurements.

When particles of matter were found to make tracks through supersaturated vapor, and in a photographic plate, the atomic theory seemed to be confirmed. However, no mechanics in which electrons, protons, or neutrons traced out individual mathematical curves proved to be satisfactory. A revolution in physical science was necessary. This developed slowly. In order to explain black-body radiation, Planck had to assume that a solid transferred energy to the electromagnetic field in discrete amounts, quanta. Einstein suggested that such quanta persisted in the field as photons and was thus able to explain the photoelectric effect. The young de Broglie saw that Einstein's equation implied a relationship between wavelength and particle momentum. He suggested that this might also apply to particles with a rest mass. Finally, Heisenberg and Schrödinger developed a suitable mechanics for such particles.

Now, all these theories are induced from experimental observations. Consequently, they appear only as a way of explaining the data on which they are based. New results, at a deeper level, may very well require modifications or additions – even a new revolution. Study the following material with these reservations in mind.

Chapter 1

Quantization of Translatory Motion

1.1. Background Remarks on Time and Space

Each of us learns about the physical world through (a) experiencing various processes in one's own body, (b) interacting with external objects close at hand and far away, (c) constructing and manipulating devices, (d) observing and measuring reactions of the resulting instruments, and (e) studying accounts of the experiences, manipulations, and measurements of others.

A person's mind, first of all, notes an *order* in impressions that are recorded. Each perception is recognized as occurring either before, simultaneously with, or after each other one. The senses also recognize mechanisms that repeat a certain process again and again. The cycles of such a process occur in sequence and can be counted. Furthermore, an observer can associate any given stimulus with a particular cycle, as long as the mechanism is operating. Since parameters in the machine may presumably be altered to make the length of a cycle as small as the observer wishes, there appears to be no limit to the precision with which he can thus pinpoint a given event.

An engineer can design a cyclic mechanism to be nearly free of secular changes and split each of its cycles into parts that are apparently equivalent in duration. The resulting machine is called a clock. Experimenters find that a good mechanism does not seem to be affected by a change in position alone. A constant times the number of cycles, or fractional cycles, executed by the clock within a given interval is taken as the *time* to be associated with the interval.

A second clock that behaves as the first one can be constructed. This can be synchronized with the first clock in the neighborhood of a given body. An experimenter can then move the second clock slowly to a different body to establish a time scale there corresponding to that already existing on the first body. Reducing the sizes of the bodies probably does not destroy the possibility of making this correspondence; in our calculations we will presume that it can be done.

The mind also receives evidence of the *coexistence* of things through the eyes, ears, and tactile nervous system. Furthermore, it finds that the sources of the stimuli map onto a three-dimensional *space* at any given time. Measuring rods, compasses, and protractors can be constructed and manipulated to study this space. Insofar as an observer can tell, such devices are not altered by a change in position alone, or by a change in time. Measurements with such tools show that the Pythagorean theorem holds to a high degree of accuracy in macroscopic situations. So for these at least, the space is Euclidean and Cartesian coordinate frames are adequate.

A person can keep such a frame in uniform rectilinear motion with respect to three or more noncolinear force-free bodies that maintain constant separations from each other. The frame is then said to be *inertial*. Each reference frame that we will employ will not differ in the pertinent properties from an inertial system during the significant intervals of time.

With available rays, an observer can only survey macroscopic and microscopic parts of space. The submicroscopic elements making up any discernible region need not be divisible into an arbitrary number of parts and may contain deviations from the Pythagorean theorem, as long as the elements fit together to yield the apparently Euclidean regions that are found. Similarly, the subchronometric elements of time add to give an apparently smooth uniform flow of the variable t; but each such small element need not be infinitely divisible and may fluctuate from the corresponding element at another location in some unknown manner. Since these deviations from uniformity and homogeneity are not directly observable, we will assume that they are not present. However, they might have to be considered in a more comprehensive theory.

A macroscopic object can be broken into smaller parts, each of the parts divided into still smaller parts, and so on, at any given time. At some stage in the process, however, one reaches seemingly indivisible units that can be associated with separated points in space. These basic constituents of matter are called *particles*. Similar to an object, a particle is characterized by a mass and a charge. Furthermore, it may possess a property analogous to angular momentum – a spin – and classifying attributes such as hypercharge.

1.2. The Statistical Nature of Position, Velocity, and Momentum

In classical mechanics, it is assumed that each particle in a given system moves *as a point* through a space that is locally Euclidean and through a time that is uniform. A definite smooth *curve* is traced out by the particle at a determinable varying rate. This curve is called the path or trajectory of the particle.

No device can locate a particle and determine its velocity and momentum at any given time with exactness. Not only are there errors in transforming the interactions with the particle to numbers, but the interactions themselves introduce errors. Classically, the uncertainties were attributed to the measurement process. Calculations of various observable effects proceeded. Many of these failed, however.

The interchange of energy between vibrational modes of a solid and the electromagnetic field did not follow the classical laws. Interactions of the electromagnetic field with electrons in a molecule or in a condensed phase resulted in discontinuous changes. The existence of stable atomic and molecular states could not be explained. Little could be done with nuclear states. The behavior of such a simple system as a homogeneous beam of particles was inexplicable.

On the other hand, considering some of the uncertainty to be an essential attribute of the particles leads to a viable theory. A person does violence to a system of particles when he assumes that each follows a definite trajectory.

In our analysis, we will assume that the space and time of the laboratory can be infinitely subdivided, as in elementary calculus, and imposed on the system under consideration. And, if a particle were at a certain point in this space at a certain time with a given velocity, it is presumed to possess a kinetic energy T and a potential energy V calculated in the same way as the corresponding classical quantities. An isolated set of particles arranged in a particular way with given velocities similarly exhibits a T and V equivalent to the classical kinetic and potential energies that such a set would have. The sum of these energies is the total energy E for the system.

If a person prepares a set of equivalent potential fields and employs an instrument to introduce an identical particle into each in the same manner, he always obtains a statistical distribution of initial positions and initial velocities. The distributions persist over time. A set of such systems that is large enough so that adding more members does not appreciably alter the statistical weight of any pertinent value of a property is said to form an *ensemble*. An observer can presumably study an ensemble and determine the probability that the particle is in a given small volume $d^3\mathbf{r}$ of the reference space at a chosen time t. Here, radius vector $\mathbf{r}$ is drawn from the origin to the center of the differential volume. Dividing the probability by the volume yields the *probability density* ρ for the particle. A system of particles, and ensembles of such a system, can be considered similarly. The probability that the first particle is in volume $d^3\mathbf{r}_1$, the second in $d^3\mathbf{r}_2$, . . . , the nth in $d^3\mathbf{r}_n$ may be determined and the corresponding probability densities calculated.

Also determinable are the statistical kinetic energy and momentum of a particle in a beam. Since these properties are independent of the density ρ at the point of measurement, but are related to particle movements in the beam, one needs an additional real function.

Essentially, two things have to be represented: distribution of the density and unidirectional movements of the particles. However, a single complex function can describe what one can know of both attributes very simply. The probability density is related directly to the square of the absolute value of the function, while the particle propagation is embodied in the phase angle. Indeed, we will find that the relationship

$$\rho = (\text{constant})\ \Psi^* \Psi \tag{1.1}$$

serves except when Einstein's relativity needs to be taken into account. Then, four complex functions are needed, rather than only one.

Symmetry considerations will enable us, in principle, to complete the formulation. In particular, we will consider how $d\Psi$ depends on Ψ, on coordinates, and on time. Except in special circumstances where another choice is convenient, the constant in (1.1) will be set equal to 1. We call Ψ the *state function*, or *wave function*, for the given particle (or particles).

The principle of *continuity*, that small causes produce small effects, is applied, not to each individual particle, but to the probability density and to the state function. So whenever the influences that act on a particle vary smoothly with the coordinates and time, the corresponding probability density varies smoothly. And, the state function is analytic wherever the relevant V is analytic.

Since observable properties are determined by the way the pertinent particles are distributed in space, on the average, and by how they propagate, these properties are derivable from state functions. Indeed, we have the theorem of *wholeness* or *completeness*: The function Ψ for a particle (or particles) represents the state of the corresponding system to the extent that this state can be determined.

From the standpoint of probability, the simplest system imaginable is one in which free movement occurs in a single direction with a constant ρ. Employing a source with constant intensity and accelerating energy, or velocity selection, one can also assume that the propagation properties are as uniform as possible.

Since ρ would be constant along such a beam, the magnitude of Ψ would be constant. The phase angle, however, would vary to represent the propagation. This variation must be introduced in a way consistent with the prevailing symmetries, as we will see in the next section.

In general, we have to consider particles subject to a varying potential V. But as long as the variations are not abrupt, we may assume that the motion across an infinitesimal element is effectively at constant V, and the results obtained for the homogeneous beam and its reflection may be applied within the element.

1.3. A State Function Governing Translation

Freely moving particles do not distinguish between the different points traversed in space or time; each point is equivalent in its average effect on a particle. Furthermore, a homogeneous beam, in which the particles move freely with the same momentum, possesses a Ψ made up of uniform parts. The resulting symmetries enable us to construct a credible state function.

Let us consider a system of equivalent particles traveling freely in one direction at one velocity (or momentum). Furthermore, let us suppose that the probability density ρ is constant throughout the region under consideration. Following the discussion in Section 1.2, we assume that all behavior of a typical particle is described by the factor of ρ labeled Ψ. This function Ψ presumably varies smoothly, so formal differentiation leads to the result

$$d\Psi = \frac{\partial\Psi}{\partial x}dx + \frac{\partial\Psi}{\partial y}dy + \frac{\partial\Psi}{\partial z}dz + \frac{\partial\Psi}{\partial t}dt \tag{1.2}$$

in which x, y, z are Cartesian coordinates of an inertial frame and t is the time.

For the pure motion we are studying, there is no reason to expect any part of Ψ to exert a different effect on $d\Psi$ than any other equivalent part. The simplest way to incorporate this symmetry is to make $d\Psi$ homogeneously linear in Ψ. Furthermore, each point in the given region and time interval is assumed to be like any other point in the region and interval. So the change in Ψ on going from one point to the next needs to be independent of the initial point. But the change must depend on

$$dx, \quad dy, \quad dz, \quad dt. \tag{1.3}$$

By symmetry, multiplying each of these small changes by a factor should multiply the effects on $d\Psi$ by the factor. The infinitesimal $d\Psi$ therefore depends linearly on each of the infinitesimals in set (1.3).

For simplicity, however, let us go to axes X', Y', Z' for which the direction of the X' axis is that of the motion. Then the only spatial coordinate affecting Ψ is x' in the region of interest. And by the arguments just given, the variation in the state function is linear in dx', linear in dt, and homogeneously linear in Ψ. We thus have

$$d\Psi = \kappa\Psi\, dx' + \gamma\Psi\, dt \tag{1.4}$$

where κ and γ are parameters to be identified. Our argument allows κ and γ to be imaginary or complex, so we may alternatively consider

$$d\Psi = ik\Psi\, dx' - i\omega\Psi\, dt. \tag{1.5}$$

Let us separate variables in (1.5)

$$\frac{d\Psi}{\Psi} = ik\, dx' - i\omega\, dt \tag{1.6}$$

and integrate

$$\Psi = A\, e^{ikx'}\, e^{-i\omega t}. \tag{1.7}$$

Imposing (1.1) with the constant taken equal to 1, as we commonly do in normalizing functions, yields the probability density

$$\rho = \Psi^* \Psi = A^*(e^{ikx'})^*(e^{-i\omega t})^* A\, e^{ikx'}\, e^{-i\omega t}. \tag{1.8}$$

This reduces to an expression independent of x' and t *only* if k and ω are real (κ and γ imaginary). Since such independence exists in the homogeneous propagating beam, we assume that k and ω are real in the beam. Then (1.8) reduces to

$$\rho = A^*A. \tag{1.9}$$

A general displacement may be written as

$$d\mathbf{r} = dx\hat{\mathbf{x}} + dy\hat{\mathbf{y}} + dz\hat{\mathbf{z}}. \tag{1.10}$$

Element dx' is the component of $\mathbf{dr}$ in the direction of motion. If $\mathbf{e}$ is the unit vector pointing in this direction and $k\mathbf{e}$ is written as $\mathbf{k}$, then

$$k\,dx' = k\mathbf{e}\cdot d\mathbf{r} = \mathbf{k}\cdot d\mathbf{r} \tag{1.11}$$

and

$$kx' = k\mathbf{e}\cdot\mathbf{r} = \mathbf{k}\cdot\mathbf{r}. \tag{1.12}$$

The expression governing pure translation (1.7) now becomes

$$\Psi = A\,e^{i\mathbf{k}\cdot\mathbf{r}}\,e^{-i\omega t} = A\,e^{i(\mathbf{k}\cdot\mathbf{r}-\omega t)}. \tag{1.13}$$

A given phase of this Ψ travels with a given value of the angle

$$kx' - \omega t = \mathbf{k}\cdot\mathbf{r} - \omega t = a. \tag{1.14}$$

Indeed, the corresponding coordinate x' obeys the equation

$$x' = \frac{\omega}{k}t + \frac{a}{k}, \tag{1.15}$$

which yields

$$\frac{dx'}{dt} = \frac{\omega}{k} = w \tag{1.16}$$

for the *phase velocity* w. Parameter ω is called the *angular frequency* with respect to time t; and parameter $\mathbf{k}$, the *wavevector* for the motion.

In the unprimed Cartesian coordinate system, parameter $\mathbf{k}$ has the components

$$\mathbf{k} = k_x\hat{\mathbf{x}} + k_y\hat{\mathbf{y}} + k_z\hat{\mathbf{z}}. \tag{1.17}$$

Since

$$\mathbf{k}\cdot\mathbf{r} = k_x x + k_y y + k_z z, \tag{1.18}$$

formula (1.13) expands to

$$\Psi = A\,e^{ik_x x}\,e^{ik_y y}\,e^{ik_z z}\,e^{-i\omega t} \tag{1.19}$$

and (1.5) to

$$d\Psi = ik_x\Psi\,dx + ik_y\Psi\,dy + ik_z\Psi\,dz - i\omega\Psi\,dt. \tag{1.20}$$

Equation (1.20) describes how Ψ varies within the homogeneous beam, for a general orientation of axes; Equation (1.19) describes the resulting coherent wave function.

Example 1.1. What is the periodicity of the exponential function?

The exponential function is related to trigonometric functions by the identity

$$e^{i\alpha} \equiv \exp i\alpha = \cos\alpha + i\sin\alpha.$$

Since both the cosine and the sine pass through one cycle when α increases by 2π, the exponential function $e^{i\alpha}$ (that is, exp $i\alpha$) does also.

Example 1.2. Show that when

$$\Psi = A\,e^{i(kx' - \omega t)},$$

as in (1.7), an increase in coordinate x' by $2\pi/k$ causes the function to go through one cycle.

At a given time t, an increase in the phase angle

$$kx' - \omega t = \alpha$$

appears as

$$k\,\Delta x' = \Delta\alpha.$$

From Example 1.1, an increase in α by 2π carries Ψ through one cycle. We then have

$$k\,\Delta x' = 2\pi$$

and

$$\frac{2\pi}{k} = \Delta x' = \lambda.$$

The distance λ that is required for a cosine, a sine, or an exponential function to go through one cycle at a given time is called its *wavelength*. We see that the wavelength of state function (1.7) is $2\pi/k$.

After the man who first applied the wave concept to all particles, in the early 1920s, the entity described by Ψ is called a *de Broglie wave.*

1.4. Diffraction by Neighboring Parallel Slits

Direct evidence for the validity of the wave functions we have formulated comes from diffraction experiments. Indeed, one finds that a homogeneous beam of particles is split into the same kind of diffraction pattern as a monochromatic beam of light, on passing a regular obstruction or grating. Since the intensity at an observation point is measured by the number of particles reaching the immediate neighborhood of the point per unit area per unit time, it is proportional to the probability density ρ there, and equals (constant) $\Psi^*\Psi$.

A simple setup is one in which equivalent particles emerge from a source with a given kinetic energy, travel to a wall containing two close parallel slits, and pass through these to form fringes on a screen. See Figure 1.1.

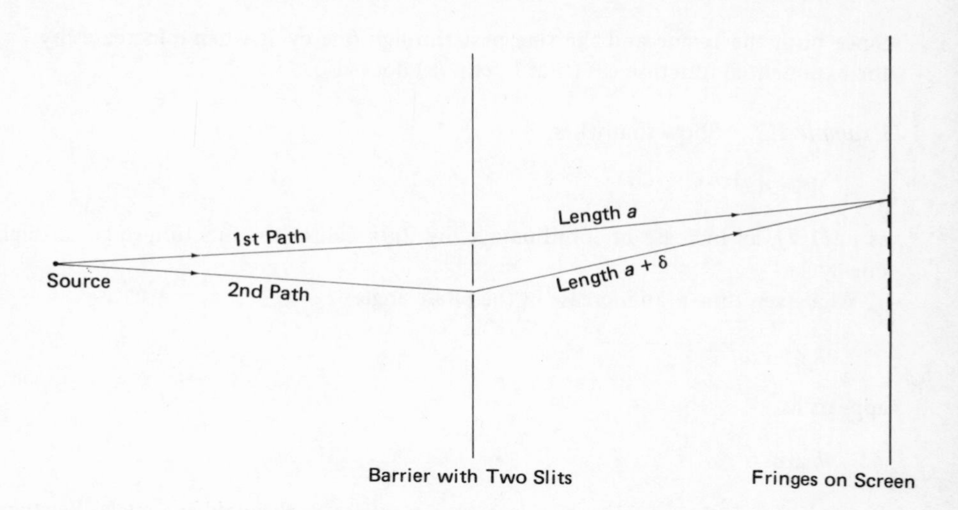

Fig. 1.1. Splitting of a beam by neighboring slits and its recombination at a point on a screen.

We consider the de Broglie wave to be scattered through a range of angles by each slit. The resulting Ψ found at an arbitrary point on the screen is not pure. There is a contribution from the part traveling along the first path of the form

$$\Psi_1 = A_1 e^{ika} e^{-i\omega t} \tag{1.21}$$

if (1.7) is valid and a is the distance traversed. Similarly, from the part traveling along the second path, we have

$$\Psi_2 = A_2 e^{ik(a+\delta)} e^{-i\omega t} \tag{1.22}$$

if $a + \delta$ is the distance traversed. When the slits are equivalent, equidistant from the source, and far enough from the screen so that the emerging rays are nearly parallel, the wavelets are of nearly equivalent intensities and

$$A_1 \simeq A_2 = A. \tag{1.23}$$

On the screen, particle densities and intensities corresponding to Ψ_1 and Ψ_2 individually are not observed. Only a combined effect, a single definite intensity is seen. This intensity is found to equal (constant) $\Psi^* \Psi$ if

$$\Psi = \Psi_1 + \Psi_2, \tag{1.24}$$

and the resulting wave function is the algebraic sum of its parts. Thus, at each observation point on the screen, we have

$$\Psi = A e^{ika} e^{-i\omega t} (1 + e^{ik\delta}). \tag{1.25}$$

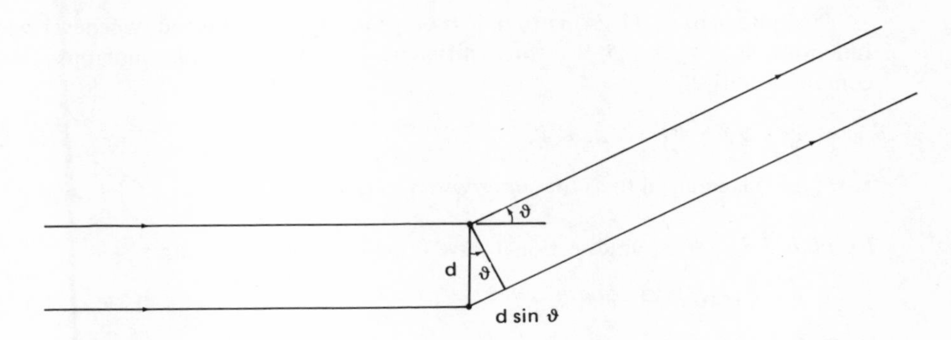

Fig. 1.2. Extra distance along the second path obtained as a projection of the distance between the slits.

The distance δ is obtained from Figure 1.2 as

$$\delta = d \sin \theta \qquad (1.26)$$

where d is the distance between the slits and θ is the angle of deflection from the initial rectilinear path. Coefficient A in general varies with θ. If it varies slowly, maxima in Ψ, and in the resulting intensity

$$I = (\text{constant})\ \Psi * \Psi \qquad (1.27)$$

occur approximately where

$$e^{ik\delta} = 1 ; \qquad (1.28)$$

that is, where

$$k\delta = 2\pi n \qquad (n \text{ is an integer}). \qquad (1.29)$$

In an experiment, a person can determine d and θ's for well-defined fringes. The order n for each of these may be found and k computed. Furthermore, the momentum p of a typical particle in the incident beam can be obtained from its energy. The wavevector k is then found to vary linearly with the momentum p:

$$p = \hbar k. \qquad (1.30)$$

But, the measurements are not very accurate. From consequences of the use of (1.30), it is found that

$$\hbar = \frac{h}{2\pi} = 1.05459 \times 10^{-34} \text{ J s}. \qquad (1.31)$$

Parameter h is called *Planck's constant*. Since the momentum **p** has the same direction as wavevector **k**, we also have

$$\mathbf{p} = \hbar \mathbf{k}. \qquad (1.32)$$

Relationship (1.32) is called *de Broglie's equation*.

The statement in (1.24) is found to be generally valid. Indeed, whenever wave functions $\Psi_1, \Psi_2, \ldots, \Psi_n$ for n different alternative motions superpose, they combine additively:

$$\Psi = \Psi_1 + \Psi_2 + \ldots + \Psi_n. \tag{1.33}$$

Law (1.33) is referred to as the *superposition principle.*

Example 1.3. What unidirectional wave functions combine to form

$$\Psi = C(\sin |k|x)\, e^{-i\omega t}?$$

The identity in Example 1.1 tells us that

$$\sin \alpha = \frac{e^{i\alpha} - e^{-i\alpha}}{2i}.$$

Substituting this into the given function yields

$$\Psi = C\frac{e^{i|k|x} - e^{-i|k|x}}{2i}\, e^{-i\omega t}$$

$$= \frac{C}{2i}\, e^{i|k|x}\, e^{-i\omega t} - \frac{C}{2i}\, e^{-i|k|x}\, e^{-i\omega t}.$$

The first term represents movement in the positive direction along the x axis; the second term represents movement in the negative direction, with the same probability.

In forming the Ψ function for a mixed state, the functions for the various independent contributing states are weighted and phased as their sources require, then added, following the superposition principle.

1.5. Diffraction by Molecules

A molecule consists of atoms vibrating about equilibrium positions. These positions act as scattering centers for particles from an incident beam. A collection of such molecules provides sufficient scattering to produce an observable diffraction pattern on a suitably placed screen. Consider the scheme in Figure 1.3, where a beam of approximately monoenergetic electrons passes through a stream of randomly oriented molecules to form circular fringes around an intense central dot on the screen.

The needed free electrons are produced by heating a metal filament. Those that emerge properly are accelerated by a voltage V to the kinetic energy

$$T = eV \tag{1.34}$$

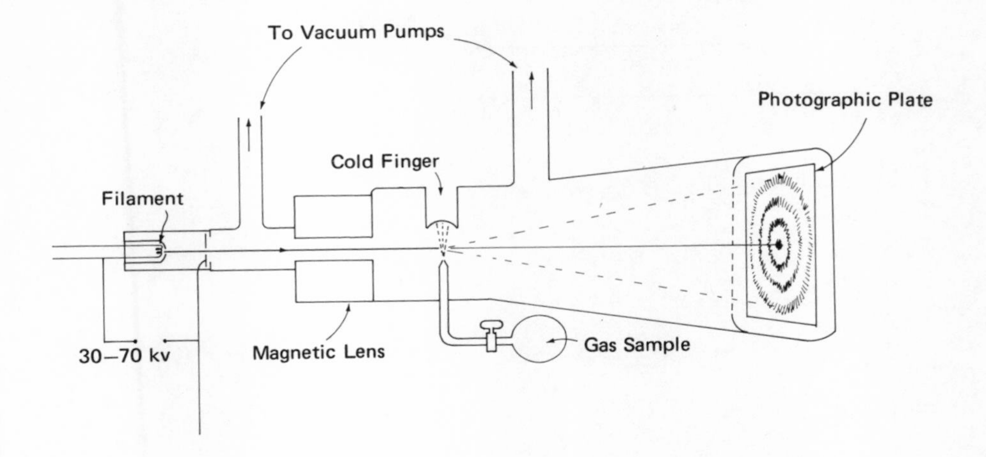

Fig. 1.3. Setup for studying the diffraction of electrons by randomly oriented molecules.

with e the charge on an electron. The Newtonian expression

$$T = \frac{p^2}{2m} \tag{1.35}$$

combines with (1.34) to give the momentum

$$p = (2meV)^{1/2}, \tag{1.36}$$

where m is the mass of an electron.

For simplicity, let us suppose that the gas in the intersecting stream is pure, with each molecule composed of two identical atoms. Also, let us consider two rays, 1 and 2, striking atoms A and B in a particular molecule and being deflected through angle θ in some longitudinal direction φ. See Figure 1.4.

The diffracted rays are governed by such functions as (1.21) and (1.22):

$$\Psi_1 = A\, e^{ika}\, e^{-i\omega t}, \tag{1.37}$$

$$\Psi_2 = A\, e^{ik(a+\delta)}\, e^{-i\omega t}. \tag{1.38}$$

Amplitude A is proportional to the amplitude of the incident wave. Furthermore, A varies with the nature of the scattering atom and with the deflection angle θ. Quantities a and $a + \delta$ are the effective distances from a reference plane at a given phase in the incident wave to the target spot on the screen which the rays strike. At this spot, the wavelets superpose to form

$$\Psi = \Psi_1 + \Psi_2 = A\, e^{ika}\, e^{-i\omega t}(1 + e^{ik\delta}). \tag{1.39}$$

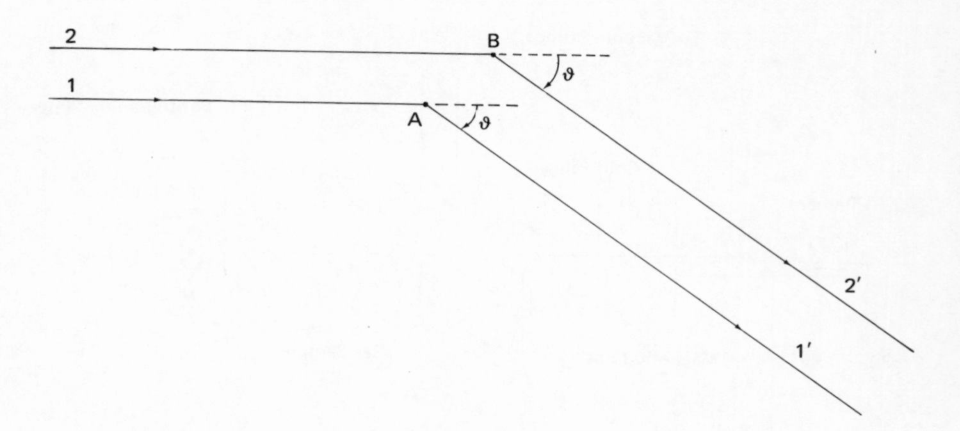

Fig. 1.4. Rays scattered at spherical-coordinate angle θ by atoms A and B in a molecule.

If we choose the constant in (1.1) so $\Psi * \Psi$ is the number of particles reaching the target spot per unit cross-sectional area per unit time, the intensity at this spot is

$$I = \Psi * \Psi = A*A(1 + e^{-ik\delta})(1 + e^{ik\delta})$$
$$= 2A*A(1 + \cos k\delta). \tag{1.40}$$

To complete the calculation, one would have to average (1.40) over all possible orientations of the molecule. The result can be expressed in the form

$$I = 2A*A\left(1 + \frac{\sin sr}{sr}\right) \tag{1.41}$$

where

$$s = 2k \sin \tfrac{1}{2}\theta \tag{1.42}$$

and r is the distance between A and B in the diffracting molecule.

Experimental patterns are described by Equation (1.41). In each instance, kr has to be properly chosen. Then if the internuclear distance r is known, k itself can be calculated and related to the momentum of a particle in the incident beam. De Broglie's equation is usually found to be satisfied within the limits of error.

Example 1.4. How are the wavevector and wavelength of a low-energy electron related numerically to the voltage that accelerated the electron?

Solve (1.30), the de Broglie equation, for k and eliminate p with (1.36):

$$k = \frac{p}{\hbar} = \frac{(2meV)^{1/2}}{\hbar}.$$

Introduce accepted values of the fundamental constants

$$\frac{(2me)^{1/2}}{\hbar} = \frac{[2(9.1095 \times 10^{-31}\ \text{kg})(1.6022 \times 10^{-19}\ \text{C})]^{1/2}}{1.0546 \times 10^{-34}\ \text{J s}}$$

$$= 5.1231 \times 10^{9}\ \text{m}^{-1}\ \text{volt}^{-1/2}$$

and obtain

$$k = 5.1231 \times 10^{9}\ V^{1/2}\ \text{m}^{-1}$$

or

$$k = 5.1231 \times 10^{7}\ V^{1/2}\ \text{cm}^{-1}.$$

From Example 1.2, the corresponding wavelength is

$$\lambda = \frac{2\pi}{k}.$$

Substituting in the k from above yields

$$\lambda = 12.264 \times 10^{-8}\ V^{-1/2}\ \text{cm},$$

whence

$$\lambda = \left(\frac{150.41}{V}\right)^{1/2}\ \text{Å}.$$

1.6. Dependence of the Function for a Definite Energy State on Time

The state function for a particle does not generally satisfy (1.5) or (1.20) because the motion may not be uniform and unidirectional throughout the region of interest, and the potential experienced by the particle may vary with position. As a consequence, the coefficients of dx, dy, dz would differ from the simple forms.

However, the movement through time is unidirectional and at a constant rate, insofar as anyone can tell. Therefore, when the particle is in a given energy state, we expect the dependence on time to be the same as that appearing in formula (1.20). At a fixed position, located by radius vector **r**, we have

$$d\Psi = -i\omega\Psi\, dt, \tag{1.43}$$

whence

$$\frac{\partial\Psi}{\partial t} = -i\omega\Psi. \tag{1.44}$$

Integrating this partial differential equation yields

$$\Psi = \psi(\mathbf{r})\, e^{-i\omega t} \equiv \psi(\mathbf{r})T(t). \tag{1.45}$$

Thus, the wave function for a given energy state factors into a spatial part $\psi(\mathbf{r})$ and a temporal part $T(t)$.

The probability density in the given state is

$$\rho = \Psi^* \Psi = \psi^* e^{i\omega t} \psi e^{-i\omega t} = \psi^* \psi \tag{1.46}$$

when the constant from (1.1) is chosen to be 1. Integrating this density over a volume yields the probability for finding the particle in the volume. Thus

$$\int_{\substack{\text{volume}\\ \text{containing}\\ \text{1 particle}}} \psi^* \psi \, d^3\mathbf{r} = 1. \tag{1.47}$$

A wave function may satisfy a governing differential equation, such as (1.5), without satisfying (1.47). When its amplitude is chosen so that it does satisfy (1.47), the wave function is said to be *normalized*.

Multiplying k in (1.5) by $\hbar$ yields the momentum p of a particle in the beam, according to the de Broglie equation. Multiplying ω in (1.5) by $\hbar$ gives an energy. By analogy, we may suppose that this is the energy E of the particle:

$$E = \hbar\omega. \tag{1.48}$$

Photoelectric measurements show that (1.48) relates the energy of a photon to the angular frequency of the associated electromagnetic wave. (Recall Einstein's pioneering work in 1905.) In fact, these measurements provide an accurate method of determining $\hbar$. On superposing Ψ's with varying ω's, we can show that (1.48) gives the energy of the associated particle. Since we do not need (1.48) in our detailed calculations at present, this demonstration will be deferred.

1.7. Independent Movements within a State

Certain potentials allow orthogonal coordinates to be constructed such that a change in one does not affect the results of changes in the others. Movement over this coordinate is then described by an independent factor in the wave function.

As a simple example, translation of a particle in any given direction in free space does not influence how the particle moves in perpendicular directions. These independences are embodied in differential (1.20). Indeed, integrating (1.20) at constant y, z, t yields the x variation of the Ψ; integration at constant z, t, x yields the y variation; integration at constant t, x, y yields the z variation; and integration at constant x, y, z yields the t variation. Each of the functional forms appears as a factor in the complete integral, (1.19) or

$$\begin{aligned}\Psi &= B\,e^{ik_x x}\, C\,e^{ik_y y}\, D\,e^{ik_z z}\, e^{-i\omega t}\\ &= X(x)Y(y)Z(z)T(t)\end{aligned} \tag{1.49}$$

where A has been replaced by BCD.

The product of the complex conjugate of a factor with the factor itself is interpreted as the probability density over the corresponding coordinate, or coordinates. Thus, in (1.49) we consider

$$X^*(x)X(x)\,\mathrm{d}x \tag{1.50}$$

to be the probability that a particle is between x and $x + \mathrm{d}x$,

$$Y^*(y)Y(y)\,\mathrm{d}y \tag{1.51}$$

to be the probability that a particle is between y and $y + \mathrm{d}y$,

$$Z^*(z)Z(z)\,\mathrm{d}z \tag{1.52}$$

to be the probability that a particle is between z and $z + \mathrm{d}z$.

Now, if a first choice can be made in n_1 equivalent ways, a second choice in n_2 equivalent ways, and so on, the number of ways in which all the choices can be made jointly is the product $n_1\, n_2\, \ldots$. As a consequence, the probability that independent events all occur equals the *product* of their respective probabilities.

Multiplying probabilities (1.50), (1.51), and (1.52) according to this theorem yields the result

$$(X^*X\,\mathrm{d}x)\,(Y^*Y\,\mathrm{d}y)\,(Z^*Z\,\mathrm{d}z) = (XYZ)^*(XYZ)\,\mathrm{d}x\,\mathrm{d}y\,\mathrm{d}z. \tag{1.53}$$

For the final expression to equal

$$\psi^*\,\psi\,\mathrm{d}x\,\mathrm{d}y\,\mathrm{d}z = \rho\,\mathrm{d}x\,\mathrm{d}y\,\mathrm{d}z, \tag{1.54}$$

we must have

$$\psi = XYZ \qquad \text{and} \qquad \Psi = XYZT, \tag{1.55}$$

as in (1.49).

A separate state function can be constructed for each particle of a multiparticle system. In the approximation that the motion of a particle is independent of the motions of any others, the function for that motion depends only on its coordinates.

Thus, a two-particle system might be described by

$$\Psi_a(\mathbf{r}_1, t) \qquad \text{and} \qquad \Psi_b(\mathbf{r}_2, t) \tag{1.56}$$

where $\mathbf{r}_1$ and $\mathbf{r}_2$ locate the first and second particles, respectively. The probability for the first particle to be in element $\mathrm{d}^3\mathbf{r}_1$ is $\Psi_a^*(\mathbf{r}_1, t)\,\Psi_a(\mathbf{r}_1, t)\,\mathrm{d}^3\mathbf{r}_1$, while the probability for the second particle to be in element $\mathrm{d}^3\mathbf{r}_2$ is $\Psi_b^*(\mathbf{r}_2, t)\,\Psi_b(\mathbf{r}_2, t)\,\mathrm{d}^3\mathbf{r}_2$. The fundamental law for combining probabilities tells us that the overall probability for both these conditions to be met is

$$\begin{aligned}&\Psi_a^*(\mathbf{r}_1, t)\,\Psi_a(\mathbf{r}_1, t)\,\mathrm{d}^3\mathbf{r}_1\;\;\Psi_b^*(\mathbf{r}_2, t)\,\Psi_b(\mathbf{r}_2, t)\,\mathrm{d}^3\mathbf{r}_2\\ &= [\Psi_a(\mathbf{r}_1, t)\,\Psi_b(\mathbf{r}_2, t)]^*\,\Psi_a(\mathbf{r}_1, t)\,\Psi_b(\mathbf{r}_2, t)\,\mathrm{d}^3\mathbf{r}_1\;\mathrm{d}^3\mathbf{r}_2.\end{aligned} \tag{1.57}$$

But if Ψ is the wave function for the system, this probability must also be given by

$$\Psi^* \Psi \, d^3\mathbf{r}_1 \, d^3\mathbf{r}_2 . \tag{1.58}$$

Expressions (1.57) and (1.58) are the same if

$$\Psi(\mathbf{r}_1, \mathbf{r}_2, t) = \Psi_a(\mathbf{r}_1, t) \, \Psi_b(\mathbf{r}_2, t). \tag{1.59}$$

We have here assumed that the particles are distinguishable. When they are indistinguishable, expressions representing the other possibilities must be added to the right side of (1.59), and the sum renormalized.

When effects of changes in one variable, or in more than one variable, are not altered by changes in any other independent variable, the effects are governed by an independent function depending only on the first variable, or variables. This state function combines multiplicatively with the state function for the other variables because of the way independent probabilities combine.

Example 1.5. A homogeneous beam, for which

$$\psi = A \, e^{ikx},$$

bombards a small section of screen with I particles per unit area per unit time. What is A if A is real?

From the kinetic energy of a typical particle, one could calculate its velocity v. Since all particles a distance v from the screen reach the target area in unit time, the volume containing I particles equals v. The volume per particle is v/I. Then (1.47) tells us that

$$\int_{\substack{\text{volume}\\\text{containing}\\\text{1 particle}}} \rho \, d^3\mathbf{r} = \int_{v/I} A \, e^{-ikx} A \, e^{ikx} \, dx \, dy \, dz = A^2 \frac{v}{I} = 1,$$

whence

$$A = \sqrt{\frac{I}{v}}$$

Example 1.6. If a single particle is confined between $x = 0$ and $x = a$ in the state for which

$$X = B \sin 2\pi \frac{x}{a},$$

what is B? Assume that B is real.

From the interpretation given to expression (1.50), the probability of finding the particle in a distance dx is $X^*X\,\mathrm{d}x$. The integral of this probability over the whole range of the particle must equal 1:

$$1 = \int_0^a X^*X\,\mathrm{d}x = \int_0^a B^2 \sin^2 2\pi\frac{x}{a}\,\mathrm{d}x$$

$$= B^2 \int_0^a \left(\frac{1}{2} - \frac{1}{2}\cos 4\pi\frac{x}{a}\right)\mathrm{d}x$$

$$= B^2 \left(\frac{1}{2}x - \frac{a}{8\pi}\sin 4\pi\frac{x}{a}\right)\Big|_0^a = B^2\,\frac{a}{2}.$$

Therefore,

$$B = \sqrt{\frac{2}{a}}\,.$$

1.8. The Continuum of Translational States

In an ideal gas, or in a system of free electrons in a metal, each particle may move freely along any line. Consequently, the argument in Section 1.3 applies to its movement with each momentum **p**. The pertinent wavevector **k** is related to this **p** by de Broglie's equation.

When the size of the system is large enough so that adding more equivalent material at the same density and temperature has no appreciable effect on the properties of interest, one may presume that such material has been added until all space is filled uniformly. All traces of the initial boundaries would be obliterated.

Interference phenomena then impose *no* conditions on the wavevector **k**. All magnitudes and directions consistent with the given energy density occur. Since **k** may vary continuously, so may momentum **p** and the energy E of a particle. A continuum of translational states is observed.

These states can be superposed in various ways. Indeed, if the motion described by

$$e^{i(\mathbf{k}\cdot\mathbf{r} - \omega t)} \tag{1.60}$$

occurs with the amplitude

$$\frac{1}{(2\pi)^{3/2}}\,\phi(\mathbf{k})\,\mathrm{d}^3\mathbf{k} \tag{1.61}$$

where $d^3\mathbf{k}$ is the appropriate differential volume in **k**-space, the complete state function is

$$\Psi(\mathbf{r}, t) = \frac{1}{(2\pi)^{3/2}} \int \phi(\mathbf{k})\, e^{i(\mathbf{k}\cdot\mathbf{r} - \omega t)}\, d^3\mathbf{k}$$

$$= \frac{1}{(2\pi)^{3/2}} \int \Phi(\mathbf{k}, t)\, e^{i\mathbf{k}\cdot\mathbf{r}}\, d^3\mathbf{k}, \tag{1.62}$$

where

$$\Phi(\mathbf{k}, t) = \phi(\mathbf{k})\, e^{-i\omega t}. \tag{1.63}$$

The function of wavevector and time $\Phi(\mathbf{k}, t)$ governs the behavior of the particle and describes its state as well as $\Psi(\mathbf{r}, t)$, since $\Phi(\mathbf{k}, t)$ determines $\Psi(\mathbf{r}, t)$. We call Ψ the *coordinate representation* and Φ the *wavevector representation* for the mixed state.

Function $\phi(\mathbf{k})$ may be constructed so that the exponential wave functions interfere with each other and reduce Ψ to a very small value except in a small region of physical space, at a given time. For simplicity here, an explicit demonstration will be deferred.

1.9. Periodic, Rectangularly Symmetric, Free Motion

The system of independent, freely moving particles may be considered differently. Instead of letting all wavevectors exist after adding the new material, one may simply require the new sections to duplicate what could be present in the first region. The resulting assembly is then periodic in three independent directions.

As before, we consider a system of free translators extensive enough so that adding more of the same type and comparably expanding the volume has no appreciable effect on the properties to be studied. We first arrange the given material in rectangular form, with the boundaries shown in Figure 1.5. Then to this unit of material we add similar units, side by side, until all space is filled, without making the system uniform in any direction. The conditions in the first block are then *repeated* in each successive block in each direction perpendicular to a face. *Periodic boundary conditions* are said to prevail.

In the composite system, a given particle experiences no changes in potential, wherever it moves. The environment presented to the particle is like that in the homogeneous beam. Consequently, the wave function for a pure state is the one in (1.19) and (1.49).

But the assumed periodicity makes the de Broglie wave repeat itself in each neighboring unit. An edge, $\mathscr{A}$, $\mathscr{B}$, $\mathscr{C}$, must then be an integral number of

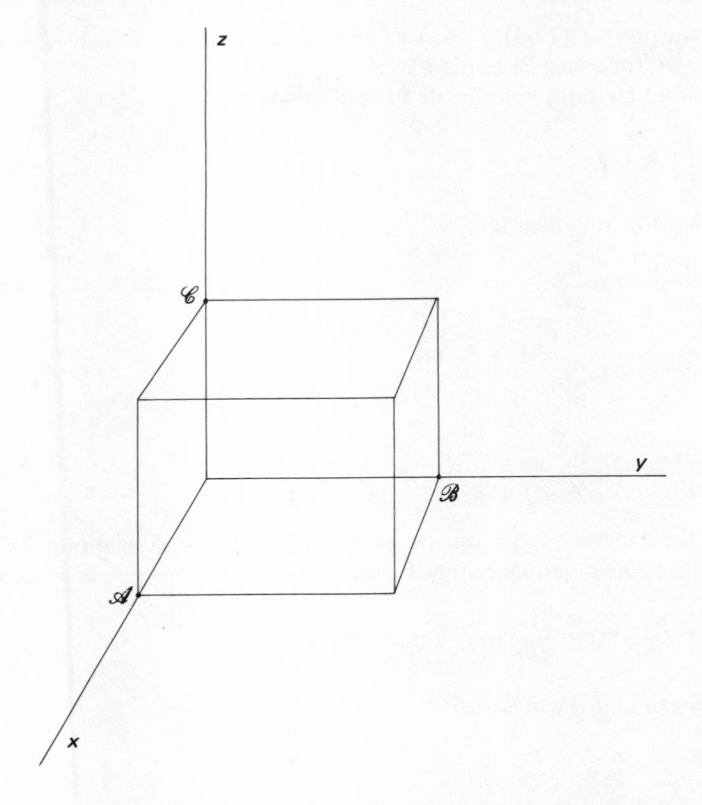

Fig. 1.5. A rectangular unit of material with appropriate coordinate axes.

wavelengths. If λ_x is the wavelength of Ψ for change in coordinate x, λ_y the wavelength for change in y, and λ_z the wavelength for change in z, we have

$$\mathscr{A} = n_x \lambda_x = \frac{2\pi n_x}{k_x}, \tag{1.64}$$

$$\mathscr{B} = n_y \lambda_y = \frac{2\pi n_y}{k_y}, \tag{1.65}$$

$$\mathscr{C} = n_z \lambda_z = \frac{2\pi n_z}{k_z}, \tag{1.66}$$

where

$$n_x = \ldots, -2, -1, 0, 1, 2, \ldots, \tag{1.67}$$

$$n_y = \ldots, -2, -1, 0, 1, 2, \ldots, \tag{1.68}$$

$$n_z = \ldots, -2, -1, 0, 1, 2, \ldots. \tag{1.69}$$

In the second equality in (1.64), (1.65), (1.66), we have replaced each λ by the corresponding $2\pi/k$, following Example 1.2.

Combining these equations with the de Broglie equation

$$\mathbf{p} = \hbar\mathbf{k} = \frac{h}{2\pi}\mathbf{k} \tag{1.70}$$

yields the components of momentum

$$p_x = \frac{h}{2\pi}\frac{2\pi n_x}{\mathscr{A}} = h\frac{n_x}{\mathscr{A}}, \tag{1.71}$$

$$p_y = \frac{h}{2\pi}\frac{2\pi n_y}{\mathscr{B}} = h\frac{n_y}{\mathscr{B}}, \tag{1.72}$$

$$p_z = \frac{h}{2\pi}\frac{2\pi n_z}{\mathscr{C}} = h\frac{n_z}{\mathscr{C}}. \tag{1.73}$$

Assuming that the kinetic energy can be calculated from the momentum, as in Section 1.5, and that the potential energy V is zero leads to the energy expression

$$E = T + V = \frac{p^2}{2m} + 0 = \frac{1}{2m}(p_x{}^2 + p_y{}^2 + p_z{}^2) \tag{1.74}$$

which (1.71) through (1.73) convert to

$$E = \frac{h^2}{2m}\left(\frac{n_x{}^2}{\mathscr{A}^2} + \frac{n_y{}^2}{\mathscr{B}^2} + \frac{n_z{}^2}{\mathscr{C}^2}\right). \tag{1.75}$$

Integers that label an allowed state of a system, as n_x, n_y, n_z do, are called *quantum numbers*. Quantity h is Planck's constant, m is the mass of the particle, and $\mathscr{A}, \mathscr{B}, \mathscr{C}$ are dimensions of the original rectangular unit.

These dimensions can be chosen so that intrinsically different state functions (those with differing sets of quantum numbers) yield the same energy E. Whenever such equivalence occurs, the pertinent level is said to be degenerate. The number of independent state functions needed to describe a level is called the *degeneracy* of the level.

Example 1.7. What does $h^2/2m$ equal when the translating particle is an electron?

Introduce accepted values of the fundamental constants

$$\frac{h^2}{2m} = \frac{(6.6262 \times 10^{-34}\ \text{J s})^2}{2(9.1095 \times 10^{-31}\ \text{kg})} = 2.4099 \times 10^{-37}\ \text{J m}^2$$

and change units to obtain

$$\frac{h^2}{2m} = \frac{2.4099 \times 10^{-37}\ \text{J m}^2}{1.6022 \times 10^{-19}\ \text{J eV}^{-1}} = 1.5041 \times 10^{-18}\ \text{eV m}^2$$

or

$$\frac{h^2}{2m} = 150.41 \text{ eV Å}^2 .$$

Note how this quantity checks with the last number in Example 1.4.

1.10. Enumerating and Filling Translational States

The restrictions that periodic boundary conditions introduce enable one to count translational states and determine how they are filled in a Fermi gas at low temperatures. The valence electrons in a metal and the neutrons and protons in the body of a nucleus behave approximately as such a gas.

Let us consider a large number of equivalent independent particles moving randomly, back and forth, through a region in which each particle's potential is constant at $-V_0$. A unit of the resulting material is taken, as shown in Figure 1.5. We assume that conditions within the parallelepiped are repeated in the x, y, and z directions with periods $\mathscr{A}$, $\mathscr{B}$, and $\mathscr{C}$. The de Broglie wavelengths of the translators then satisfy formulas (1.64) through (1.69).

From these equations, the components of the *wave number*, defined as

$$\bar{\mathbf{k}} = \frac{\mathbf{k}}{2\pi}, \tag{1.76}$$

with magnitude $1/\lambda$, are

$$\bar{k}_x = \frac{k_x}{2\pi} = \frac{n_x}{\mathscr{A}}, \tag{1.77}$$

$$\bar{k}_y = \frac{k_y}{2\pi} = \frac{n_y}{\mathscr{B}}, \tag{1.78}$$

$$\bar{k}_z = \frac{k_z}{2\pi} = \frac{n_z}{\mathscr{C}}. \tag{1.79}$$

Quantum numbers n_x, n_y, and n_z are restricted to integral values, as (1.67)–(1.69) specify.

Plotting the allowed wave numbers in a three-dimensional Euclidean space produces a rectangular lattice of points with a cell size

$$\frac{1}{\mathscr{A}\,\mathscr{B}\,\mathscr{C}}. \tag{1.80}$$

Each lattice point can be associated with the cell for which the point is the greatest $\bar{k}_x$, $\bar{k}_y$, and $\bar{k}_z$. When

$$\mathscr{A} = \mathscr{B} = \mathscr{C} = 1, \tag{1.81}$$

the size of each cell is unity and any given translational state is associated with a separate unit volume in the $\mathbf{k}$ plot. These unit volumes fill all $\mathbf{k}$-space.

Fundamental particles exhibit spin, as well as movement through physical space. The spin of an electron, neutron, or proton may be either $(+\frac{1}{2})\hbar$ or $(-\frac{1}{2})\hbar$ with respect to a given axis. The corresponding quantum numbers are $+\frac{1}{2}$ and $-\frac{1}{2}$. In a multiparticle system, it is found that no two fractional $\hbar$ spin particles of a given kind can have the same set of quantum numbers. This law, which we will here consider as empirical, is called the *Pauli exclusion principle*. As a consequence, no more than two such particles can occupy a translational state for which n_x, n_y, and n_z are given. The corresponding gas is called a *Fermi gas*, and the particles are called *fermions*.

Let us assume that unit volume of the given system contain $\mathcal{N}$ Fermi-gas particles of a given type. Let us also suppose that the temperature of the system is low, so that thermal agitation cannot excite any appreciable number of these particles out of their lowest $\mathcal{N}/2$ translational states. Since the kinetic energy of each particle is

$$E = \frac{p^2}{2m} = \frac{h^2 k^2}{2m}, \tag{1.82}$$

these states have the smallest k's. The $(\frac{1}{2}\mathcal{N})$th smallest wave number, the highest utilized k at 0 K, is labelled $k_{\max}$.

When the particle density $\mathcal{N}$ is large, cell size (1.80) is relatively small, and the number of these filled states equals the volume within a sphere of radius $k_{\max}$ in $\mathbf{k}$-space, to a good approximation. We have

$$\int_0^{k_{\max}} 4\pi k^2 \, \mathrm{d}k = \frac{4}{3}\pi k_{\max}^3 = \frac{\mathcal{N}}{2}. \tag{1.83}$$

Solving for the highest employed wave number gives us

$$k_{\max} = \left(\frac{3\mathcal{N}}{8\pi}\right)^{1/3} = \frac{1}{2}\left(\frac{3\mathcal{N}}{\pi}\right)^{1/3}. \tag{1.84}$$

The kinetic energy of a particle at this level is

$$E_{\mathrm{F}} = \frac{h^2 k_{\max}^2}{2m} = \frac{h^2}{8m}\left(\frac{3\mathcal{N}}{\pi}\right)^{2/3}. \tag{1.85}$$

We call E_{F} the *Fermi energy* for the particle.

The potential of a valence electron outside a metal, and the potential of a nucleon outside its nucleus, are taken to be 0. When moving well inside the metal, or nucleus, the potential is approximately constant at V_0. The most easily removed particle needs to acquire energy W, its *work function*, to raise its total energy to 0, the energy of a free particle at rest. When the temperature of the system is low enough so that no appreciable excitation has occurred, the work function plus the Fermi energy yield $-V_0$ and subscript 0 is added to W. (See Figure 1.6 and Table 1.1.)

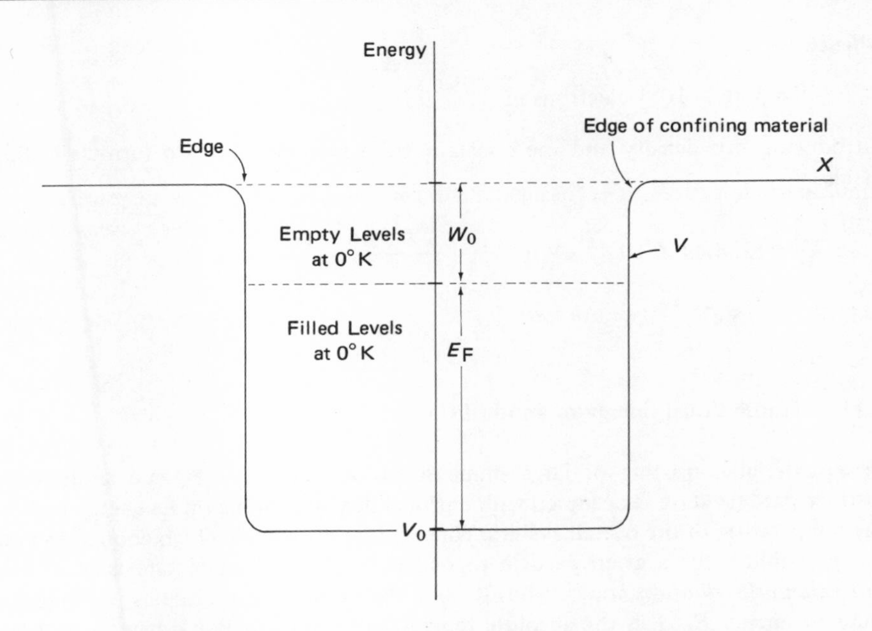

Fig. 1.6. Relationship of the potential depth $-V_0$ to the Fermi energy E_F and work function W_0, for a system of confined translating fermions.

TABLE 1.1.
Selected work functions and Fermi energies

Metal	W_0, in eV	E_F, in eV
Ag	4.7	5.5
Au	4.8	5.5
Ca	3.2	4.7
Cu	4.1	7.1
K	2.1	2.1
Li	2.3	4.7
Na	2.3	3.1

Example 1.8. If the density of silver is 10.5 g cm^{-3} and its atomic weight 107.9 u, what is its Fermi energy?

Since the predominant valence of silver is +1, we assume that there is one free electron for each atom. Consequently, the number of such electrons in unit volume is

$$\mathcal{N} = \frac{(10.5 \text{ g cm}^{-3})(6.02 \times 10^{23} \text{ electrons mole}^{-1})}{107.9 \text{ g mole}^{-1}}$$

$$= 5.86 \times 10^{22} \text{ electrons cm}^{-3},$$

whence

$$\mathcal{N} = 5.86 \times 10^{28} \text{ electrons m}^{-3}.$$

Introducing this density and the constant from Example 1.7 into formula (1.85) yields

$$E_F = (1.5041 \times 10^{-18} \text{ eV m}^2)\frac{1}{4}\left(\frac{3 \times 5.86 \times 10^{28} \text{ m}^{-3}}{3.1416}\right)^{2/3}$$

$$= 5.5 \text{ eV}.$$

1.11. Translational Energy of an Ideal Gas

In a material consisting of large numbers of particles of the same kinds, each distinct particle state is occupied with a probability depending on its energy and on the temperature of the overall system. When this temperature is high enough so that the probability for a given particle to occupy each significant state is small, the *Boltzmann distribution law* is valid. If N_j is the number of molecules occupying a state of energy E_j, T is the absolute temperature, and k is Boltzmann's constant, we then have

$$N_j = A\,e^{-E_j/kT}. \tag{1.86}$$

The internal energy of a system is the sum of energies of the particles from which the system is formed. For the total translational energy, we have

$$E = \sum_{\substack{\text{all}\\ \text{states}}} N_j E_j, \tag{1.87}$$

as long as E_j is the energy of the jth translational state. Letting E_j in (1.86) be the same energy and substituting the result into (1.87) yields

$$E = A \sum E_j\,e^{-E_j/kT}. \tag{1.88}$$

If we also let

$$Z \equiv \sum e^{-E_j/kT}, \tag{1.89}$$

then

$$\left(\frac{\partial Z}{\partial T}\right)_V = \frac{1}{kT^2} \sum E_j\,e^{-E_j/kT}, \tag{1.90}$$

where the partial derivative is calculated at constant total volume V. Expression Z is called the *state sum*.

The number of particles equals the sum of particles in each state:

$$N = \Sigma N_j = A\,\Sigma\, e^{-E_j/kT}$$
$$= AZ, \tag{1.91}$$

whence

$$A = \frac{N}{Z}. \tag{1.92}$$

Combining (1.90), (1.92) with (1.88) leads to

$$E = \frac{N}{Z} kT^2 \left(\frac{\partial Z}{\partial T}\right)_V = NkT^2 \left(\frac{\partial \ln Z}{\partial T}\right)_V. \tag{1.93}$$

Let us consider an ideal gas, for which the quantization in (1.75) is valid. We then have

$$Z_{\text{tr}} = \Sigma \exp\left[-\frac{h^2}{2mkT}\left(\frac{n_x^2}{\mathscr{A}^2} + \frac{n_y^2}{\mathscr{B}^2} + \frac{n_z^2}{\mathscr{C}^2}\right)\right]$$
$$= \Sigma \exp\left(-\frac{n_x^2 h^2}{2m\mathscr{A}^2 kT}\right) \Sigma \exp\left(-\frac{n_y^2 h^2}{2m\mathscr{B}^2 kT}\right) \Sigma \exp\left(-\frac{n_z^2 h^2}{2m\mathscr{C}^2 kT}\right)$$
$$\equiv Z_x Z_y Z_z. \tag{1.94}$$

If we set

$$\alpha = \frac{h^2}{2m\mathscr{A}^2 kT}, \tag{1.95}$$

we reduce the first factor to

$$Z_x = \sum_{-\infty}^{\infty} e^{-\alpha n_x^2}. \tag{1.96}$$

Except at low temperatures, α is relatively small and the sum can be approximated by the integral

$$Z_x = \int_{-\infty}^{\infty} e^{-\alpha n^2}\, dn = \left(\frac{\pi}{\alpha}\right)^{1/2}. \tag{1.97}$$

So

$$Z_x = \left(\frac{\pi 2m\mathscr{A}^2 kT}{h^2}\right)^{1/2}$$
$$= (2\pi mkT)^{1/2} \frac{\mathscr{A}}{h}. \tag{1.98}$$

Similarly,

$$Z_y = (2\pi mkT)^{1/2} \frac{\mathcal{B}}{h}, \tag{1.99}$$

$$Z_z = (2\pi mkT)^{1/2} \frac{\mathcal{C}}{h}, \tag{1.100}$$

and the translational state sum is

$$Z_{\text{tr}} = (2\pi mkT)^{3/2} \frac{V}{h^3}. \tag{1.101}$$

Taking the logarithm,

$$\ln Z_{\text{tr}} = \frac{3}{2} \ln T + \ln V + \text{constant}, \tag{1.102}$$

and substituting into (1.93) leads to

$$E_{\text{tr}} = NkT^2 \frac{3}{2}\frac{1}{T} = \frac{3}{2} NkT. \tag{1.103}$$

When N is Avogadro's number, Nk is the gas constant R and

$$E_{\text{tr}} = \frac{3}{2} RT. \tag{1.104}$$

The energy associated with one translational coordinate is $\frac{1}{2}kT$ per molecule. The energy associated with all three translational coordinates is $\frac{3}{2}kT$ per molecule, or $\frac{3}{2}RT$ per mole of gas.

1.12. Standing-Wave Translational Functions

When the beam described by (1.19) or (1.49) strikes a reflecting wall, its normal component is reversed. When the beam is introduced between confining reflecting walls, the reversals occur periodically. The complete wave consists of the sum of the initial wave, its reflection, its rereflection if different from the initial wave, and so on. The net result may be either nothing or a standing wave.

For simplicity in our discussion, let us suppose that the walls are perpendicular to the coordinate axes. Passage between the walls is presumed to be free, at constant potential. The movements are then described by wavevectors with a given magnitude $|\mathbf{k}|$.

The movement along the x axis in the positive direction is governed by the function

$$X_1 = \frac{1}{\sqrt{a}} e^{i|k_x|x} \tag{1.105}$$

with wavevector $|k_x|$; the movement in the negative direction by

$$X_2 = \frac{1}{\sqrt{a}} e^{-i|k_x|x} \tag{1.106}$$

with wavevector $-|k_x|$. Let us multiply both (1.105), (1.106) by $1/\sqrt{2}$ and add, following the superposition principle:

$$X_3 = \frac{1}{\sqrt{2a}} e^{i|k_x|x} + \frac{1}{\sqrt{2a}} e^{-i|k_x|x}$$

$$= \sqrt{\frac{2}{a}} \frac{e^{i|k_x|x} + e^{-i|k_x|x}}{2} = \sqrt{\frac{2}{a}} \cos |k_x|x. \tag{1.107}$$

An independent function is obtained on multiplying (1.105) by $1/i\sqrt{2}$, (1.106) by $-1/i\sqrt{2}$ and adding:

$$X_4 = \frac{1}{i\sqrt{2a}} e^{i|k_x|x} - \frac{1}{i\sqrt{2a}} e^{-i|k_x|x}$$

$$= \sqrt{\frac{2}{a}} \frac{e^{i|k_x|x} - e^{-i|k_x|x}}{2i} = \sqrt{\frac{2}{a}} \sin |k_x|x. \tag{1.108}$$

The multiplying factors are chosen with a magnitude $1/\sqrt{2}$ so that the normalizations of (1.105), (1.106) and (1.107), (1.108) are the same. (See Examples 1.5 and 1.6.) Note that while (1.105) and (1.106) describe traveling waves, (1.107) and (1.108) describe standing waves.

The general sinusoidal standing wave with wavevector $|k_x|$ is the superposition

$$X = B \sin |k_x|x + E \cos |k_x|x. \tag{1.109}$$

Similarly, the general sinusoidal standing wave for the y direction is

$$Y = C \sin |k_y|y + F \cos |k_y|y, \tag{1.110}$$

while the general one for z variation is

$$Z = D \sin |k_z|z + G \cos |k_z|z. \tag{1.111}$$

An ideal wall is conceived as a barrier that does not let any particles pass. Beyond it, the probability density ρ, its factor Ψ, and factors X, Y, Z are presumably zero. A person must also consider that Ψ and ψ go to zero as the ideal wall is approached because such a barrier can be formed by raising the potential beyond a certain surface without limit. During the process, the potential inside, here constant, is connected smoothly to that beyond the surface. Spatial factor ψ then varies smoothly across the transition region. When the rise becomes very steep and high, ψ drops to a very small number a short distance into this region, and remains small at greater distances. Then, ψ would have to be small just inside the boundary. In the limit, when the potential outside became infinite, ψ would become zero all the way up to the boundary from the outside. Thus, at confining walls,

$$\psi = 0. \tag{1.112}$$

1.13. Confined Rectangularly Symmetric Motion

Let us consider a particle confined to a rectangular box with edges *a, b*, and *c* units in length. For convenience, let us choose Cartesian axes oriented as illustrated, in Figure 1.5.

The potential of the particle is constant, within the box, and infinite outside. Each pure state is therefore governed by a given wavevector magnitude. The wave function itself drops to zero at the walls, as boundary condition (1.112) states.

The projection of the particle's movement on the x axis is governed by (1.109). To make ψ vanish at the first wall, at $x = 0$, we set

$$E = 0. \tag{1.113}$$

To make ψ vanish at the opposite wall, at $x = a$, we also set

$$|k_x|a = n_x\pi \qquad \text{with} \quad n_x = 1, 2, 3, \ldots. \tag{1.114}$$

Quantum number n_x cannot be zero because X must be different from zero somewhere for the particle to be in the box.

Rearranging (1.114) yields

$$|k_x| = \frac{n_x\pi}{a}. \tag{1.115}$$

This wavevector, together with condition (1.113), reduces (1.109) to

$$X = B \sin \frac{n_x\pi}{a} x. \tag{1.116}$$

In like manner, we obtain the factors

$$Y = C \sin \frac{n_y\pi}{b} y \qquad \text{with} \quad n_y = 1, 2, 3, \ldots \tag{1.117}$$

and

$$Z = D \sin \frac{n_z\pi}{c} z \qquad \text{with} \quad n_z = 1, 2, 3, \ldots. \tag{1.118}$$

The product of these is

$$\psi = A \left(\sin \frac{n_x\pi}{a} x\right) \left(\sin \frac{n_y\pi}{b} y\right) \left(\sin \frac{n_z\pi}{c} z\right). \tag{1.119}$$

For each factor to be normalized as Example 1.6 indicates, we must have

$$A = \sqrt{\frac{2}{a}}\sqrt{\frac{2}{b}}\sqrt{\frac{2}{c}} = \sqrt{\frac{8}{abc}} = \frac{2\sqrt{2}}{\sqrt{V}} \tag{1.120}$$

where V is the volume of the box.

The assumption that the kinetic energy can be calculated from the momentum, as in Section 1.5, and that the potential energy is zero inside the box leads to the energy expression

$$E = \frac{p^2}{2m} + 0 = \frac{1}{2m}\,(p_x{}^2 + p_y{}^2 + p_z{}^2). \tag{1.121}$$

Each component of momentum is related to a wavevector component by de Broglie's equation

$$\mathbf{p} = \mathbf{k}\hbar. \tag{1.122}$$

Substituting relationship (1.115) into (1.122) yields

$$|p_x| = \frac{n_x\pi}{a}\,\frac{h}{2\pi} = \frac{n_x h}{2a}. \tag{1.123}$$

Similarly,

$$|p_y| = \frac{n_y h}{2b}, \qquad |p_z| = \frac{n_z h}{2c}. \tag{1.124}$$

With these momenta, Equation (1.121) becomes

$$E = \frac{h^2}{8m}\left(\frac{n_x{}^2}{a^2} + \frac{n_y{}^2}{b^2} + \frac{n_z{}^2}{c^2}\right). \tag{1.125}$$

Note that E is the energy of the particle, h is Planck's constant, m is the mass of the particle, while a, b, and c are the dimensions of the confining box, and n_x, n_y, and n_z are positive integers, quantum numbers describing the state.

Decreasing a dimension of the box shortens the corresponding wavelength and raises E, while increasing the dimension lowers E. A generalization of this result is that increasing the confinement of a particle without altering its quantum numbers raises its energy, while decreasing its confinement without altering its quantum numbers lowers its energy.

1.14. Attenuated Motion

In describing the molecular, atomic, and subatomic behavior of systems, we have assumed that each submicroscopic particle moves through the Euclidean space and uniform time that classical laboratory measurements establish. Furthermore, we have related the potential and kinetic energies of a particle to its position with respect to the other involved particles and to the particle momenta according to classical laws.

However, a particle cannot be placed in a definite position with a definite momentum at any time. And no observer can determine the position and momentum of any particle with absolute accuracy at any time. Some uncertainty always exists. So we have been led to make this uncertainty an integral part of the theory. We have assumed that the state of each particle is described by a function Ψ related to the probability density ρ by Equation (1.1). Normalizing Ψ by choosing the constant to be 1 reduces (1.1) to

$$\rho = \Psi^* \Psi. \tag{1.126}$$

Considerations of simplicity and symmetry enabled us to construct relationships (1.2), (1.5), and (1.20). Along a path of equivalent points, Equation (1.20) integrates to

$$\Psi = A\, e^{i\mathbf{k}\cdot\mathbf{r}}\, e^{-i\omega t}. \tag{1.127}$$

We also assumed that wavelets from different paths combine additively where they meet.

These conditions and equations are found to govern the diffraction of particles in a uniform beam by parallel slits and by arrays of scattering centers, if wavevector $\mathbf{k}$ is related to momentum $\mathbf{p}$ of a typical particle in the beam by de Broglie's equation

$$\mathbf{p} = \hbar \mathbf{k}. \tag{1.128}$$

The magnitude of $\mathbf{p}$ is obtained from the kinetic energy T in the conventional manner:

$$T = \frac{p^2}{2m}. \tag{1.129}$$

The direction of $\mathbf{p}$ is given by the direction in which the beam moves.

The particles in many beams are accelerated by a decrease in potential energy. The total energy E of each particle satisfies the equation

$$E = T + V, \tag{1.130}$$

where V is the potential energy and T the kinetic energy. Eliminating T from Equations (1.129) and (1.130) leads to

$$p^2 = 2m(E - V). \tag{1.131}$$

An increase in potential energy, on the other hand, decelerates the particles. However, an increase to or above E does not stop a particle immediately, if our equations are true, because we can still set up a suitable wave function.

Consider a uniform region where $E < V$. Rewrite (1.131) in the form

$$p^2 = -2m(V - E); \tag{1.132}$$

take the square root

$$p = i[2m(V - E)]^{1/2}, \tag{1.133}$$

and combine with de Broglie's equation:

$$k = \frac{p}{\hbar} = i\,\frac{[2m(V - E)]^{1/2}}{\hbar} = i\kappa. \tag{1.134}$$

The expression κ defined by the last equality is called the *attenuation constant*. With the x' axis oriented in the direction of motion, Equation (1.127) becomes

$$\Psi = A\ e^{ikx'}\ e^{-i\omega t} = A\ e^{-\kappa x'}\ e^{-i\omega t}. \tag{1.135}$$

In contrast to conditions in a homogeneous beam, the particle density ρ now decreases with distance. Indeed, we find that

$$\rho = \Psi^* \Psi = A^* A\ e^{-2\kappa x'}. \tag{1.136}$$

The probability that a particle penetrates distance x' into the high-potential region decreases exponentially with x'. In distance $1/(2\kappa)$, it drops to $1/e$ of its initial value. Penetration of a high-potential region can be observed when V falls again below E, as in Figure 1.7.

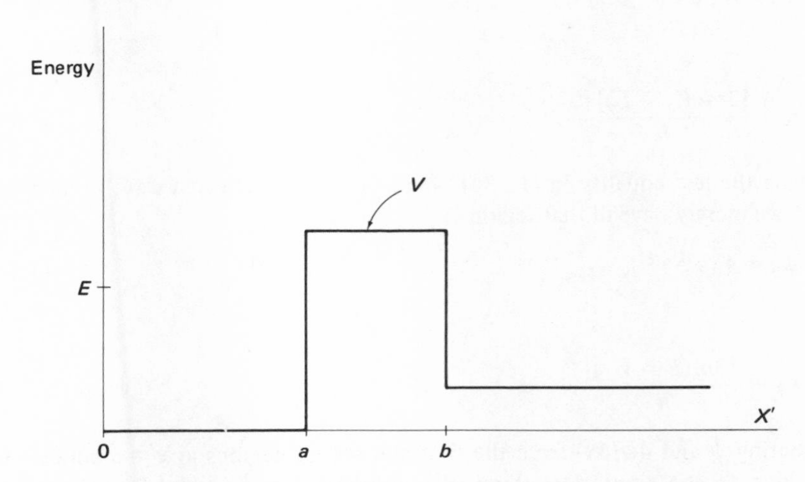

Fig. 1.7. Piecewise constant potential energy barrier for a beam of particles.

1.15. Joining Regions of Differing Potentials

When a region where the potential of a particle is a constant V_1 is connected by a transition zone to a region where the potential is a constant V_2, the variation in potential causes the particle density ρ to vary. Other things being equal, ρ will be lowest where the potential V is highest.

If V of the particle varies smoothly through the transition zone, then so would ρ and its factor Ψ. Then not only Ψ, but also its first and higher derivatives, would be continuous through the zone. Decreasing the thickness of the transition zone would not destroy these continuities. However, it would cause the higher derivatives to change more abruptly. In the limit, when the thickness becomes zero, we will find that we can still take not only Ψ continuous, as we did at a confining wall, but also its first derivative normal to the boundary.

Let us consider the barrier in Figure 1.7 being struck by a homogeneous beam of particles from the left. Such a beam is partially reflected at $x' = a$. It is attenuated on moving to $x' = b$ and is again partially reflected. However, some of the beam does emerge and travels on.

In the region where $x' < a$, both a forward and a backward traveling wave exist:

$$\psi_1 = A_1\,e^{ik_1x'} + B_1\,e^{-ik_1x'}. \tag{1.137}$$

The magnitude of the wavevector is given by (1.134) with V zero:

$$k_1 = \frac{(2mE)^{1/2}}{\hbar}. \tag{1.138}$$

For the region $a < x' < b$, we similarly have

$$\psi_2 = A_2\,e^{-\kappa x'} + B_2\,e^{\kappa x'} \tag{1.139}$$

and

$$\kappa = \frac{[2m(V_2 - E)]^{1/2}}{\hbar}, \tag{1.140}$$

following the last equality in (1.134). Since there is no reflected wave in the region $x' > b$, we merely have in that region

$$\psi_3 = A_3\,e^{ik_3x'} \tag{1.141}$$

with

$$k_3 = \frac{[2m(E - V_3)]^{1/2}}{\hbar}. \tag{1.142}$$

Equating ψ and $d\psi/dx'$ from the first and second regions at $x' = a$ imposes two conditions on the amplitudes. Similarly, equating ψ and $d\psi/dx'$ from the second and third regions at $x' = b$ imposes two more conditions. Since there are five amplitudes, one may then be chosen arbitrarily. This may be A_1, which is determined by the intensity of the incident beam. From the physics of the situation, we know that this can be chosen arbitrarily, so no difficulty arises.

A nonzero A_1 leads to a nonzero A_3, regardless of the value of V in the intermediate region. When there is a beam incident on the barrier, some of the beam passed on through. Classically, such transmission could not occur when V became greater than E unless a tunnel was constructed through the barrier. As a consequence, the peculiar quantum mechanical transmission is called *tunneling*.

The rate at which it occurs increases markedly as the thickness of the barrier is decreased and as the height of the barrier is decreased towards E. Indeed, when the potential barrier against escape of an alpha particle rises only moderately above the energy of the particle before falling, on increasing its distance from the center of the nucleus, alpha radioactivity is observed. A tunnel diode contains a barier around 100 Å thick between two heavily doped semiconductors. A small forward bias causes a current to flow; but when the band of conducting electrons on one side is raised so that an appreciable part of it coincides with the forbidden region on the other, the current drops with increasing bias. This negative-resistance behavior serves to amplify signals.

Discussion Questions

1.1. What kind of reasoning does a person employ in formulating physical concepts and laws? How are the laws tested?

1.2. How does a person discover and develop his concepts of space and time?

1.3. Why is it permissible to impose a Euclidean space and a uniform flow of time on the submicroscopic world?

1.4. What is a particle? What conceivable properties of a particle are uncertain?

1.5. Are the uncertainties accidental and eliminatable, or fundamental and ever present?

1.6. How are the potential and kinetic energies of a particle defined?

1.7. How does the probability density ρ arise? What properties can ρ govern?

1.8. What part of state function Ψ does ρ fix? What kind of arguments are used to determine the rest of Ψ?

1.9. Why do we expect Ψ to be an analytic function?

1.10. Why do we expect all properties of a submicroscopic particle to be derivable from its Ψ?

1.11. What is a homogeneous beam of particles? How is such a beam produced?

1.12. Explain how $\mathrm{d}\Psi$ within a homogeneous beam varies with (a) Ψ, (b) x', (c) t, (d) $\mathrm{d}x'$, (e) $\mathrm{d}t$.

1.13. Integrate the differential equation from Question 1.12. Show how the resulting Ψ yields a constant ρ along the propagation path.

1.14. How is a homogeneous beam diffracted by (a) two parallel slits, (b) a jet of homonuclear diatomic molecules?

1.15. What is the superposition principle? What direct evidence do we have for its validity?

1.16. How did we introduce wavevector **k** and its components, angular frequency ω, wavelength λ, wave number **ƛ** and its components? How are these continuum properties related to the discrete properties particle momentum **p** and particle energy E?

1.17. How does Ψ depend on time t when the system is in a given energy state?

1.18. What is normalization?

1.19. What circumstances permit the movements over a variable to be governed by an independent wave function?

1.20. Why do wave functions describing independent motions in a system combine multiplicatively?

1.21. To what extent is the behavior of a multiparticle system governed by a single three-dimensional wave function? How is a $3N$-dimensional wave function for an N-particle system constructed?

1.22. Why is there a continuum of translational states?

1.23. How is the wavevector representation related to the coordinate representation of a system?

1.24. Explain how the wave function for rectangularly symmetric translatory motion factors.

1.25. Justify the use of periodic boundary conditions in calculating translatory properties.

1.26. How are the corresponding translatory energy levels obtained?

1.27. What is the Pauli exclusion principle?

1.28. Describe a Fermi gas. Why do the valence electrons in a metal behave approximately as a Fermi gas?

1.29. How does one obtain the number of translational states having a wave number magnitude less than $ƛ_{max}$?

1.30. How is the Fermi energy related to the density of a Fermi gas?

1.31. Explain how a person calculates the translational energy of an ideal gas obeying Boltzmann statistics.

1.32. How do standing-wave translational wave functions arise?

1.33. Why does ψ drop to zero at a confining wall?

1.34. Show how the energy levels of the particle in a box are constructed.

1.35. How does increasing confinement alter the energy levels of a particle?

1.36. Can de Broglie's equation be considered as a relationship between the wavevector and the total energy minus the potential energy?

1.37. How does a particle move through a region in which its total energy is less than its potential energy?

1.38. Why can both Ψ and the first normal derivative of Ψ be kept continuous through a jump in the potential energy V?

1.39. What is quantum mechanical tunneling? Where is it observed?

Problems

1.1. Show that the number of cycles that a translational state function goes through in unit distance is wavevector k divided by 2π.

1.2. A homogeneous beam is being diffracted by two parallel slits 1000 Å apart. How far from the slits would the screen have to be located for a 1.00 Å wave to produce fringes with maxima 1.00 mm apart?

1.3. Resolve the wave function $(A \cos ax)\, e^{-ibt}$ into undirectional translational functions.

1.4. Calculate the wavevector k and wavelength λ for a beam of electrons accelerated from rest by a potential increment of 240 V.

1.5. On the average, 10.0 electrons with a de Broglie wavelength of 1.00 Å pass through a 1.00 $Å^2$ cross section of a homogeneous beam each second. Write down Ψ for the beam and choose the coefficient (normalize) so that $\Psi^* \Psi$ yields (a) the intensity at a stationary point, and (b) the probability density ρ.

1.6. In a certain system, the effects of changes in a generalized coordinate q are not influenced by changes in any coordinate measuring motion orthogonal to that measured by q throughout a region. If changing q alone anywhere in the region takes the model particle over equivalent points, how does Ψ depend on q?

1.7. In the Fermi gas of electrons in a metal, there are 0.0086 free translators per cubic ångström. What is the kinetic energy of the highest filled state at O K?

1.8. What is the density over wave number of the translational states when the unit cell in physical space has volume V? What is the corresponding density over energy? How many particles can occupy each state?

1.9. For a particle confined at constant potential within a cubic box, find the quantum numbers of (a) the lowest excited state that is not degenerate, and (b) the lowest energy level with a degeneracy of 6. Neglect spin.

1.10. Each electron in a homogeneous beam striking a barrier of height 10.0 eV has 1.00 eV kinetic energy. If the barrier is essentially infinite in depth, at what distance is the particle density reduced to $1/e$ of its value at the surface?

1.11. Show that the number of cycles of the de Broglie wave, for a definite energy state, passing a given point in unit time is angular frequency ω divided by 2π.

1.12. Construct wave function ψ for a homogeneous beam of 10.0 keV protons.

1.13. Obtain expressions for wavevector k and angular frequency ω at a point in a region of varying potential where $\Psi = A \exp i\, f(x, t)$.

1.14. Show that in a metal the average translational energy of a conduction electron is $\frac{3}{5} E_F$.

1.15. At what temperature is the energy associated with one translational coordinate of an ideal gas 0.500 eV per molecule, on the average?

1.16. A particle moves back and forth between planes $x = -a/2$ and $x = a/2$ in its fourth excited state. If its potential energy is constant, what is the x-dependent factor in its wave function?

1.17. For a particle confined at constant potential within a cubic box, find the quantum numbers associated with the lowest level described by wave functions that cannot all be reorientations of each other.

1.18. Resolve the wave function $(A \cosh ax)\, e^{-ibt}$ into unidirectional translational functions.

1.19. A homogeneous freely traveling beam strikes a barrier of constant height $V > E$. How are the coefficients for the transmitted wave and the reflected wave related to that for the incident wave?

1.20. In a crude model for the hydrogen atom, the orbital electron moves in a cubic box of edge $5r$ about the proton as center, the kinetic energy being

$$T = \frac{h^2}{8m} \left[\frac{1}{(5r)^2} + \frac{1}{(5r)^2} + \frac{1}{(5r)^2} \right].$$

Since the appropriate average distance of the electron from the center is r, the potential energy is taken to be $V = -e^2/(4\pi\epsilon_0 r)$. Construct the corresponding total energy E and vary r to minimize E, obtaining an expression for the energy of the ground state.

References

Books

Bastin, T. (ed.): 1971, *Quantum Theory and Beyond*, Cambridge University Press, London, pp. 1–40.

Jammer, M.: 1966, *The Conceptual Development of Quantum Mechanics*, McGraw-Hill, New York, pp. 1–382.

Kuhn, T. S.: 1978, *Black-Body Theory and the Quantum Discontinuity, 1894–1912*, Oxford University Press, New York, pp. 1–356.

Park, D.: 1974, *Introduction to the Quantum Theory*, 2nd edn, McGraw-Hill, New York, pp. 3–59.

Petersen, A.: 1968, *Quantum Physics and the Philosophical Tradition*, M.I.T. Press, Cambridge, Mass., pp. 1–190.

Ziock, K.: 1969, *Basic Quantum Mechanics*, Wiley, New York, pp. 1–22.

Articles

Andrews, M.: 1981, 'Matching Conditions on Wave Functions at Discontinuities of the Potential', *Am. J. Phys.* **49**, 281–282.

Ballentine, L. E.: 1972, 'Einstein's Interpretation of Quantum Mechanics', *Am. J. Phys.* **40**, 1763–1771.

Beam, J. E.: 1970, 'Multiple Reflection in Potential-Barrier Scattering', *Am. J. Phys.* **38**, 1395–1401.

Benioff, P.: 1978, 'A Note on the Everett Interpretation of Quantum Mechanics', *Found. Phys.* **8**, 709–720.

Bohm, D. J., and Hiley, B. J.: 1975, 'On the Intuitive Understanding of Nonlocality as Implied by Quantum Theory', *Found. Phys.* **5**, 93–109.

Branson, D.: 1979, 'Continuity Conditions on Schrödinger Wave Functions at Discontinuities of the Potential', *Am. J. Phys.* **47**, 1000–1003.

Christoudouleas, N. D.: 1975, 'Particles, Waves, and the Interpretation of Quantum Mechanics', *J. Chem. Educ.* **52**, 573–575.

Corben, H. C.: 1979, 'Another Look Through the Heisenberg Microscope', *Am. J. Phys.* **47**, 1036–1037.

Cummings, F. E.: 1977, 'The Particle in a Box is Not Simple', *Am. J. Phys.* **45**, 158–160.

Draper, J. E.: 1979, 'Use of $|\Psi|^2$ and Flux to Simplify Analysis of Transmission past Rectangular Barriers or Wells', *Am. J. Phys.* **47**, 525–530.

Draper, J. E.: 1980, 'Quantal Tunneling through a Rectangular Barrier Using $|\Psi|^2$ and Flux', *Am. J. Phys.* **48**, 749–751.

El'yashevich, M. A.: 1977, 'From the Origin of Quantum Concepts to the Establishment of Quantum Mechanics', *Soviet Phys. Uspekhi* **20**, 656–682.

Freundlich, Y.: 1977, 'Two Views of an Objective Quantum Theory', *Found. Phys.* **7**, 279–300.

Goodman, M.: 1981, 'Path Integral Solution to the Infinite Square Well', *Am. J. Phys.* **49**, 843–847.

Heisenberg, W.: 1975, 'Development of Concepts in the History of Quantum Theory', *Am. J. Phys.* **43**, 389–394.

Johnson, E. A., and Williams, H. T.: 1982, 'Quantum Solutions for a Symmetric Double Square Well', *Am. J. Phys.* **50**, 239–243.

Jönsson, C. (translated by Brandt, D., and Hirschi, S.): 1974, 'Electron Diffraction at Multiple Slits', *Am. J. Phys.* **42**, 4–11.

Kiang, D.: 1974, 'Multiple Scattering by a Dirac Comb', *Am. J. Phys.* **42**, 785–787.

Levy-Leblond, J.-M.: 1981, 'Classical Apples and Quantum Potatoes', *Eur. J. Phys.* **2**, 44–47.

MacKinnon, E.: 1976, 'De Broglie's Thesis: A Critical Retrospective', *Am. J. Phys.* **44**, 1047–1055.

Maxwell, N.: 1982, 'Instead of Particles and Fields: A Micro Realistic Quantum "Smearon" Theory', *Found. Phys.* **12**, 607–631.

Mermin, N. D.: 1981, 'Bringing Home the Atomic World: Quantum Mysteries for Anybody', *Am. J. Phys.* **49**, 940–943.

Newton, R. G.: 1980, 'Probability Interpretation of Quantum Mechanics', *Am. J. Phys.* **48**, 1029–1034.

Patsakos, G.: 1976, 'Classical Particles in Quantum Mechanics', *Am. J. Phys.* **44**, 158–165.

Peres, A.: 1975, 'A Single System Has No State', *Am. J. Phys.* **43**, 1015–1016.

Peres, A.: 1979, 'Proposed Test for Complex versus Quaternion Quantum Theory', *Phys. Rev. Letters* **42**, 683–686.

Powell, R. A.: 1978, 'Photoelectric Effect: Back to Basics', *Am. J. Phys.* **46**, 1046–1051.

Rogers, D. W.: 1972, 'Symmetry and Degeneracy: The Particle in a "Squeezed Box" ', *J. Chem. Educ.* **49**, 501–502.

Schreiber, H. D., and Spencer, J. N.: 1971, 'Contour Functions for the Particle in a 3-D Box', *J. Chem. Educ.* **48**, 185–187.

Stapp, H. P.: 1972, 'The Copenhagen Interpretation', *Am. J. Phys.* **40**, 1098–1116.

Wigner, E. P.: 1971, 'Rejoinder', *Am. J. Phys.* **39**, 1097–1098.

Chapter 2

Quantization of Rotatory Motion

2.1. Separation of Different Modes of Motion from Each Other

Composite submicroscopic particles, such as molecules, atoms, or nuclei, exhibit not only motion from place to place (translation), but also internal motions that may be classified as rotations and vibrations. Each kind of movement is presumably governed by a state function whose form depends on the energy associated with the movement. When this energy is independent of the energies associated with the other motions, the corresponding state function is independent.

Since the energy associated with free translation is independent of the energy associated with internal motions, the translational motion can be considered by itself, as we have done in Chapter 1. The energy associated with free rotation is independent of that associated with vibrations in the approximation that all dimensions of the composite system can be approximated by fixed numbers. Then the rotational motion can also be considered by itself, analogous to the translational motion.

The rotational energy of a linear molecule is represented by the energy of a particle of appropriate mass moving on a sphere in an inertial system. A suitable model will be constructed in Section 2.2. The rotation of a cylindrically symmetric nonlinear molecule can be related to rotations of linear systems. Furthermore, the rotational levels of an asymmetric molecule can be correlated with those of related cylindrically symmetric molecules.

Other motions in a composite system can be resolved into nearly independent vibrational modes. A model for such a mode will be developed in the next chapter.

To the approximation that each kind of movement, each mode, is independent, each is governed by a separate factor in the overall wave function, just as the movement of a particle along a single coordinate is described by a separate factor when the translation is independent of the motions along the other orthogonal coordinates.

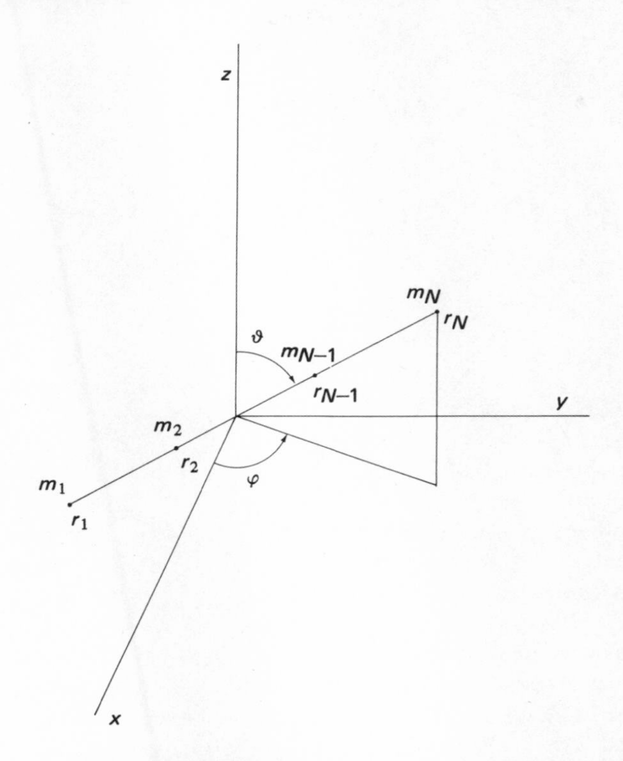

Fig. 2.1. Coordinates for the atoms in a rotating linear molecule. Positive r_j's lie on the side to which angles θ and φ are drawn, negative r_j's on the opposite side, of the origin.

2.2. One-Particle Model for a Linear Rotator

The turning motion of two or more particles bound together along a straight line is represented by the revolution of a single particle about a fixed line in space.

Suppose that the masses of the constituent particles are

$$m_1, m_2, \ldots, m_N \tag{2.1}$$

and that these are at positions

$$r_1, r_2, \ldots, r_N \tag{2.2}$$

along the straight line at time t. Let the origin on this axis be located at the center of mass, so

$$m_1 r_1 + m_2 r_2 + \ldots + m_N r_N = 0. \tag{2.3}$$

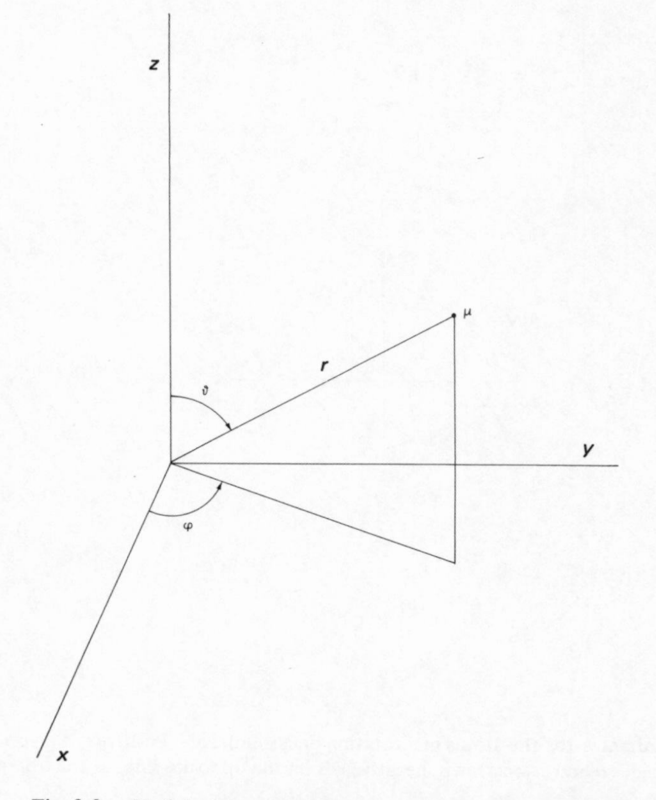

Fig. 2.2. Model whose movement represents rotation of a linear molecule.

Also, suppose the axis is oriented with respect to a nonrotating coordinate frame at the time t, as Figure 2.1 shows.

By the Pythagorean theorem, the square of an element of path traced out by m_j in time $\mathrm{d}t$ is

$$\mathrm{d}s_j^2 = \mathrm{d}r_j^2 + r_j^2\, \mathrm{d}\theta^2 + r_j^2 \sin^2\theta\, \mathrm{d}\varphi^2. \tag{2.4}$$

The corresponding velocity squared is

$$v_j^2 = \dot{r}_j^2 + r_j^2\, \dot{\theta}^2 + r_j^2 \sin^2\theta\, \dot{\varphi}^2 \tag{2.5}$$

if

$$\dot{r}_j = \frac{\mathrm{d}r_j}{\mathrm{d}t}, \quad \dot{\theta} = \frac{\mathrm{d}\theta}{\mathrm{d}t}, \quad \dot{\varphi} = \frac{\mathrm{d}\varphi}{\mathrm{d}t}. \tag{2.6}$$

Let us consider the system to be turning freely, with a potential energy against rotation of zero. The kinetic energy associated with the changing angles is then

$$E = \Sigma \frac{1}{2} m_j v_j^2 = \frac{1}{2} \Sigma m_j r_j^2 [\dot{\theta}^2 + (\sin^2 \theta) \dot{\varphi}^2]$$

$$= \frac{I}{2} [\dot{\theta}^2 + (\sin^2 \theta) \dot{\varphi}^2], \tag{2.7}$$

with the *moment of inertia* the sum

$$I = m_1 r_1{}^2 + m_2 r_2{}^2 + \ldots + m_N r_N{}^2. \tag{2.8}$$

A possible model consists of a single particle of mass μ moving with respect to an inertial frame as Figure 2.2 indicates. Let the spherical coordinates locating this particle be r, θ, φ, while its mass μ is given by

$$\frac{1}{\mu} = \frac{1}{m_1} + \frac{1}{m_2} + \ldots + \frac{1}{m_N}. \tag{2.9}$$

Calculating the velocity v of the particle as in (2.5), with r constant, and substituting the result into the kinetic energy expression, yields

$$E = \frac{1}{2} \mu v^2 = \frac{1}{2} \mu r^2 [\dot{\theta}^2 + (\sin^2 \theta) \dot{\varphi}^2]. \tag{2.10}$$

To make (2.10) agree with (2.7), we choose r so that

$$\mu r^2 = I, \tag{2.11}$$

the moment of inertia of the model equals that of the system. Parameter μ is called the *reduced mass*.

Any change in particle positions that causes I to vary causes r to vary, following Equation (2.11). For a two-particle system, r is the interparticle distance and the total potential V depends uniquely on r. For multiparticle systems, V also involves other coordinates.

Since the rotational properties of a system are determined by how its rotational energy depends on the coordinates, the one-particle system does simulate the pertinent rotational behavior. Whenever we are interested in such behavior, we may consider the system to be a particle of reduced mass, μ, moving on a sphere of radius r about the center of mass.

Example 2.1. Show that for a two-particle system, distance r is the interparticle distance.

When N is 2, Equations (2.8) and (2.9) become

$$I = m_1 r_1{}^2 + m_2 r_2{}^2$$

and

$$\frac{1}{\mu} = \frac{1}{m_1} + \frac{1}{m_2}.$$

Multiplying each side of the first equation by the corresponding side of the second yields

$$\frac{I}{\mu} = \frac{1}{m_1}(m_1 r_1{}^2 + m_2 r_2{}^2) + \frac{1}{m_2}(m_1 r_1{}^2 + m_2 r_2{}^2)$$

$$= r_1{}^2 + \frac{m_2}{m_1} r_2{}^2 + \frac{m_1}{m_2} r_1{}^2 + r_2{}^2.$$

Since the origin has been placed at the center of mass, we have

$$m_1 r_1 = -m_2 r_2,$$

and

$$\frac{m_2 r_2{}^2}{m_1} = -\frac{m_1 r_1 r_2}{m_1} = -r_1 r_2,$$

$$\frac{m_1 r_1{}^2}{m_2} = -\frac{m_2 r_2 r_1}{m_2} = -r_2 r_1.$$

These formulas convert the relationship for I/μ to

$$\frac{I}{\mu} = r_1{}^2 - r_1 r_2 - r_2 r_1 + r_2{}^2 = (r_2 - r_1)^2$$

or

$$I = \mu(r_2 - r_1)^2.$$

Comparing this expression with (2.11) yields

$$r = r_2 - r_1.$$

Since r_1 is negative, coordinate r equals the distance between mass m_1 and mass m_2.

2.3. Variation of Ψ in a Spherically Symmetric Field

When the potential energy of a linear system depends only on the interparticle separations, the space of the system exhibits spherical symmetry about the center of mass. Placing the origin of a coordinate frame at the center of symmetry, and drawing from this origin an axis, allows a rotatory motion about the axis to be defined which is particularly simple.

In our development, let us consider the model system of Figure 2.2. Since a state function is analytic wherever there are no singularities in the potential V, the change in Ψ caused by infinitesimal changes in coordinates r, θ, φ and in time t is

$$\mathrm{d}\Psi = \frac{\partial \Psi}{\partial r}\,\mathrm{d}r + \frac{\partial \Psi}{\partial \theta}\,\mathrm{d}\theta + \frac{\partial \Psi}{\partial \varphi}\,\mathrm{d}\varphi + \frac{\partial \Psi}{\partial t}\,\mathrm{d}t. \tag{2.12}$$

When the actual particles of the system are bound together by an interaction that varies with I, and so with r, potential V depends on r. Moving the model particle along a radial line drawn from the origin then subjects it to a changing V. Points along such a path are not similar.

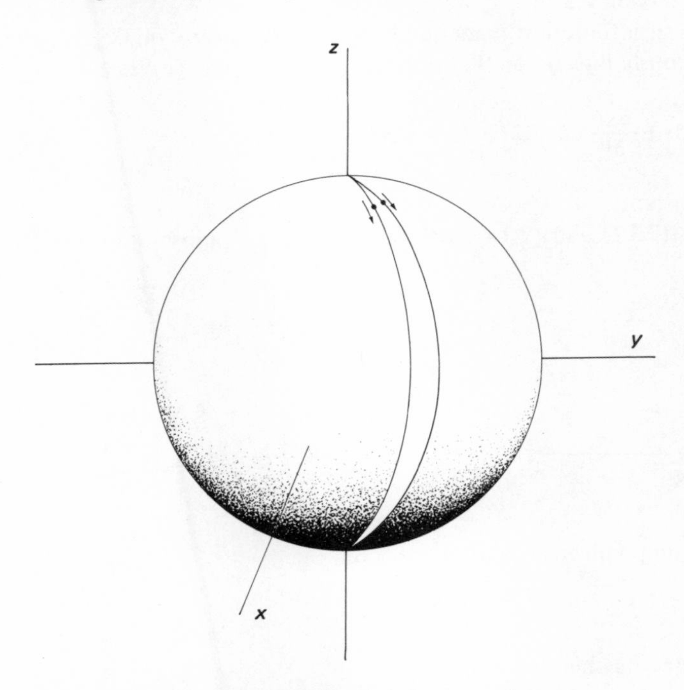

Fig. 2.3. Neighboring longitudinal circles on a sphere.

On the other hand, moving the model particle along a longitudinal circle, varying θ alone, does not alter its potential energy. However, the corresponding point then continuously alters its distance from a comoving point on a neighboring longitudinal circle, as Figure 2.3 shows. Successive positions along such a path are not equivalent with respect to the comoving point. Hence, the effect of a given $\mathrm{d}\theta$ on $\mathrm{d}\Psi$ must vary with θ, even though V does not change.

Moving the model particle along a latitudinal circle varies only coordinate φ. Since the particle then remains at a constant distance from a comoving point on each neighboring latitudinal circle, points separated by changes in φ alone are equivalent as long as

$$V = V(r). \tag{2.13}$$

By symmetry, the effect of a small variation in φ on a variation in Ψ is then independent of the initial φ. Also, by symmetry, multiplying $d\varphi$ by a factor merely multiplies the effect of the infinitesimal change by the factor.

When the motion at constant r and θ is unidirectional, with a definite angular momentum around the z axis, each equivalent part of Ψ presumably produces the same effect on $d\Psi$ as in Section 1.3. Furthermore, the result of a change in t can be represented as in Section 1.3.

Consequently, an infinitesimal variation in Ψ depends linearly on $d\varphi$, linearly on dt, and homogeneously linearly on Ψ. Formula (2.12) then reduces to

$$d\Psi = \frac{\partial \Psi}{\partial r}\, dr + \frac{\partial \Psi}{\partial \theta}\, d\theta + iM\Psi\, d\varphi - i\omega\Psi\, dt \tag{2.14}$$

with M and ω constant.

On comparing (2.12) and (2.14), we see that for the pure motion being considered

$$\frac{\partial \Psi}{\partial \varphi} = iM\Psi \tag{2.15}$$

and

$$\frac{\partial \Psi}{\partial t} = -i\omega\Psi. \tag{2.16}$$

Integrating these simultaneous equations leads to

$$\Psi = F(r, \theta)\, e^{iM\varphi}\, e^{-i\omega t} = F(r, \theta)\, \Phi(\varphi)\, T(t). \tag{2.17}$$

The last two factors together

$$\Phi(\varphi)\, T(t) = e^{i(M\varphi - \omega t)} \tag{2.18}$$

describe a wave traveling around the z axis.

2.4. The Rotational State Described by the Traveling Wave

Constant ω in function (2.18) is the energy of the model particle divided by $\hbar$, according to formula (1.48). Constant M, on the other hand, needs further interpretation.

From Equation (1.5), the change of Ψ in distance $\mathrm{d}s$ along a straight line of equivalent points is

$$\mathrm{d}\Psi = ik\Psi\,\mathrm{d}s - i\omega\Psi\,\mathrm{d}t, \tag{2.19}$$

where k is the wavevector related to the corresponding linear momentum p by de Broglie's equation

$$p = \hbar k. \tag{2.20}$$

An infinitesimal change in φ alone moves the model particle along the element

$$\mathrm{d}s = r \sin\theta\,\mathrm{d}\varphi, \tag{2.21}$$

which is essentially straight. The corresponding change in the state function is

$$\mathrm{d}\Psi = i\frac{M}{r\sin\theta}\Psi r\sin\theta\,\mathrm{d}\varphi \tag{2.22}$$

by Equation (2.14) and

$$\mathrm{d}\Psi = ik\Psi r\sin\theta\,\mathrm{d}\varphi \tag{2.23}$$

according to (2.19) and (2.21).

On comparing Equations (2.22) and (2.23), we see that

$$k = \frac{M}{r\sin\theta}, \tag{2.24}$$

whence

$$\hbar k r \sin\theta = M\hbar. \tag{2.25}$$

The de Broglie relationship reduces this equation to

$$pr\sin\theta = M\hbar. \tag{2.26}$$

The expression on the left is the angular momentum $\mathscr{M}_z$ of the particle about the z axis; we have

$$\mathscr{M}_z = M\hbar. \tag{2.27}$$

Hence, quantity M in the φ-dependent factor of Ψ,

$$\Phi(\varphi) = \mathrm{e}^{iM\varphi}, \tag{2.28}$$

is the number of $\hbar$ units in the angular momentum around the z axis. Analogous to the classical law that angular momentum about an axis is conserved is the fact, coming from symmetry, that

$$M = \text{constant}. \tag{2.29}$$

2.5. Quantization of the Simple Rotator

The diffraction experiments of Sections 1.4 and 1.5 indicate that the various wavelets reaching a given point on a screen combine additively. Even when distinct wavelets can be identified by the different paths they follow, they do not produce distinct effects. We are thus led to assume that any given wave function has a single value at each point in space and time.

But the correspondence between coordinates and physical points can be many-to-one. Variation of the coordinates over only a part of their complete range then causes the point to move over all space. With spherical coordinates, a single sweep results when r varies from 0 to ∞, θ from 0 to π, and φ from 0 to 2π.

For a function of such coordinates to be a suitable state function, the function must yield the same result at all coordinate values corresponding to the same physical point. In the model system that we have been considering, coordinates

$$r = a, \qquad \theta = b, \qquad \varphi = c + 2\pi n, \qquad t = d, \tag{2.30}$$

where n is an integer, correspond to the same point as coordinates

$$r = a, \qquad \theta = b, \qquad \varphi = c, \qquad t = d. \tag{2.31}$$

Consequently, both sets have to yield the same Ψ. Factor $\Phi(\varphi)$ must repeat itself with each increase of φ by 2π.

Since the period of the exponential function is $2\pi i$, the constant M in (2.28) must be an integer. Because there is no essential difference between the positive and negative directions around the axis, the upper and lower limits on M, for a given amount of rotational energy, have the same magnitude. Thus

$$M = -J, \ldots, -1, 0, 1, \ldots, J. \tag{2.32}$$

The magnitude of the limits, J, is called the *rotational quantum number*. Since the magnetic moment generated by a separation of charge in the chemical bonds of the rotating system is proportional to $\mathscr{M}_z$, and to M, M is called the *magnetic quantum number*.

Because state function Ψ is required to be smoothly varying with θ as well as with φ, the pure motion we have been discussing cannot exist by itself. Instead, it has to be accompanied by some up-down-and-back motion through the plane at $\theta = 90°$. Since this motion produces no net circulation about the center of mass, the net angular momentum continues about the z axis.

But in the general pure rotational state, the angular momentum is around a different axis, which we label z', which does pass through the center of mass. Accompanying this are back-and-forth motions through the corresponding equatorial plane. The moment of inertia is still

$$I = \mu r^2, \tag{2.33}$$

the angular momentum for the pure motion is

$$\mathcal{M}_{z'} = M'\hbar, \tag{2.34}$$

where M' is the magnetic quantum number for the z' axis, and the related energy is

$$E' = \frac{\mathcal{M}_{z'}{}^2}{2I} = \frac{M'^2 \hbar^2}{2I}. \tag{2.35}$$

This formula describes the rotational energy of one part of a molecule turning with respect to another part around a single bond, to the approximation that the rotation is free. The accompanying bending of the bond is classified as a vibration. For a freely turning linear rotator, however, the energy associated with the back-and-forth rocking movement is also considered to be rotational. Since the net effect of such movement is to introduce standing waves orthogonal to the traveling wave representing the pure motion around the imposed axis, we have to investigate the nature of such waves.

2.6. Standing-Wave Rotational Functions

In Chapter 1 we described two different kinds of translational motion: (a) movement of particles in one direction, as in a homogeneous beam, and (b) movement of particles with equal probability in opposing directions, as in a confined gas. A traveling de Broglie wave governs the former; a standing de Broglie wave, the latter.

Similarly, there are two kinds of rotational motion. In the first kind, the configuration of particles turns in one direction about some axis in space. In the second kind, the array rotates with equal probability in both directions about the axis. Any quantum mechanical system turning in one direction also exhibits some motion of the second kind about axes perpendicular to the unique axis. The first kind of motion is described by a traveling wave; the second kind, by a standing wave.

Rotation in the positive direction about the z axis is represented by the φ-dependent factor

$$\Phi_+ = e^{i|M|\varphi}; \tag{2.36}$$

rotation in the negative direction about the z axis, by the factor

$$\Phi_- = e^{-i|M|\varphi}. \tag{2.37}$$

The corresponding complete rotational state functions arise when these are multiplied by the same allowed function of r, θ, and t.

From the superposition principle, the Φ's may be added in any proportions with any relative phasing. When they are added in equal amounts, the rotator turns with equal probability in either direction about the axis. Thus, we may multiply both (2.36) and (2.37) by $1/\sqrt{2}$ and add to obtain

$$\Phi_1 = \frac{1}{\sqrt{2}}\,\Phi_+ + \frac{1}{\sqrt{2}}\,\Phi_- = \sqrt{2}\,\frac{e^{i|M|\varphi} + e^{-i|M|\varphi}}{2}$$

$$= \sqrt{2}\cos|M|\varphi. \tag{2.38}$$

As in Section 1.12, choosing the magnitudes of the two multipliers to be $1/\sqrt{2}$ ensures that the normalization is not altered. We obtain (2.38) rotated by one fourth cycle on multiplying (2.36) by $1/i\sqrt{2}$, (2.37) by $-1/i\sqrt{2}$, and adding:

$$\Phi_2 = \frac{1}{i\sqrt{2}}\,\Phi_+ - \frac{1}{i\sqrt{2}}\,\Phi_- = \sqrt{2}\,\frac{e^{i|M|\varphi} - e^{-i|M|\varphi}}{2i}$$

$$= \sqrt{2}\sin|M|\varphi. \tag{2.39}$$

The general standing wave for a given $|M|$ is the general linear combination of the independent functions Φ_1 and Φ_2.

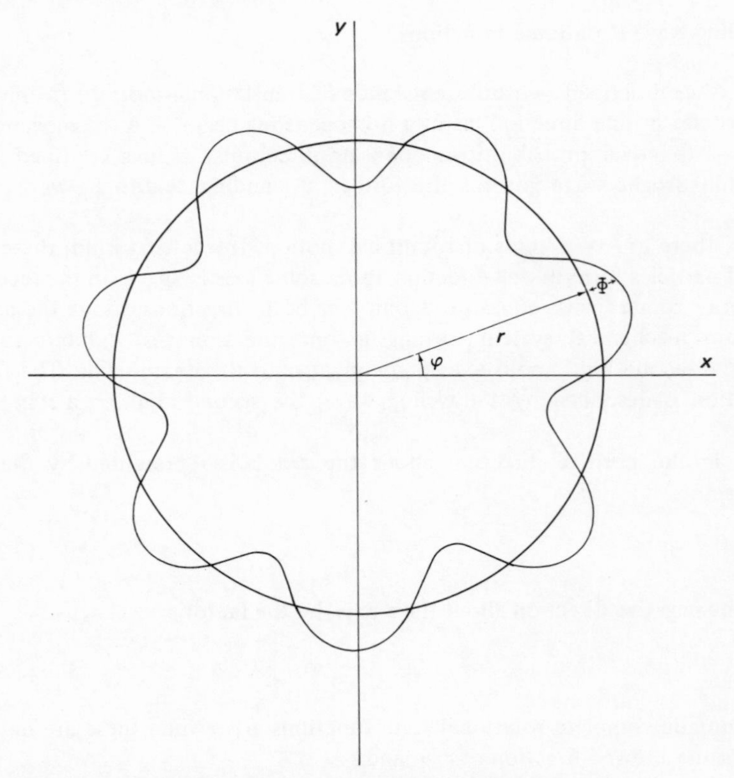

Fig. 2.4. A one-dimensional standing rotational wave.

But for such a combination to be suitable, it must yield a single value at each physical angle. An increase of φ by 2π must cause the trigonometric functions to go through an integral number of cycles. A suitable wave is sketched in Figure 2.4.

Consequently, $|M|$ must be a positive integer or zero. Since two independent functions correspond to each positive $|M|$, and one,

$$\Phi = e^0 = 1, \tag{2.40}$$

exists for $|M|$ equal to zero, there are $2J + 1$ independent standing waves for a given J, just as there are $2J + 1$ independent traveling waves. We have $2J + 1$ independent states for the given J.

Imposing a magnetic field introduces a unique direction, that of the field, about which rotation of the first kind can occur. When the rotating array is polar, because of separation of charges, such rotation leads to a circulation of net charge around the axis pointing in the unique direction. Each distinct angular momentum then generates a different magnetic moment and a different interaction with the field. As a consequence, the field separates the energies of states corresponding to different M's.

2.7. Energy Levels for a Two-Dimensional Rotator

The square of angular momentum and the energy associated with the back-and-forth movement through angle $\theta = \pi/2$ can be deduced from the square of angular momentum and the energy of the rotation along angle φ. Considerations of symmetry and simplicity are sufficient.

We have seen how a two-dimensional rotator is represented by a mass μ moving at constant potential V on the surface of a sphere of radius r. Rotation about the z axis with a definite angular momentum $\mathscr{M}_z$ is governed by function (2.28). The accompanying back-and-forth motion through the xy plane is governed by a standing-wave function $\Theta(\theta)$, which we have not yet constructed.

But from (2.27) and (2.32), we have

$$\mathscr{M}_z{}^2 = M^2\hbar^2 \qquad \text{with} \quad M = -J, -J+1, \ldots, J. \tag{2.41}$$

The square of the quantized angular momentum depends only on the absolute value of M. This value is the same for both contributions to (2.38) and (2.39); so Equation (2.41) for $\mathscr{M}_z{}^2$ applies to each standing-wave state by itself.

When both Θ and Φ are standing waves, there is no essential difference between directions and each of the independent states in a complete set for the given rotational quantum number J occurs with the same probability. To get the average $\mathscr{M}_z{}^2$, we need only add the $M^2\hbar^2$ for each of these states and divide by the number of states.

Thus, we have

$$\overline{\mathcal{M}_z^2} = \frac{\sum_{M=-J}^{M=J} M^2\hbar^2}{2J+1} = \frac{0 + 2\sum_{M=1}^{M=J} M^2\hbar^2}{2J+1} = \frac{2\hbar^2 \sum_1^J M^2}{2J+1}. \tag{2.42}$$

Introducing the identity

$$\sum_1^J M^2 = \frac{J(J+1)(2J+1)}{6}, \tag{2.43}$$

proved in Example 2.3, leads to

$$\overline{\mathcal{M}_z^2} = J(J+1)\frac{1}{3}\hbar^2. \tag{2.44}$$

In a classical rotator, the total angular momentum squared is

$$\mathcal{M}_x^2 + \mathcal{M}_y^2 + \mathcal{M}_z^2. \tag{2.45}$$

In our problem, we need to calculate the average

$$\overline{\mathcal{M}_x^2} + \overline{\mathcal{M}_y^2} + \overline{\mathcal{M}_z^2}. \tag{2.46}$$

But if the system presents a spherically symmetric mean image, as we assume, each of the terms in (2.46) must equal the same quantity; consequently, the total angular momentum squared is three times the average around the z axis:

$$J(J+1)\hbar^2. \tag{2.47}$$

Designating the total angular momentum $\mathcal{M}$, we find that

$$\mathcal{M} = \sqrt{J(J+1)}\,\hbar. \tag{2.48}$$

The corresponding energy is obtained on dividing (2.47) by twice the moment of inertia:

$$E = \frac{\mathcal{M}^2}{2I} = J(J+1)\frac{\hbar^2}{2I}. \tag{2.49}$$

In the *vector model*, angular momenta and possible quantum relationships among them are described by vectors. The magnitude of the total angular momentum for a linear molecule is given by (2.48). The possible projections on a given axis are obtained from (2.34). These differ from 0 by an integral number of $\hbar$ units. The magnitude of a projection cannot be greater than the magnitude of the total angular momentum; so a maximum and a minimum projection exist.

When J is 2, the total angular momentum is a vector $\sqrt{2(3)}\,\hbar$ long. This momentum may have any orientation that yields the projections $2\hbar$, $\hbar$, 0, $-\hbar$, $-2\hbar$ on a given axis, as Figure 2.5 illustrates.

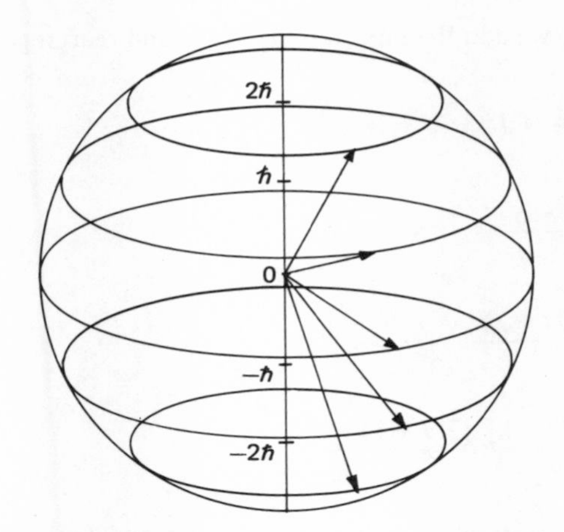

Fig. 2.5. Permissible orientations of the total angular momentum $\sqrt{2(3)}\,\hbar$ according to the vector model.

Example 2.2. Verify the formula

$$\sum_{M=1}^{M=J} M^2 = \frac{J(J+1)(2J+1)}{6}$$

by mathematical induction.

When J is 1, the left side reduces to

$$\sum_{1}^{1} M^2 = 1,$$

while the right side is

$$\frac{J(J+1)(2J+1)}{6} = \frac{1 \cdot 2 \cdot 3}{6} = 1.$$

When J is 2, we have

$$\sum_{1}^{2} M^2 = 1 + 4 = 5$$

and

$$\frac{J(J+1)(2J+1)}{6} = \frac{2 \cdot 3 \cdot 5}{6} = 5.$$

The formula is thus valid when $J = 1, 2$. Suppose it has been proved for a certain J:

$$\sum_{1}^{J} M^2 = \frac{J(J+1)(2J+1)}{6}$$

To obtain the summation to $J + 1$, we add the next term $(J + 1)^2$ and rearrange:

$$\sum_{1}^{J+1} M^2 = \frac{J(J+1)(2J+1)}{6} + J^2 + 2J + 1$$

$$= \frac{2J^3 + 9J^2 + 13J + 6}{6}$$

$$= \frac{(J+1)(J+2)(2J+3)}{6}$$

$$= \frac{[(J+1)]\ [(J+1)+1]\ [2(J+1)+1]}{6}$$

The result has $J + 1$ replacing J everywhere in the formula. So if the relationship holds when the upper limit is J, it must hold when the limit is $J + 1$. But we know it is true when J is 1 and J is 2. Hence it has to be valid when the upper limit is $2 + 1 = 3, 3 + 1 = 4, 4 + 1 = 5$, and so on, through all the positive integers.

2.8. Rotational Spectrum of a Linear Molecule

A turning of a linear system around an axis (the first dimension of its rotation), the accompanying wobbling motion (the second dimension of its rotation), and variations in the moment of inertia are represented by movements of particle μ of Figure 2.2. When the axis for the turning motion is the z axis and the magnitude of the angular momentum thereabout is definite, function Φ of azimuthal angle φ factors out of Ψ and has the form we have found, depending on M or $|M|$. The variation over colatitude θ is described by a function Θ to be determined later. This function depends on both $|M|$ and J.

A molecule with definite rotational energy need not exhibit a definite M or $|M|$. Each of the M's consistent with the prevailing J may be present with a certain probability. Transitions to differing mixtures, and to differing J's, occur during collisions and interactions with neighboring molecules.

When the rotating system is polar, or polarizable, it can also interact with the electromagnetic field, absorbing or emitting photons, or altering photons that happen to be passing by. Let us here consider the limitations restricting the absorption or emission of single photons, and describe the resulting spectrum.

A separation of the center of negative charge from the center of positive charge gives the system a dipole moment. This dominant part of the charge distribution complex is represented by placing the appropriate fractional charge on model particle μ and an opposite charge at the origin. On describing any possible rotation of the physical system, the model particle undergoes acceleration toward the center. From classical theory, such an accelerating charge would radiate energy and the particle would spiral inward, losing angular momentum. For angular momentum to be conserved, a radiated wave would have to carry it away.

In the corresponding quantum picture, the lost momentum is associated with radiated photons. All evidence supports the assumption that only one unit, $\hbar$, is carried by a single photon. Furthermore, the energy of each photon is related to its frequency ν by the Einstein equation

$$E = \hbar\omega = \frac{h}{2\pi} 2\pi\nu = h\nu. \tag{2.50}$$

Usually, one and only one photon is absorbed or emitted at a time, and no angular momentum is associated with relative motion of the photon and the molecule.

Therefore, in the elementary process of radiation, the angular momentum of the molecule jumps by $\hbar$ with respect to some axis, and the limiting $\mathcal{M}_{z'}$ about this axis is shifted by one unit, requiring

$$\Delta J = \pm 1. \tag{2.51}$$

When a unique direction is imposed by an external field, the z axis is made to point in that direction. The change in total angular momentum is then accompanied by either no change or a quantized change about the axis:

$$\Delta M = 0 \qquad \text{or} \qquad \Delta M = \pm 1. \tag{2.52}$$

Since the energy of any state having rotational quantum number J is

$$E_{\text{rot}} = J(J+1)\frac{\hbar^2}{2I}, \tag{2.53}$$

according to (2.49), the changes in energy allowed by selection rule (2.51) are

$$\Delta E = (J_0 \pm 1)(J_0 \pm 1 + 1)\frac{\hbar^2}{2I} - J_0(J_0 + 1)\frac{\hbar^2}{2I}$$

$$= \frac{\hbar^2}{2I}(\pm 2J_0 \pm 1 + 1) = Bhc(\pm 2J_0 \pm 1 + 1), \tag{2.54}$$

with

$$B = \frac{\hbar}{4\pi Ic}. \tag{2.55}$$

In these equations, h is Planck's constant, $\hbar$ is $h/2\pi$, c is the speed of light, J_0 the initial value of quantum number J, and I the moment of inertia of the molecule. The corresponding photon wave number is

$$k = \frac{1}{\lambda} = \frac{h\nu}{hc} = \frac{\pm\Delta E}{hc} = \pm B(\pm 2J_0 \pm 1 + 1). \tag{2.56}$$

When a molecule absorbs a photon rotationally, its quantum number J goes from J_0 to $J_0 + 1$ and

$$\bar{k} = B(+2J_0 + 1 + 1) = 2B(J_0 + 1); \tag{2.57}$$

in emission, J goes from J_0 to $J_0 - 1$ and

$$\bar{k} = -B(-2J_0) = 2BJ_0. \tag{2.58}$$

The wave numbers of consecutive lines in the rotational spectrum of a linear molecule appear at $2B, 4B, 6B, 8B, \ldots$, as Figure 2.6 shows.

This calculation assumes that the bond distances and angles are fixed. Actually, rotation introduces centrifugal forces that stretch and distort the molecule and increase its moment of inertia. The net effect is a reduction in spacing with increasing J. The correction depends on the rigidity of the molecule and will not be estimated here.

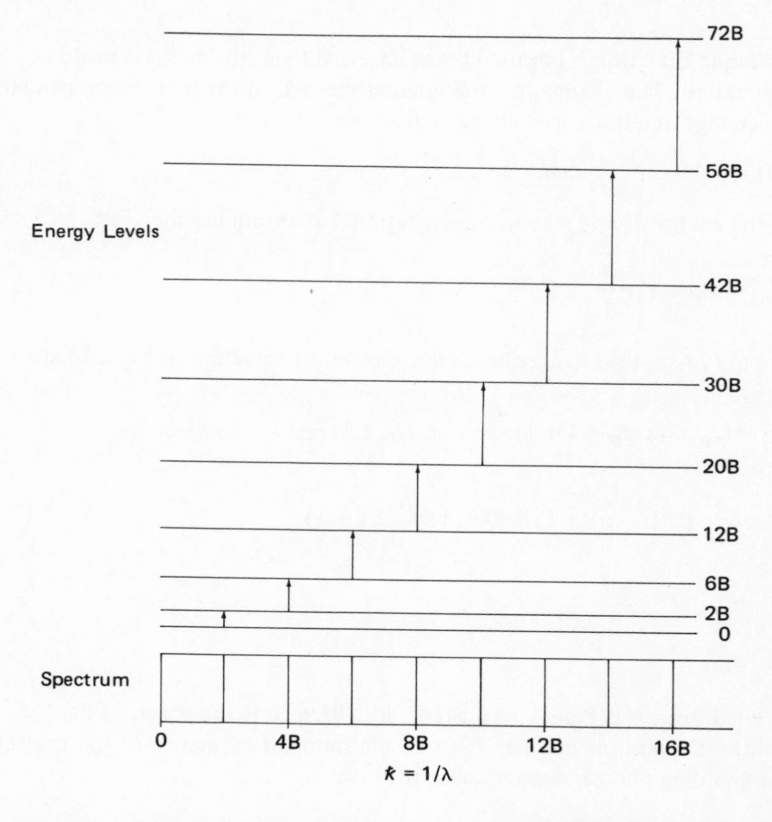

Fig. 2.6. The lower rotational energy levels of a rigid linear molecule and the allowed transitions that occur between these in absorption.

Example 2.3. Evaluate the constant

$$\frac{\hbar}{4\pi c};$$

that is, the coefficient of $1/I$ in Equation (2.55).

Substitute accepted values of the fundamental constants into the expression and carry out the indicated operations:

$$\frac{\hbar}{4\pi c} = \frac{1.0546 \times 10^{-27}\ \text{g cm}^2\ \text{s}^{-1}}{4(3.1416)\,(2.9979 \times 10^{10}\ \text{cm s}^{-1})}$$

$$= 2.7993 \times 10^{-39}\ \text{g cm}.$$

Example 2.4. The absorption spectrum of HCl has maxima spaced about 20.68 cm^{-1} apart in the far infrared. What is the corresponding moment of inertia I?

In the approximation that the moment of inertia does not change with J, Equation (2.57) applies, the spacing equals $2B$, and

$$B = 10.34\ \text{cm}^{-1}.$$

Solving (2.55) for I and employing the number obtained in Example 2.3 leads to

$$I = \frac{\hbar}{4\pi cB} = \frac{2.7993 \times 10^{-39}\ \text{g cm}}{10.34\ \text{cm}^{-1}}$$

$$= 2.707 \times 10^{-40}\ \text{g cm}^2.$$

Example 2.5. What is the bond length in HCl?

Let N be 2 in (2.9), solve for μ, and substitute the atomic weights of H and Cl for m_1 and m_2:

$$\mu = \frac{m_1 m_2}{m_1 + m_2} = \frac{(1.00797\ \text{u})\,(35.453\ \text{u})}{36.461\ \text{u}} = 0.9801\ \text{u}.$$

Then convert to grams using Avogadro's number:

$$\mu = \frac{0.9801\ \text{u}}{6.0221 \times 10^{23}\ \text{u g}^{-1}} = 1.628 \times 10^{-24}\ \text{g}.$$

Solve Equation (2.11) for r and insert the values of I and μ.

$$r = \left(\frac{I}{\mu}\right)^{1/2} = \left(\frac{2.707 \times 10^{-40}\ \text{g cm}^2}{1.628 \times 10^{-24}\ \text{g}}\right)^{1/2} = 1.29 \times 10^{-8}\ \text{cm}.$$

2.9. Broadening of Spectral Lines

A given transition is never observed at just one wave number. Instead, it always appears spread over a range of wave numbers, as a distribution, for the following reasons.

First of all, *experimental errors* arise because of defects in the observer and in the instrument being used. Some of these are random, others systematic. Among the former are errors caused by misjudging the position of a hand or curve, by shaking the apparatus, by random fluctuations in the line voltage. Among the latter are errors arising from the width of a slit, unwanted diffraction effects, expansion or contraction of parts, graininess in the receiver, and so on. As a result, a particular wave number in an incident beam is recorded over a distribution of wave numbers.

Secondly, *interaction broadening* occurs. A molecule by itself exhibits certain properties. But when it interacts with neighboring molecules and with fields, those properties are perturbed by amounts depending on the quantum numbers. Because of continual changes in the configurations, fluctuations in the perturbations appear and persist. Consequently, the energy shift associated with a given change in a quantum number vacillates. The energy of the involved photon (or photons) varies with the distribution around the emitting or absorbing molecule, with the distances to each perturbing neighbor, and with the nature of any added species. For a given gas, the resulting line width varies directly with the pressure at a given temperature, over a considerable range. Similarly in a solution, the broadening varies markedly with the concentration of each solute. Thus, one may speak of pressure broadening or concentration broadening as taking place.

Thirdly, the so-called *temperature broadening* occurs. Molecules in a gas, liquid, or solid exist in potential fields. With each particle confined, the de Broglie wavelength about the minimum potential point (or points) is finite, even when the particle has its minimum energy. So at absolute zero, the motion with respect to an external receiver does not cease. However, this motion does increase with increasing temperature. The movement with respect to an observing instrument at the time of emission produces a Doppler effect. This varies from emitting molecule to emitting molecule. A line width results which increases with temperature.

Fourthly, a *natural width* exists. Each emission or absorption by a molecule involves the formation, or annihilation, of an effectively small, compact, electromagnetic wave: a photon. The packet produced, or destroyed, does not exhibit a unique wave number or frequency. Instead, a Fourier analysis would always reveal it as a superposition of sinusoidal waves, with a certain width over wave numbers. A spectroscope designed to separate different wave numbers would reveal this natural distribution if the preceding effects could be made small enough. The size of the packet, and so the distribution over wave numbers, varies with the mean life of the excited state. A relationship will be obtained in Chapter 6.

2.10. Angular Momenta and the Resultant Energies for a Nonlinear Rotator

The rotational motion of a nonlinear system with respect to arbitrary nonrotating axes may be highly complicated. However, a simple relationship between energy and angular momenta emerges when these axes coincide with the principal axes. Furthermore, making two of the principal moments of inertia equal causes the energy expression to reduce to a function only of the angular momentum about the unique axis and the total angular momentum, both of which can be quantized easily.

Let us consider three or more particles bound together to form a nonlinear system. From the relative positions of the masses, we can calculate the moments and products of inertia with respect to a Cartesian coordinate system erected on the center of mass. Let the Cartesian axes be nonrotating and oriented, at a given instant, so that the products of inertia vanish. The coordinate axes then coincide with the principal axes of the body; the corresponding moments of inertia are called the principal moments.

Let us number the coordinate axes so that these moments satisfy the conditions

$$I_1 \leqslant I_2 \leqslant I_3. \tag{2.59}$$

Now, if the instantaneous angular velocities of the body around the first, second, and third axes are ω_1, ω_2, ω_3, the rotational kinetic energy is

$$\begin{aligned} E &= \frac{1}{2} I_1 \omega_1{}^2 + \frac{1}{2} I_2 \omega_2{}^2 + \frac{1}{2} I_3 \omega_3{}^2 \\ &= \frac{\mathscr{M}_1{}^2}{2I_1} + \frac{\mathscr{M}_2{}^2}{2I_2} + \frac{\mathscr{M}_3{}^2}{2I_3}. \end{aligned} \tag{2.60}$$

In the second equality, we have introduced the components of angular momentum:

$$I_1 \omega_1 = \mathscr{M}_1, \qquad I_2 \omega_2 = \mathscr{M}_2, \qquad I_3 \omega_3 = \mathscr{M}_3. \tag{2.61}$$

In a cylindrically symmetric rotator, two of the principal moments are equal. When these are the larger two,

$$I_2 = I_3, \tag{2.62}$$

the rotator is said to be *prolate*. Eliminating I_3 from (2.60) then yields

$$\begin{aligned} E &= \frac{\mathscr{M}_1{}^2}{2I_1} + \frac{1}{2I_2} (\mathscr{M}_2{}^2 + \mathscr{M}_3{}^2) \\ &= \frac{\mathscr{M}^2}{2I_2} + \mathscr{M}_1{}^2 \left(\frac{1}{2I_1} - \frac{1}{2I_2} \right), \end{aligned} \tag{2.63}$$

where

$$\mathcal{M}^2 = \mathcal{M}_1{}^2 + \mathcal{M}_2{}^2 + \mathcal{M}_3{}^2. \tag{2.64}$$

If I_2 is eliminated instead of I_3, with (2.62), the equivalent form

$$E = \frac{\mathcal{M}^2}{2I_3} + \mathcal{M}_1{}^2\left(\frac{1}{2I_1} - \frac{1}{2I_3}\right) \tag{2.65}$$

is obtained.

When the smaller moments are equal,

$$I_1 = I_2, \tag{2.66}$$

the rotator is said to be *oblate*. Then Equation (2.60) reduces to

$$E = \frac{\mathcal{M}_3{}^2}{2I_3} + \frac{1}{2I_1}(\mathcal{M}_1{}^2 + \mathcal{M}_2{}^2)$$

$$= \frac{\mathcal{M}^2}{2I_1} + \mathcal{M}_3{}^2\left(\frac{1}{2I_3} - \frac{1}{2I_1}\right) \tag{2.67}$$

or

$$E = \frac{\mathcal{M}^2}{2I_2} + \mathcal{M}_3{}^2\left(\frac{1}{2I_3} - \frac{1}{2I_2}\right). \tag{2.68}$$

As before, $\mathcal{M}$ is the total angular momentum.

Figure 2.7 illustrates the nature of the two kinds of bodies. A prolate body is extended along its unique axis, while an oblate body is flattened at the poles (like the earth).

2.11. Quantizations of Prolate and Oblate Rotators

The linear rotator, whose behavior has been discussed, is a prolate rotator with no I_1 or $\mathcal{M}_1$; I_1 in effect vanishes. Allowing I_1 to increase from zero, keeping $I_2 = I_3$, introduces an additional angular momentum $\mathcal{M}_1$, which must be quantized. At the stage where I_1 becomes equal to I_2 and to I_3, the labeling is changed; the unique axis becomes the third axis for more increases in this moment.

Now, motion of each element of the body around its unique axis is like motion of the model particle around the z axis. Consequently, we expect a similar quantization. Just as angular momentum of the particle appears in units of $\hbar$ about the z axis, so also should the angular momentum of all elements of the body appear in units of $\hbar$ about the unique line. In the prolate rotator, the first principal axis, or line, is unique and we have

$$\mathcal{M}_1 = K\hbar \qquad \text{with} \quad K = -J, \ldots, -1, 0, 1, \ldots, J. \tag{2.69}$$

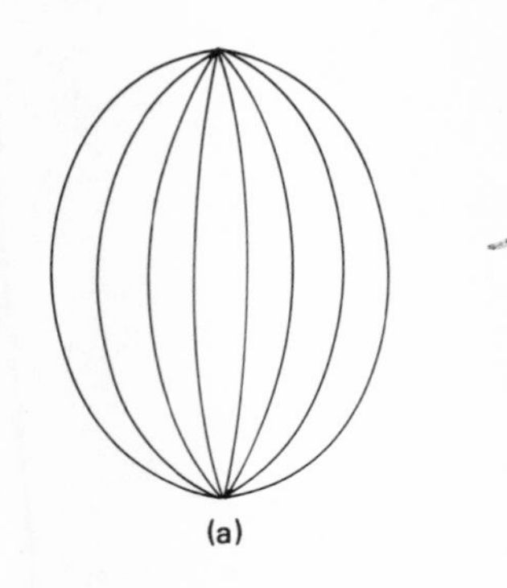

(a)

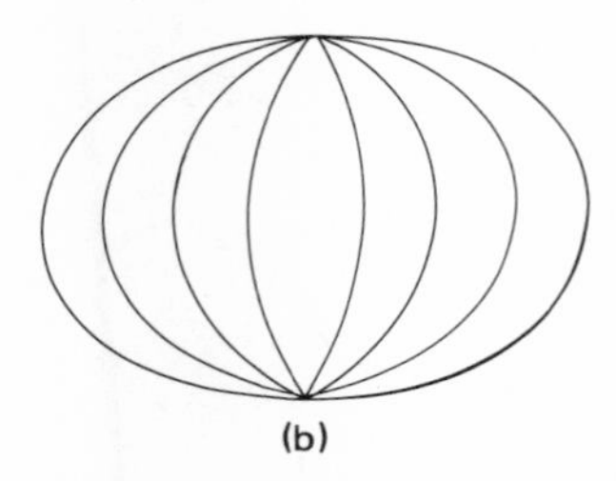

(b)

Fig. 2.7. Spheroids of (a) prolate and (b) oblate forms, with their unique axes vertical.

In the oblate rotator, the third principal axis is unique and we have

$$\mathscr{M}_3 = K\hbar \qquad \text{with} \quad K = -J, \ldots, -1, 0, 1, \ldots, J. \tag{2.70}$$

Furthermore, treating the total angular momentum as in (2.48) yields

$$\mathscr{M} = \sqrt{J(J+1)}\,\hbar \qquad \text{with} \quad J = 0, 1, 2, \ldots. \tag{2.71}$$

Expressions (2.69) and (2.71) reduce (2.63) for the prolate rotator to

$$\begin{aligned} E &= J(J+1)\,\frac{\hbar^2}{2I_2} + K^2\hbar^2\left(\frac{1}{2I_1} - \frac{1}{2I_2}\right) \\ &= hc\,[J(J+1)B + K^2(A - B)]. \end{aligned} \tag{2.72}$$

The *rotational constants A, B, C* have been introduced as

$$A = \frac{\hbar}{4\pi I_1 c}, \qquad B = \frac{\hbar}{4\pi I_2 c}, \qquad C = \frac{\hbar}{4\pi I_3 c}. \tag{2.73}$$

Dividing (2.72) by hc yields the energy in reciprocal centimeters:

$$E_{\text{rot}} = J(J+1)B + K^2(A - B). \tag{2.74}$$

Since $B = C$, we also have

$$E_{\text{rot}} = J(J+1)C + K^2(A - C). \tag{2.75}$$

For the oblate rotator, we similarly obtain

$$E_{\text{rot}} = J(J+1)B + K^2(C - B) \tag{2.76}$$

or

$$E_{\text{rot}} = J(J+1)A + K^2(C - A) \tag{2.77}$$

when the energy is in reciprocal centimeters. The nature of these levels is illustrated by Figure 2.8.

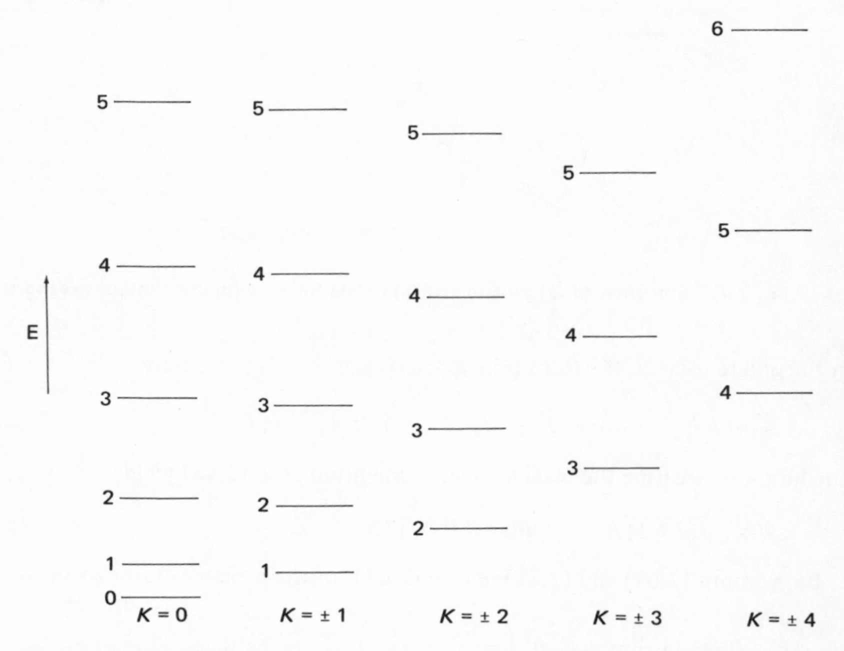

Fig. 2.8. Energy levels of an oblate rotator for which $A = 2C$. The number at left of each line is the J for the level.

Transitions among the various rotational states are induced by collisions and interactions with neighboring molecules. They may also be caused by the electromagnetic field when certain conditions are met.

The systems we are considering in this section possess a rotational symmetry about a unique axis. As a consequence, they do not exhibit any unbalanced shift of charge away from this axis. Any dipole moment the given system has must therefore lie along this axis and rotation about the axis is like the rotation of a nonpolar linear molecule. The electromagnetic field cannot interact through the dipole moment on it and we have the selection rule

$$\Delta K = 0. \tag{2.78}$$

When the molecule does exhibit a dipole moment along the axis, the total angular momentum can be changed by the field as described in Section 2.8. So we have the selection rule

$$\Delta J = \pm 1. \tag{2.79}$$

Combining both rules with either Equation (2.74) or (2.76) yields

$$k = 2B(J_0 + 1) \tag{2.80}$$

for absorption and

$$k = 2BJ_0 \tag{2.81}$$

for emission.

When two principal moments are equal, a molecule rotates as a cylindrically symmetric top; when all three principal moments are equal, it rotates as a spherically symmetric top. Pyramidal NH_3 is an example of a molecule with two moments equal; tetrahedral CH_4 and octahedral SF_6 are examples of molecules with all three moments equal.

Example 2.6. Obtain the energy levels of a molecule for which $I_1 = I_2 = I_3$.

A spherical-top molecule can be considered as a special cylindrical-top molecule. Indeed, we obtain the desired energy levels from (2.77) on letting

$$C = A.$$

Then

$$E_{\text{rot}} = J(J + 1)A,$$

where

$$A = \frac{\hbar}{4\pi Ic}$$

and energy E_{rot} is in reciprocal centimeters.

2.12. The Asymmetric Rotator

A general nonlinear rotator behaves in a complicated manner that we will not analyze in detail. Instead, we will relate its states to those of similar prolate and oblate rotators.

Consider a multiparticle system bound together so that its principal moments, labeled as in (2.59), are all different. Let the rotational constants A, B, C be related to these moments by Equations (2.73), as before. Note that there is now no unique axis about which the quantization determined by K can exist. But since there is still a total angular momentum, the rotational quantum number J is still pertinent.

Furthermore, the asymmetric system may be altered to make its intermediate moment I_2 equal to its smallest moment I_1. Rotational constant B would then be increased to A and the levels would move up to those of an oblate symmetric top. Alternatively, one may distort the molecule to make I_2 equal to the largest moment I_3. Then B would be decreased to C and the levels would move down to those of a prolate symmetric top.

Each state of the asymmetric rotator is distinguished by its J and by the $|K|$'s of the prolate and oblate states to which it is thus related. (See Figure 2.9 for an example.) Note how the $|K|$'s are placed as subscripts on the J in the label for a state. Thus, the 3_{21} level in the example correlates with the prolate $J = 3$, $|K| = 2$ level and with the oblate $J = 3$, $|K| = 1$ level.

The wave function for a general rotational state depends on the three angles that measure the orientation of the molecule in space. We will not construct such a Ψ or the related energy levels here, but merely quote the results in Table 2.1.

An asymmetric molecule usually exhibits some dipole moment. This moment interacts with an electromagnetic field, leading to absorption or emission of photons. In each step, the angular momentum $\hbar$ of the photon involved either increases

TABLE 2.1.
Energy levels of an asymmetric rotator

J_{po}	E_{rot}
0_{00}	0
1_{01}	$B + C$
1_{11}	$A + C$
1_{10}	$A + B$
2_{02}	$2A + 2B + 2C - 2[(B - C)^2 + (A - C)(A - B)]^{1/2}$
2_{12}	$A + B + 4C$
2_{11}	$A + 4B + C$
2_{21}	$4A + B + C$
2_{20}	$2A + 2B + 2C + 2[(B - C)^2 + (A - C)(A - B)]^{1/2}$
3_{03}	$2A + 5B + 5C - 2[4(B - C)^2 + (A - B)(A - C)]^{1/2}$
3_{13}	$5A + 2B + 5C - 2[4(A - C)^2 - (A - B)(B - C)]^{1/2}$
3_{12}	$5A + 5B + 2C - 2[4(A - B)^2 + (A - C)(B - C)]^{1/2}$
3_{22}	$4A + 4B + 4C$
3_{21}	$2A + 5B + 5C + 2[4(B - C)^2 + (A - B)(A - C)]^{1/2}$
3_{31}	$5A + 2B + 5C + 2[4(A - C)^2 - (A - B)(B - C)]^{1/2}$
3_{30}	$5A + 5B + 2C + 2[4(A - B)^2 + (A - C)(B - C)]^{1/2}$

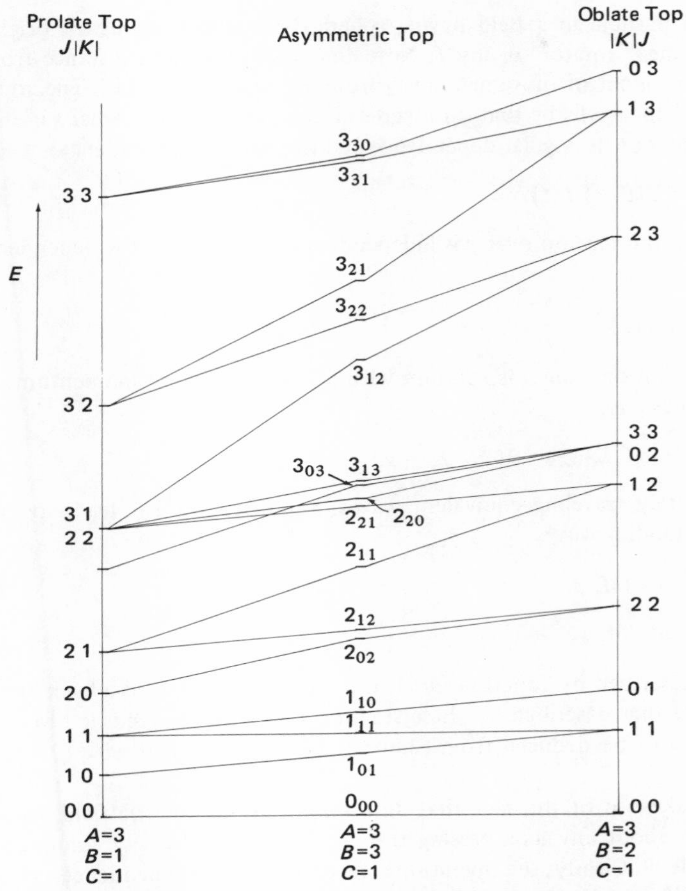

Fig. 2.9. Correlation of low-lying symmetric-rotator energy levels with levels of corresponding prolate and oblate rotators.

or decreases J by 1, or merely acts to rearrange the total angular momentum, of the molecule. Thus, we have the selection rule

$$\Delta J = \pm 1 \qquad \text{and} \qquad \Delta J = 0. \tag{2.82}$$

Rules also exist for combining the subscripts, but these will not be considered.

2.13. Dependence of Ψ on θ in a Spherically Symmetric Field

Let us now return to the one-particle model for a *linear rotator* and complete construction of the angular part of its state function. How this part varies with colatitude θ will be obtained, through symmetry arguments, from how it varies with azimuthal angle φ, in representative pure states.

Consider a particle in a field having spherical symmetry about the origin, the model for a linear rotator of any I. Note that motion a given distance from the center is independent of movement away from or toward the center. The argument in Section 1.7 then tells us that, in a state at given energy, the radial variation of Ψ is independent of its angular dependence and its temporal dependence:

$$\Psi = R(r)\, Y(\theta, \varphi)\, T(t). \tag{2.83}$$

If, in addition, the motion over φ is independent of that over θ, the dependence on φ factors out:

$$Y = \Theta(\theta)\, \Phi(\varphi). \tag{2.84}$$

When this motion over φ is rotation with a definite angular momentum, Equation (2.28) holds and

$$\Phi = e^{iM\varphi}, \qquad \text{where} \quad M = -J, \ldots, -1, 0, 1, \ldots, J. \tag{2.85}$$

Mixing oppositely traveling equivalent waves, as in Section 2.6, leads to the two independent standing waves

$$\Phi = \sqrt{2} \cos |M| \varphi, \tag{2.86}$$

$$\Phi = \sqrt{2} \sin |M| \varphi. \tag{2.87}$$

The motion described by function $\Theta(\theta)$, for a given J and $|M|$, is a back-and-forth movement like that described by these standing waves. We are thus led to consider how its form can be deduced from (2.86) and (2.87), together with (2.85) when $M = 0$.

We will make use of the fact that, in a spherically symmetric field, rotating a state function about any axis passing through the center by any amount always yields a suitable Ψ. Clearly, the operation affects the angular dependences. However, the rotated $Y(\theta, \varphi)$ must be a linear combination (a superposition) of the original Y's for the same J, because mixtures of them describe all possibilities.

The pertinent rotations can be carried out readily in rectangular coordinates. Through Figure 2.2, we see that these coordinates are obtained from the spherical coordinates on projecting line r on the three axes:

$$x = r \sin\theta \cos\varphi, \tag{2.88}$$

$$y = r \sin\theta \sin\varphi, \tag{2.89}$$

$$z = r \cos\theta. \tag{2.90}$$

Dividing (2.90) by r and taking the inverse cosine of both sides yields

$$\theta = \cos^{-1} \frac{z}{r}. \tag{2.91}$$

Consequently, a function of θ is a function of z at a given r. From (2.88) and (2.89), we also have

$$\cos\varphi = \frac{x}{r\sin\theta}, \tag{2.92}$$

$$\sin\varphi = \frac{y}{r\sin\theta}. \tag{2.93}$$

When $J = 0$, M has to be zero and Equation (2.85) reduces to

$$\Phi = 1. \tag{2.94}$$

The corresponding (2.84) is at most a function of θ alone:

$$Y_a = \Theta_{00}(\theta) = F_0\left(\cos^{-1}\frac{z}{r}\right). \tag{2.95}$$

Rotating (2.95) about any axis inclined to the z axis at the origin leads to a function depending on φ as well as on θ. But the only standard form when $J = 0$ is (2.95) itself, which does not depend on φ. Therefore, F_0 must be constant. If the constant is chosen so that the integral of Y^2 over the solid angle about the center is 4π, we have

$$Y_s = 1 \cdot 1 = 1, \tag{2.96}$$

where subscript s indicates that $J = 0$.

When $J = 1$, M can be zero. Then Equations (2.85) and (2.84) yield

$$Y_a = \Theta_{10}(\theta) = F_0\left(\cos^{-1}\frac{z}{r}\right) \tag{2.97}$$

where the subscripts on Θ are the J and $|M|$ for the function, while the subscripts on Y and F are merely temporary labels. When $J = 1$ and $|M| = 1$, Equations (2.86), (2.87), and (2.84) yield

$$Y_b = \sqrt{2}\,\Theta_{11}(\theta)\cos\varphi = \sqrt{2}\,\Theta_{11}(\theta)\frac{x}{r\sin\theta} = F_1\left(\cos^{-1}\frac{z}{r}\right)\frac{x}{r}, \tag{2.98}$$

$$Y_c = \sqrt{2}\,\Theta_{11}(\theta)\sin\varphi = \sqrt{2}\,\Theta_{11}(\theta)\frac{y}{r\sin\theta} = F_1\left(\cos^{-1}\frac{z}{r}\right)\frac{y}{r}. \tag{2.99}$$

Rotating function (2.98) by replacing x with y, y with z, and z with x leads to

$$Y_d = F_1\left(\cos^{-1}\frac{x}{r}\right)\frac{y}{r}, \tag{2.100}$$

a form containing y/r as a factor. Since this factor is only present in (2.99) out of the set for $J = 1$, we must have

$$Y_d = aY_c \tag{2.101}$$

or

$$F_1\left(\cos^{-1}\frac{x}{r}\right)\frac{y}{r} = aF_1\left(\cos^{-1}\frac{z}{r}\right)\frac{y}{r} \tag{2.102}$$

where a is some constant.

Equation (2.102) can be true only if function F_1 is a constant. Letting the constant be A and representing Y_b as Y_{p_x}, we write

$$Y_{p_x} = A\frac{x}{r}. \tag{2.103}$$

Since F_1 in (2.99) is the same function, we rewrite Y_c in the form

$$Y_{p_y} = A\frac{y}{r}. \tag{2.104}$$

A rotation that replaces y by z converts (2.104) to an explicit form for (2.97):

$$Y_{p_z} = A\frac{z}{r}. \tag{2.105}$$

The p part of the subscript on each Y indicates that $J = 1$; the variable part, x, y, z, indicates the form of the polynomial in the numerator of the expression. For the integral of Y^2 over the solid angle about the center to be 4π, we must have

$$A = \sqrt{3}. \tag{2.106}$$

When $J = 2$ and $M = 0$, Equations (2.85) and (2.84) combine to give

$$Y_a = \Theta_{20}(\theta) = F_0\left(\cos^{-1}\frac{z}{r}\right). \tag{2.107}$$

Again, the subscripts on Θ are the J and $|M|$ for the state, while the subscripts on Y and F now indicate a different set of functions.

When $J = 2$ and $|M| = 1$, Equations (2.86), (2.87), and (2.84) combine to form

$$Y_b = \sqrt{2}\,\Theta_{21}(\theta)\cos\varphi = \sqrt{2}\,\Theta_{21}(\theta)\frac{x}{r\sin\theta} = F_1\left(\cos^{-1}\frac{z}{r}\right)\frac{x}{r}, \tag{2.108}$$

$$Y_c = \sqrt{2}\,\Theta_{21}(\theta)\sin\varphi = \sqrt{2}\,\Theta_{21}(\theta)\frac{y}{r\sin\theta} = F_1\left(\cos^{-1}\frac{z}{r}\right)\frac{y}{r}; \tag{2.109}$$

when $J = 2$ and $|M| = 2$, they yield

$$Y_d = \sqrt{2}\,\Theta_{22}(\theta)\cos 2\varphi = \sqrt{2}\,\Theta_{22}(\theta)(\cos^2\varphi - \sin^2\varphi)$$

$$= \sqrt{2}\,\Theta_{22}(\theta)\frac{x^2 - y^2}{r^2\sin^2\theta} = F_2\left(\cos^{-1}\frac{z}{r}\right)\frac{x^2 - y^2}{r^2}, \tag{2.110}$$

$$Y_e = \sqrt{2}\,\Theta_{22}(\theta)\sin 2\varphi = \sqrt{2}\,\Theta_{22}(\theta)\,2\sin\varphi\cos\varphi$$

$$= \sqrt{2}\,\Theta_{22}(\theta)\frac{2xy}{r^2\sin^2\theta} = F_2\left(\cos^{-1}\frac{z}{r}\right)\frac{2xy}{r^2}. \tag{2.111}$$

Rotating function (2.108) by replacing x with y, y with z, and z with x leads to

$$Y_f = F_1\left(\cos^{-1}\frac{x}{r}\right)\frac{y}{r}. \tag{2.112}$$

If (2.112) were interpreted as a multiple of (2.109), expression F_1 would be constant and (2.108) would be Y_{p_x}. Since we have already taken care of this possibility (it occurs with $J = 1$), we choose the alternative. We note that Y_e is the only function in the set that contains y/r times a function of x/r and take

$$Y_f = aY_e, \tag{2.113}$$

whence

$$F_1\left(\cos^{-1}\frac{x}{r}\right)\frac{y}{r} = aF_2\left(\cos^{-1}\frac{z}{r}\right)\frac{2xy}{r^2}. \tag{2.114}$$

Equation (2.114) can be true only if function F_2 is a constant,

$$F_2 = B, \tag{2.115}$$

and $F_1(\cos^{-1} x/r)$ a constant times x/r. To be consistent, we take

$$F_1\left(\cos^{-1}\frac{x}{r}\right) = 2B\frac{x}{r}. \tag{2.116}$$

Then

$$F_1\left(\cos^{-1}\frac{z}{r}\right) = 2B\frac{z}{r} \tag{2.117}$$

and Equations (2.111), (2.109), (2.108), (2.110) yield

$$Y_{d_{xy}} = 2B\frac{xy}{r^2}, \tag{2.118}$$

$$Y_{d_{yz}} = 2B\frac{yz}{r^2}, \tag{2.119}$$

$$Y_{d_{zx}} = 2B\,\frac{zx}{r^2}, \tag{2.120}$$

$$Y_{d_{x^2-y^2}} = B\,\frac{x^2 - y^2}{r^2}. \tag{2.121}$$

An explicit form for (2.107) is constructed as follows. Rotate (2.121) by replacing x with y, y with z, z with x, obtaining

$$Y_{d_{y^2-z^2}} = B\,\frac{y^2 - z^2}{r^2} = B(\sin^2\theta\,\sin^2\varphi - \cos^2\theta). \tag{2.122}$$

Also, replace x with z, y with x, z with y, obtaining

$$Y_{d_{z^2-x^2}} = B\,\frac{z^2 - x^2}{r^2} = B(\cos^2\theta - \sin^2\theta\,\cos^2\varphi). \tag{2.123}$$

Note that from neither of these Y's does a $\Phi(\varphi)$ factor. But a constant times the difference (2.123) minus (2.122) is a Y with no dependence on φ:

$$Y_{d_{3z^2-r^2}} = C\,\frac{2z^2 - x^2 - y^2}{r^2} = C\,\frac{3z^2 - r^2}{r^2}. \tag{2.124}$$

Set (2.118)–(2.121) and (2.124) is complete because it contains a function for $M = 0$ and two independent functions for each nonzero $|M|$ up to and including $|M| = J$. The d part of the subscript on each Y indicates that $J = 2$; the variable part gives the form of the polynomial in the numerator of the explicit expression. For the integral of Y^2 over the solid angle about the center to be 4π, we must also have

$$B = \frac{\sqrt{15}}{2}, \tag{2.125}$$

$$C = \frac{\sqrt{5}}{2}. \tag{2.126}$$

2.14. Explicit State Functions for Angular Motion in a Central Field

A free linear rotator, molecule or hydrogen-like atom, is modeled by a single particle in a spherically symmetric field. In each pure energy state, the radial motion of the particle separates from (is independent of) the angular motion, and the spatial part of its wave function has the form

$$\psi = R(r)\,Y(\theta, \varphi). \tag{2.127}$$

In Section 2.13 we constructed complete Y's from the dependence on φ which we had found earlier. The results are summarized in Table 2.2. Substitutions (2.88), (2.89), (2.90) change each of the listed state functions to a form that factors further:

$$\psi = R(r)\,\Theta(\theta)\,\Phi(\varphi). \tag{2.128}$$

TABLE 2.2.
Low angular-quantum-number state functions for a particle in a spherically symmetric field

Quantum numbers		ψ
J or l	$\lvert M\rvert$ or $\lvert m\rvert$	
0	0	$R_{n0}(r)$
1	0	$R_{n1}(r)\sqrt{3}\,\dfrac{z}{r}$
1	1	$R_{n1}(r)\sqrt{3}\,\dfrac{x}{r}$
1	1	$R_{n1}(r)\sqrt{3}\,\dfrac{y}{r}$
2	0	$R_{n2}(r)\,\dfrac{\sqrt{5}}{2}\,\dfrac{3z^2 - r^2}{r^2}$
2	1	$R_{n2}(r)\sqrt{15}\,\dfrac{zx}{r^2}$
2	1	$R_{n2}(r)\sqrt{15}\,\dfrac{yz}{r^2}$
2	2	$R_{n2}(r)\sqrt{15}\,\dfrac{xy}{r^2}$
2	2	$R_{n2}(r)\,\dfrac{\sqrt{15}}{2}\,\dfrac{x^2 - y^2}{r^2}$

The φ-dependent part appears as (2.86) or (2.87); the θ-dependent part, a form listed in Table 2.3. By superposing these solutions, a person can construct the traveling wave functions for each M. It is found that the colatitude factor multiplying (2.85) is the one listed for the pertinent J and $\lvert M\rvert$.

The radial factor depends on the nature of $V(r)$. Explicit forms will be obtained in subsequent chapters. In general, its quantization entails the introduction of an additional quantum number, which we have here designated n. Quantum numbers J and $\lvert M\rvert$ or M are generally employed to describe rotational states of molecules, as we have seen.

For a hydrogen-like atom, the rotational quantum number J is replaced by the *azimuthal quantum number* l. The capital letter M for the *magnetic quantum number* is replaced by the small letter m. The number n associated with quantization of $R(r)$ is called the *principal quantum number*. When l is 0, 1, 2, 3, 4, . . . , we call the state function an *s, p, d, f, g, . . . orbital.*

TABLE 2.3.
Colatitude factors for a two-dimensional rotator and for a hydrogen-like atom

Quantum numbers		Θ
J or l	$\lvert M \rvert$ or $\lvert m \rvert$	
0	0	1
1	0	$\sqrt{3}\cos\theta$
1	1	$\sqrt{\frac{3}{2}}\sin\theta$
2	0	$\frac{\sqrt{5}}{2}(3\cos^2\theta - 1)$
2	1	$\sqrt{\frac{15}{2}}\cos\theta\sin\theta$
2	2	$\frac{1}{2}\sqrt{\frac{15}{2}}\sin^2\theta$

Example 2.7. Show that rotating a p_x state function through an arbitrary angle transforms it to a linear combination of p_x, p_y, and p_z functions.

State functions ψ_{p_x}, ψ_{p_y}, and ψ_{p_z} consist of (2.103), (2.104), and (2.105) multiplied by the appropriate radial factor $R_{n1}(r)$, as in Table 2.2. Let us rotate ψ_{p_x} together with an attached primed set of axes to an arbitrary orientation, obtaining

$$\psi_{p_{x'}} = Ax' \frac{R_{n1}(r)}{r}.$$

Let us also draw a radius vector from the origin to a point with the coordinates (x', y', z') and (x, y, z). We have

$$\mathbf{r} = x'\hat{\mathbf{x}}' + y'\hat{\mathbf{y}}' + z'\hat{\mathbf{z}}' = x\hat{\mathbf{x}} + y\hat{\mathbf{y}} + z\hat{\mathbf{z}}.$$

Dot multiply both sides of the last equality by $\hat{\mathbf{x}}'$,

$$x' = x\hat{\mathbf{x}}\cdot\hat{\mathbf{x}}' + y\hat{\mathbf{y}}\cdot\hat{\mathbf{x}}' + z\hat{\mathbf{z}}\cdot\hat{\mathbf{x}}',$$

and substitute the result into the rotated function:

$$\psi_{p_{x'}} = \hat{\mathbf{x}}\cdot\hat{\mathbf{x}}'\,Ax\frac{R_{n1}(r)}{r} + \hat{\mathbf{y}}\cdot\hat{\mathbf{x}}'\,Ay\frac{R_{n1}(r)}{r} + \hat{\mathbf{z}}\cdot\hat{\mathbf{x}}'\,Az\frac{R_{n1}(r)}{r}$$

$$= \hat{\mathbf{x}}\cdot\hat{\mathbf{x}}'\,\psi_{p_x} + \hat{\mathbf{y}}\cdot\hat{\mathbf{x}}'\,\psi_{p_y} + \hat{\mathbf{z}}\cdot\hat{\mathbf{x}}'\,\psi_{p_z}.$$

Example 2.8. What function is paired with

$$B(x^2 - y^2)\frac{R_{n2}(r)}{r^2}$$

to eliminate one direction of travel around the z axis?

A general standing wave with definite J and $|M|$ has the form

$$\psi_1 = R(r)\,\Theta(\theta)\sqrt{2}\cos|M|(\varphi - \alpha).$$

Rotating this by one-fourth cycle yields

$$\psi_2 = R(r)\,\Theta(\theta)\sqrt{2}\sin|M|(\varphi - \alpha).$$

Mixing ψ_1 and ψ_2 in equivalent amounts, with an appropriate relative phasing,

$$\psi = \frac{1}{\sqrt{2}}\psi_1 \pm i\frac{1}{\sqrt{2}}\psi_2,$$

leads to the traveling-wave form

$$\begin{aligned}\psi &= R(r)\,\Theta(\theta)\,[\cos|M|(\varphi - \alpha) \pm i\sin|M|(\varphi - \alpha)]\\ &= R(r)\,\Theta(\theta)\,\mathrm{e}^{\mp i|M|\alpha}\,\mathrm{e}^{\pm i|M|\varphi}\\ &= R(r)\,\Theta(\theta)\,(\text{constant})\,\mathrm{e}^{iM\varphi}\end{aligned}$$

in which the φ factor appears as in (2.85).

Substituting expressions (2.88) and (2.89) for x and y into the given function and reducing produces

$$\begin{aligned}B(x^2 - y^2)\frac{R_{n2}(r)}{r^2} &= Br^2(\sin^2\theta)(\cos^2\varphi - \sin^2\varphi)\frac{R_{n2}(r)}{r^2}\\ &= B(\sin^2\theta)(\cos 2\varphi)R_{n2}(r),\end{aligned}$$

which is ψ_1 with $|M| = 2$ and $\alpha = 0$. On replacing $\cos 2\varphi$ with $\sin 2\varphi$, we obtain the corresponding ψ_2,

$$\begin{aligned}B(\sin^2\theta)(\sin 2\varphi)R_{n2}(r) &= B(\sin^2\theta)2(\sin\varphi)(\cos\varphi)R_{n2}(r)\\ &= 2B\frac{xy}{r^2}R_{n2}(r),\end{aligned}$$

which is the desired function.

Example 2.9. Show that for the integral of Y^2 over the solid angle about the center to be 4π, coefficient A in (2.105) must be $\sqrt{3}$.

In spherical coordinates, the volume element is

$$\mathrm{d}^3\mathbf{r} = (r\,\mathrm{d}\theta)(r\sin\theta\,\mathrm{d}\varphi)\,\mathrm{d}r = r^2\,\mathrm{d}\Omega\,\mathrm{d}r.$$

Consequently, an element of solid angle is

$$d\Omega = \sin\theta \, d\theta \, d\varphi.$$

Integrating the square of (2.105) over the complete solid angle about the center leads to

$$\int_0^{2\pi}\int_0^{\pi} A^2 \cos^2\theta \sin\theta \, d\theta \, d\varphi = A^2\varphi\Big|_0^{2\pi}\int_0^{\pi}\cos^2\theta\sin\theta\, d\theta$$

$$= 2\pi A^2\left(-\frac{\cos^3\theta}{3}\right)\Big|_0^{\pi} = 2\pi A^2\left(+\frac{2}{3}\right) = 4\pi\frac{A^2}{3}.$$

When this result is set equal to 4π, we obtain

$$A = \sqrt{3}.$$

Discussion Questions

2.1. What different kinds of motion does a composite particle exhibit?

2.2. When is one mode of motion independent of the other modes? How is this independence reflected in the state function or wave function?

2.3. Why is the free translational motion of a molecule independent of its other motions? Why is the free rotational motion only approximately independent?

2.4. How does a single particle model (or represent) (a) the translation of a molecule, (b) the rotation of a linear molecule?

2.5. Why does movement of the particle of mass μ radially carry it over non-equivalent points?

2.6. Explain in what sense points along a longitudinal circle equidistant from a center of symmetry are not equivalent. Why are points on a latitudinal circle equivalent?

2.7. Explain how Ψ for a particle varies with r, θ, φ, and t when the potential is spherically symmetric around the origin and the particle is in a definite energy state with a unique angular momentum with respect to the z axis.

2.8. What wave results from integrating the relationship in Question 2.7?

2.9. Why is the coefficient of φ, iM, in the exponent of $\Phi(\varphi)$, constant? To what angular momentum is M related? Explain.

2.10. Justify the assumption that ψ has a single value at each point in physical space.

2.11. How is the assumption in Question 2.10 used to determine the allowed values for M? What is J?

2.12. How is the energy of a one-dimensional rotator quantized?

2.13. Describe the two extreme kinds of (a) translation, (b) rotation, which occur.

2.14. Why must standing-wave motion always accompany pure traveling-wave rotation?

2.15. Why should each independent state consistent with a given rotational energy occur with the same probability?

2.16. How is $\overline{\mathcal{M}_z^2}$ for a given J constructed? How are the corresponding $\overline{\mathcal{M}_x^2}$ and $\overline{\mathcal{M}_y^2}$ found?

2.17. Why is the total angular momentum a definite quantity? How is it obtained from the preceding results?

2.18. What is the corresponding rotational energy?

2.19. Describe the vector model for angular momenta. What is the significance of the model?

2.20. Why is a rotational state usually a mixture of states with various M's?

2.21. When can a molecule interact with an electromagnetic field through its rotation? Why does such interaction lead to a gain or loss of the molecule's angular momentum?

2.22. Why does J change by 1 in a rotational transition involving absorption or emission of a photon? How can M change in the process?

2.23. Explain how rotational lines are spaced in the absorption or emission spectra. What is the effect of centrifugal distortion on the spacing?

2.24. What cause spectral lines to appear as bands? Can each kind of broadening be reduced? If so, how?

2.25. How is the energy of a nonlinear rotating molecule related to its angular momenta?

2.26. Is a molecule cylindrically symmetric when two of its principal moments of inertia are equal? Is it spherically symmetric when all three principal moments are equal?

2.27. When does a molecule have a unique axis? How is angular momentum quantized about such an axis?

2.28. How is the total angular momentum quantized?

2.29. Describe the energy levels of a rotator in which two principal moments of inertia are equal.

2.30. How are the energy levels of an asymmetric rotator related to the levels of prolate and oblate rotators?

2.31. Why does the asymmetric rotator not exhibit 2_{01}, 2_{22}, or 2_{10} states?

2.32. Explain how the expressions for the 1_{11} and 1_{10} levels may be obtained from the expression for the 1_{01} level; those for 2_{11}, 2_{21} from 2_{12}; those for 3_{13}, 3_{12} from 3_{03}.

2.33. Explain why (a) the temporal dependence and (b) the angular dependence factor out of the state function for the model particle.

2.34. Explain when the angular dependence factors so that $Y = \Theta(\theta)\,\Phi(\varphi)$. Why does the Θ factor generally differ from any possible Φ factor?

2.35. Why is $\Theta = 1$ when $J = 0$ and $|M| = 0$?

2.36. Why is $\Theta = \sqrt{3}\cos\theta$ when $J = 1$ and $|M| = 0$? Why is $\Theta = \sqrt{\frac{3}{2}}\sin\theta$ when $J = 1$ and $|M| = 1$?

2.37. Explain how the standard forms for the d orbitals can be derived from the standard Φ factors.

2.38. How can vector algebra be employed in rotating an orbital?

Problems

2.1. How many cycles does the rotational wave function $e^{iM\varphi}\,e^{-i\omega t}$ execute when φ increases by 2π?

2.2. A linear molecule rotates about an axis passing through its center of mass with an angular momentum of $3\hbar$. Construct the corresponding factor in the wave function, normalized so the integral of the corresponding factor in the probability density over a complete turn is 2π.

2.3. Independent AB molecules are rotating freely at various energies. What are $\overline{\mathcal{M}_x^2}$, $(\overline{\mathcal{M}_x^2})^{1/2}$, and $\overline{\mathcal{M}_x}$ for a rotational level that is well occupied?

2.4. If the separation between successive maxima in the rotational absorption band for $^1H^{35}Cl$ is 20.68 cm^{-1}, what is the corresponding separation for $^2H^{35}Cl$? The masses of ^1H, ^2H, and ^{35}Cl are 1.007825, 2.014102, and 34.96885 u, respectively.

2.5. If the bond length in $^{127}I^{35}Cl$ is 2.32 Å, where are the first three lines in its rotational spectrum? The mass of ^{127}I is 126.9044 u.

2.6. Calculate the rotational constants for $^{11}B^{19}F_3$. In this molecule each fluorine is at the corner of an equilateral triangle, while the boron is at the center, with each B–F bond 1.29 Å long. The masses of ^{11}B and ^{19}F are 11.0093 and 18.9984 u.

2.7. Substitute the results from Problem 2.6 into the appropriate rotational energy formula. Then locate and sketch the lowest rotational levels of the molecule.

2.8. At what B does the 3_{03} level cross (a) the 2_{20} level, (b) the 2_{21} level, when $A = 3C$ in an asymmetric rotator?

2.9. Rotate the nd_{xy} state function around the z axis by an arbitrary angle and express the result as a linear combination of nd orbitals.

2.10. Find the f function to be paired with

$$E\,[x(x^2 - 3y^2)]\;\frac{R_{n3}(r)}{r^3}$$

to represent the mass μ circling the z axis in one direction only.

2.11. Calculate the reduced mass of the system in which an electron revolves around (a) a positron (as in positronium), (b) a proton (as in the hydrogen atom). The masses of e and p are 0.000548585 and 1.0072765 u, respectively.

2.12. Half the time a model particle of mass μ has angular momentum $4\hbar$ about the z axis, while the rest of the time it has angular momentum $-4\hbar$. Formulate the corresponding φ-dependent factor in the wave function and normalize it so the integral of $\Phi^*\Phi$ over a complete turn is 2π.

2.13. If, at a certain energy level, the root-mean-square angular momentum of a free rotator around an axis in space is $2\hbar$, what is its J?

2.14. If the masses of ^7Li and ^1H are 7.0160 and 1.00783 u, respectively, while the lithium–hydrogen bond length is 1.60 Å, what is the rotational constant for $^7Li^1H$? What is the formula for its rotational energy levels? Where are the first four lines in the corresponding spectrum?

2.15. If the rotational constant A for CH_3Br is 5.08 cm^{-1}, while $B = C = 0.31$ cm^{-1}, where are the lowest rotational energy levels? What transitions are allowed and what spectrum do these allowed transitions produce?

2.16. The H_2O molecule is nonlinear with an HOH angle of 104° 27′ and an O–H bond distance equal to 0.958 Å. Calculate its rotational constants. The mass of ^{16}O is 15.994915 u.

2.17. Use the results from Problem 2.16 to locate the lowest rotational levels of the H_2O molecule.

2.18. Express

$$\frac{\sqrt{5}}{2}\,\frac{3x^2 - r^2}{r^2}\,R(r)$$

as a linear combination of the standard d state functions.

2.19. Find the f orbital that a person must pair with $Y = 2D(xyz/r^3)$ to describe an electron circulating around the z axis in one direction.

2.20. Normalize the rotational wave function $Y = 2B(zx/r^2)$ to 4π.

References

Books

Finch, A., Gates, P. N., Radcliffe, K., Dickson, F. N., and Bentley, F. F.: 1970, *Chemical Applications of Far Infrared Spectroscopy*, Academic Press, New York, pp. 1–102.

Hedvig, P., and Zentai, G. (translated by Morgan, E. D.): 1969, *Microwave Study of Chemical Structures and Reactions*, Iliffe Books, London, pp. 9–140.

Wollrab, J. E.: 1967, *Rotational Spectra and Molecular Structure*, Academic Press, New York, pp. 1–113, 282–317.

Articles

Gardner, M.: 1973, 'Mathematical Games: "Look-see" Diagrams that Offer Visual Proof of Complex Algebraic Formulas', *Scient. American* **229** (4), 114–118.

Hoskins, L. C.: 1975, 'Pure Rotational Raman Spectroscopy of Diatomic Molecules', *J. Chem. Educ.* **52**, 568–572.

Levy-Leblond, J.-M.: 1976, 'Quantum Heuristics of Angular Momentum', *Am. J. Phys.* **44**, 719–722.

McGervey, J. D.: 1981, 'Obtaining L^2 and J^2 from Expectation Values of L_z and J_z', *Am. J. Phys.* **49**, 494–495.

Peach, G.: 1975, 'The Width of Spectral Lines', *Contemp. Phys.* **16**, 17–34.

Chapter 3

Quantization of Vibratory Motion

3.1. Additional Concerted Movements

When submicroscopic entities are examined sufficiently closely, many are found to be composite. Often the unit consists of relatively massive particles held together by oppositely charged, much less massive, light particles. Free movement of the center of mass of the unit is called 'translation', while free movement around the center is called 'rotation'. We have already considered such motions.

In a molecule, the mass of each of the light particles is a very small fraction of the mass of any of the heavy particles. As a consequence, every light particle executes complete orbits while the more massive particles move only slightly. A concerted displacement of the heavy particles from an equilibrium situation alters the interaction uniquely and smoothly.

A general displacement is made up of translations, rotations, and phases of nearly independent vibrations. In a typical pure vibration, the restoring force is approximately proportional to the extent of displacement from the equilibrium configuration, as measured by the pertinent generalized coordinate. The resulting oscillation in the generalized coordinate, a *mode* of motion, is modeled by movement of a particle of appropriate mass μ in a parabolic potential field with the pertinent force constant f.

In such a field neighboring points are not equivalent, except about the equilibrium position. However, a person may approximate the potential over any small region by its average, without appreciably altering any pure state of the system. The wave function outside the region would remain unaltered. Furthermore, Ψ and its spatial derivative just within each junction would equal Ψ and the derivative just outside, as in Section 1.15.

Function Ψ along the replacement segment can be broken down into a wave moving in one direction, Ψ_1, and a wave moving in the opposite direction, Ψ_2. The symmetry argument of Section 1.3 then enables us to construct the variations of Ψ_1 and Ψ_2 over this segment, where V is constant. A unique expression for the change in the derivative of the sum $\Psi_1 + \Psi_2$ emerges. Because this derivative must match the derivative of Ψ for the unaltered potential at each end point, the result does apply to the original system in which V varies smoothly.

Acceptable solutions for the resulting differential equation will be constructed for the parabolic potential. Then, vibrational spectra will be discussed.

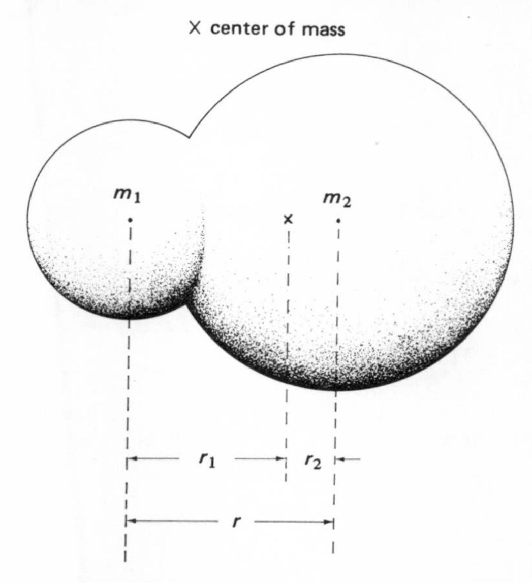

Fig. 3.1. Nuclei and the surrounding electron clouds in a two-atom system.

3.2. Vibrational Kinetic and Potential Energies for a Diatomic Molecule

The electron mass associated with an atom in a molecule is small with respect to its nuclear mass. So even if the center of the electron mass is displaced from the corresponding nuclear center as in a highly polar region, the center of the atom's mass is practically at the middle of the nucleus. Consequently, in our discussions we will consider each atom's mass to be located at the position of its nucleus.

Here we will consider the simplest oscillating molecule: one consisting of two atoms. We will determine how the internal kinetic and potential energies are represented by the kinetic and potential energies of a single particle in a one-dimensional potential dip. A picture of the molecule appears in Figure 3.1.

Let the masses of the two atoms, m_1 and m_2, be located at positions r_1 and r_2 on an axis, with r_1 negative and r_2 positive (recall Figure 2.1). Let us place the origin at the center of mass of the molecule, so that

$$m_1 r_1 + m_2 r_2 = 0 \tag{3.1}$$

or

$$r_1 = -\frac{m_2}{m_1} r_2 . \tag{3.2}$$

The derivative, with respect to time, of (3.2) is

$$\dot{r}_1 = -\frac{m_2}{m_1}\dot{r}_2, \tag{3.3}$$

where

$$\dot{r}_1 = \frac{\mathrm{d}r_1}{\mathrm{d}t} \quad \text{and} \quad \dot{r}_2 = \frac{\mathrm{d}r_2}{\mathrm{d}t}. \tag{3.4}$$

The kinetic energy associated with the oscillatory motion is

$$T = \frac{1}{2} m_1\dot{r}_1{}^2 + \frac{1}{2} m_2\dot{r}_2{}^2 = \frac{1}{2} m_1 \frac{m_2{}^2}{m_1{}^2}\dot{r}_2{}^2 + \frac{1}{2} m_2\dot{r}_2{}^2$$

$$= \frac{1}{2}\left(\frac{1}{m_1} + \frac{1}{m_2}\right) m_2{}^2\dot{r}_2{}^2 = \frac{1}{2}\mu\frac{m_2{}^2\dot{r}_2{}^2}{\mu^2} \tag{3.5}$$

if *reduced mass* μ satisfies the equation

$$\frac{1}{\mu} = \frac{1}{m_1} + \frac{1}{m_2}. \tag{3.6}$$

The internuclear distance itself is

$$r = r_2 - r_1 = r_2 + \frac{m_2}{m_1} r_2 = \left(\frac{1}{m_2} + \frac{1}{m_1}\right) m_2 r_2$$

$$= \frac{m_2 r_2}{\mu} = -\frac{m_1 r_1}{\mu}, \tag{3.7}$$

whence

$$\dot{r} = \frac{m_2\dot{r}_2}{\mu} = -\frac{m_1\dot{r}_1}{\mu} \tag{3.8}$$

and

$$T = \frac{1}{2}\mu\dot{r}^2. \tag{3.9}$$

If the equilibrium value of r is r_e, the displacement of r from its equilibrium point is

$$x = r - r_e \tag{3.10}$$

and we can rewrite (3.9) in the form

$$T = \frac{1}{2}\mu\dot{x}^2. \tag{3.11}$$

Remember, T is the kinetic energy.

Classically, a sinusoidal motion implies a restoring force proportional to the displacement from the equilibrium position and a potential that is a quadratic function of this displacement:

$$V = \frac{1}{2} fx^2. \tag{3.12}$$

The motion is then said to be *harmonic*. Coefficient f is called the *force constant*.

Note that (3.11) and (3.12) also describe the kinetic and potential energies of a particle of mass μ moving in the parabolic potential. So this particle is said to model the diatomic system. A similar model can be constructed for each normal mode of a polyatomic system.

Example 3.1. Consider the two masses, m_1 and m_2, joined by a bond as a classical harmonic oscillator with force constant f. Calculate the resultant frequency v.

Let r_1 and r_2 locate the masses as Figure 3.1 shows. The equations in Section 3.2 are then valid. Differentiating (3.12) and combining the result with the differential of (3.10) and the first equation in (3.7) leads to

$$\mathrm{d}V = \mathrm{d}\left(\frac{1}{2} fx^2\right) = fx\ \mathrm{d}x = fx\ \mathrm{d}r_2 - fx\ \mathrm{d}r_1$$

$$= -F_2\ \mathrm{d}r_2 - F_1\ \mathrm{d}r_1.$$

In the last step, the forces

$$F_2 = -\frac{\partial V}{\partial r_2} \quad \text{and} \quad F_1 = -\frac{\partial V}{\partial r_1}$$

that act on the second and first masses have been introduced.

But Newton's second law yields

$$m_2 \frac{\mathrm{d}^2 r_2}{\mathrm{d}t^2} = F_2 = -fx$$

and

$$m_1 \frac{\mathrm{d}^2 r_1}{\mathrm{d}t^2} = F_1 = fx.$$

Therefore,

$$\frac{\mathrm{d}^2 x}{\mathrm{d}t^2} = \frac{\mathrm{d}^2 r_2}{\mathrm{d}t^2} - \frac{\mathrm{d}^2 r_1}{\mathrm{d}t^2} = -\left(\frac{f}{m_2} + \frac{f}{m_1}\right) x$$

$$= -\frac{f}{\mu} x.$$

Integrating this equation leads to the formula

$$x = A \cos\left[\left(\frac{f}{\mu}\right)^{1/2} t + \alpha\right]$$

$$= A \cos(2\pi\nu t + \alpha),$$

from which the frequency of oscillation is

$$\nu = \frac{1}{2\pi}\left(\frac{f}{\mu}\right)^{1/2}.$$

3.3. Derivatives of the State Function in Regions where the Potential Energy is not Constant

Let us consider the model particle of mass μ moving where its potential energy V is a smoothly varying function of its coordinate x. For simplicity, let us suppose that the particle is in a pure energy state. (Any mixed state can be described as a superposition of such pure states.)

As before, we assume that a function Ψ determining all observable aspects of the motion, through formula (1.1), exists. This function presumably varies smoothly with all coordinates and with time. Since the given particle moves over only one coordinate x, formal differentiation yields the result

$$d\Psi = \frac{\partial \Psi}{\partial x}\, dx + \frac{\partial \Psi}{\partial t}\, dt. \tag{3.13}$$

Any increase in V with displacement introduces the probability of reflection. At each position, Ψ must consequently contain a part Ψ_1 governing motion to the right and a part Ψ_2 governing motion to the left. These constituent functions also obey (3.13).

Except at minima, maxima, and inflection points in V, neighboring points on the x axis are not equivalent. But we may replace any arbitrary infinitesimal section of V with its average over the section. From the principle of continuity, this action does not appreciably alter the overall situation. The wave function outside the section would remain unaffected.

But along the altered section, successive points are equivalent and an infinitesimal change in x in the direction of motion produces an effect that is proportional to the magnitude of the change. With Ψ_1 and Ψ_2 representing pure unidirectional motions, every equivalent part of each produces the same effect on its variation. The influence of a change in time appears as before. Thus, by symmetry, we have

$$d\Psi_1 = ik\Psi_1\,(dx) - i\omega\Psi_1\, dt \tag{3.14}$$

and

$$d\Psi_2 = ik\Psi_2\,(-dx) - i\omega\Psi_2\, dt. \tag{3.15}$$

Comparing each of these with the appropriate form of (3.13) leads to

$$\frac{\partial \Psi_1}{\partial x} = ik\Psi_1 \tag{3.16}$$

and

$$\frac{\partial \Psi_2}{\partial x} = -ik\Psi_2 . \tag{3.17}$$

Over the altered section k is constant, and differentiating (3.16) and (3.17) gives us

$$\frac{\partial^2 \Psi_1}{\partial x^2} = -k^2 \Psi_1 \qquad \text{and} \qquad \frac{\partial^2 \Psi_2}{\partial x^2} = -k^2 \Psi_2 , \tag{3.18}$$

whence

$$\frac{\partial^2 \Psi}{\partial x^2} = -k^2 \Psi , \tag{3.19}$$

where

$$\Psi = \Psi_1 + \Psi_2 . \tag{3.20}$$

Factoring out the dependence on time, through letting

$$\Psi = \psi(x)\, T(t), \tag{3.21}$$

reduces (3.19) to

$$\frac{d^2 \psi}{dx^2} = -k^2 \psi . \tag{3.22}$$

Following the argument in Section 1.15, a person matches Ψ and $\partial\Psi/\partial x$ at each end of the replacement section with Ψ and $\partial\Psi/\partial x$ just outside. The spatial rate of change of $\partial\Psi/\partial x$ over the replacement section must therefore equal that for the original section. Furthermore, there is nothing special about the location of this section. Consequently, Equations (3.19) and (3.22), which describe this rate of change, are valid for any point along a given smoothly varying V.

Wavevector k is related to the corresponding momentum p by de Broglie's equation

$$k = \frac{p}{\hbar} . \tag{3.23}$$

The momentum is related to total energy E minus potential energy V by the kinetic energy equation

$$\frac{p^2}{2\mu} = E - V . \tag{3.24}$$

Combining (3.23), (3.24) with (3.22) yields

$$\frac{d^2\psi}{dx^2} = \frac{2\mu(V-E)}{\hbar^2}\psi. \tag{3.25}$$

Equation (3.25) is a one-dimensional form of the *equation* first proposed by Erwin *Schrödinger*. Deductions from it agree very closely with experimental results wherever the refinements of Einsteinian relativity can be neglected.

Example 3.2. Solve the one-dimensional Schrödinger equation for a constant potential.

When V does not vary, $E - V$ equals a constant E' and (3.25) becomes

$$\frac{d^2\psi}{dx^2} + \frac{2\mu E'}{\hbar^2}\psi = 0.$$

This standard second-order homogeneous differential equation has the solution

$$\psi = A \exp\left[i\frac{(2\mu E')^{1/2}}{\hbar}x\right] + B\exp\left[-i\frac{(2\mu E')^{1/2}}{\hbar}x\right]$$

$$= A\,e^{ikx} + B\,e^{-ikx}.$$

3.4. The Schrödinger Equation for Simple Harmonic Motion

In this chapter we are particularly concerned with the approximately independent vibratory movements of molecules. Each of these involves a normal mode of motion of the system under consideration.

We have noted that the restoring force for such a mode is, commonly, nearly proportional to the change in a generalized coordinate from its equilibrium value. The resulting behavior is modeled by a particle of mass μ in the parabolic potential

$$V = \frac{1}{2}fx^2. \tag{3.26}$$

Now, substituting (3.26) into (3.25) yields

$$\frac{d^2\psi}{dx^2} = \frac{2\mu}{\hbar^2}\left(\frac{1}{2}fx^2 - E\right)\psi. \tag{3.27}$$

To avoid carrying along several constants, let

$$\frac{f\mu}{\hbar^2} = a^2 \tag{3.28}$$

and

$$\frac{2\mu E}{\hbar^2} = ab. \tag{3.29}$$

Then (3.27) becomes

$$\frac{d^2\psi}{dx^2} = (a^2x^2 - ab)\,\psi \tag{3.30}$$

or

$$\frac{1}{a}\frac{d^2\psi}{dx^2} = (ax^2 - b)\,\psi. \tag{3.31}$$

If we also define

$$w = a^{1/2}x, \tag{3.32}$$

then

$$\frac{d}{dw} = \frac{dx}{dw}\frac{d}{dx} = \frac{1}{a^{1/2}}\frac{d}{dx}, \tag{3.33}$$

$$\frac{d^2}{dw^2} = \frac{1}{a}\frac{d^2}{dx^2}, \tag{3.34}$$

and (3.31) simplifies to

$$\frac{d^2\psi}{dw^2} = (w^2 - b)\,\psi, \tag{3.35}$$

whence

$$\frac{d^2\psi}{dw^2} + (b - w^2)\,\psi = 0. \tag{3.36}$$

The parabolic potential rises without limit as w goes to $\pm\infty$. As long as the particle has a finite amount of energy E, the probability of finding it at $w = \pm\infty$, and the corresponding density, are zero. Consequently, ψ must go asymptotically to zero as w increases or decreases without limit. Such behavior is possible with only certain discrete b's.

3.5. Suitable State Functions for the Harmonic Oscillator

In its most stable state, the model particle has reduced its mean kinetic energy and its mean potential energy as much as possible. At each classically attainable position, its kinetic energy is as low, and its de Broglie wavelength is as large, as possible. Such a large wavelength is achieved by allowing ψ to increase from 0 at $w = -\infty$ to a flat maximum at $w = 0$, and then to decrease symmetrically to 0 at $w = \infty$.

The function, in its real form, should appear as a normal distribution curve – bell-shaped. The first excited state must exhibit a shorter de Broglie wavelength at each point. This could be achieved by introducing one node at the middle. The function ψ would vary as the derivative of the ground state function. Likewise, the ψ for the second excited state is expected to vary as the derivative of ψ for the first excited state, and so on.

We are thus led to look at

$$y = e^{-w^2} \tag{3.37}$$

and relationships involving its derivatives. In particular, let us differentiate y,

$$\frac{dy}{dw} = -2w\, e^{-w^2}, \tag{3.38}$$

combine the result with (3.37),

$$\frac{dy}{dw} + 2wy = 0, \tag{3.39}$$

and differentiate (3.39) $v + 1$ times to get

$$\frac{d^{v+2}y}{dw^{v+2}} + 2w\frac{d^{v+1}y}{dw^{v+1}} + 2(v+1)\frac{d^v y}{dw^v} = 0. \tag{3.40}$$

The order of this equation can be reduced to 2 by the substitution

$$z = \frac{d^v y}{dw^v} = \frac{d^v}{dw^v}(e^{-w^2}). \tag{3.41}$$

Thus

$$\frac{d^2 z}{dw^2} + 2w\frac{dz}{dw} + 2(v+1)z = 0. \tag{3.42}$$

To eliminate the first derivative, we set

$$u = e^{w^2/2} z. \tag{3.43}$$

Thus,

$$\frac{d^2 u}{dw^2} = e^{w^2/2}\frac{d^2 z}{dw^2} + 2w\, e^{w^2/2}\frac{dz}{dw} + e^{w^2/2} z + w^2 e^{w^2/2} z. \tag{3.44}$$

Multiplying (3.42) by $\exp(w^2/2)$ and rearranging

$$e^{w^2/2}\left[\frac{d^2 z}{dw^2} + 2w\frac{dz}{dw} + z + w^2 z + (2v + 1 - w^2)z\right] = 0, \tag{3.45}$$

then introducing (3.43) and (3.44), gives us

$$\frac{d^2 u}{dw^2} + (2v + 1 - w^2)u = 0. \tag{3.46}$$

The Schrödinger equation for the harmonic oscillator, (3.36), reduces to this derived equation, (3.46), when

$$b = 2v + 1. \tag{3.47}$$

Because $v + 1$ is the number of differentiations, introduced after (3.39), v itself is an integer. As Equations (3.36) and (3.46) are linear and homogeneous, ψ must be a constant times u, say

$$\psi_v = N(-1)^v u, \tag{3.48}$$

for b's given by (3.47). Substituting in the expression for u from (3.43) and (3.41) leads to

$$\begin{aligned} \psi_v &= N(-1)^v \, e^{w^2/2} \frac{d^v}{dw^v} e^{-w^2} \\ &= N e^{-w^2/2} (-1)^v \, e^{w^2} \frac{d^v}{dw^v} e^{-w^2} \\ &= N e^{-w^2/2} H_v(w). \end{aligned} \tag{3.49}$$

In the last step, $H_v(w)$ has been introduced by definition; it represents a polynomial of the vth degree. Since v may be any positive integer or zero, a superposition of $H_v(w)$'s can represent any polynomial over any finite range of w; and a superposition of $N\,e^{-w^2/2} H_v(w)$'s can represent any well-behaved function that goes to zero as $A\,e^{-w^2/2}$ when $w \to \pm\infty$.

Each state described by (3.49) appears at a definite energy E_{vib}. From (1.45) and (1.48), its temporal dependence is therefore

$$T(t) = \exp\left(-\frac{iE_{\text{vib}}}{\hbar}\,t\right), \tag{3.50}$$

and Ψ itself is a standing wave.

Substituting condition (3.47) and the square root of (3.28)

$$a = \frac{(f\mu)^{1/2}}{\hbar}, \tag{3.51}$$

into (3.29) and solving for the energy yields

$$\begin{aligned} E_{\text{vib}} &= \frac{\hbar^2}{2\mu} \frac{(f\mu)^{1/2}}{\hbar} (2v + 1) = \left(v + \frac{1}{2}\right)\hbar\left(\frac{f}{\mu}\right)^{1/2} \\ &= \left(v + \frac{1}{2}\right)\hbar\,\omega \qquad \text{with} \quad v = 0, 1, 2, \ldots. \end{aligned} \tag{3.52}$$

In the last step, the result for the classical angular frequency, $\omega = 2\pi\nu$, from Example 3.1, has been brought in. Since (3.52) describes how the energy is quantized in the oscillator, v is called the *vibrational quantum number*.

3.6. Properties of the Harmonic Oscillator ψ_v's

Each state function which we have found for the particle in a parabolic potential field is the product of a normalization constant N, a transcendental factor $e^{-w^2/2}$, and a polynomial factor $H_v(w)$.

The normalization constant is usually chosen so that the integral of $\psi^* \psi$ over the whole range of coordinate x, from $-\infty$ to $+\infty$ is 1. We will see later that such a result is obtained if

$$N = \frac{1}{(2^v v!)^{1/2}} \left(\frac{a}{\pi}\right)^{1/4}. \tag{3.53}$$

The transcendental factor predominates when $|w|$ is large; it imposes the proper asymptotic behavior as x increases, or decreases, without limit. The polynomial factor shortens the de Broglie wavelength at each point, as is appropriate for the given v. Particularly noticeable are the waviness and nodes which it introduces in the region where $V < E$.

The expression

$$H_v(w) = (-1)^v e^{w^2} \frac{d^v}{dw^v} e^{-w^2} \tag{3.54}$$

is called the *Hermite polynomial* of vth degree. Explicit forms for the first eleven appear in Table 3.1. Note that the $(-1)^v$ factor serves to make the highest term positive.

The harmonic oscillator exhibits symmetry about the minimum in its potential. Indeed, reflecting the expressions for its kinetic energy and its potential energy

TABLE 3.1.
Hermite polynomials

Symbol	Formula
$H_0(w)$	1
$H_1(w)$	$2w$
$H_2(w)$	$4w^2 - 2$
$H_3(w)$	$8w^3 - 12w$
$H_4(w)$	$16w^4 - 48w^2 + 12$
$H_5(w)$	$32w^5 - 160w^3 + 120w$
$H_6(w)$	$64w^6 - 480w^4 + 720w^2 - 120$
$H_7(w)$	$128w^7 - 1344w^5 + 3360w^3 - 1680w$
$H_8(w)$	$256w^8 - 3584w^6 + 13\,440w^4 - 13\,440w^2 + 1680$
$H_9(w)$	$512w^9 - 9216w^7 + 48\,384w^5 - 80\,640w^3 + 30\,240w$
$H_{10}(w)$	$1024w^{10} - 23\,040w^8 + 161\,280w^6 - 403\,200w^4 + 302\,400w^2 - 30\,240$

through this point does not alter either. Correspondingly, replacing w by $-w$ in the Schrödinger equation does not change the form of the equation. However, such a replacement can alter the phase of the state function. When we check form (3.49), we find that the reflection through the origin leaves ψ_v unchanged when v is even and changes the sign of ψ_v when v is odd.

A function is said to have *even parity*, with a quantum number P equal to $+1$, when reflection through a point of symmetry leaves it unchanged. A function is said to have *odd parity*, with P equal to -1, when reflection through the point of symmetry merely changes its sign. When v is zero, the parity of ψ_v is even. Each increase in v by 1 introduces a change in parity. The parity quantum number of the state described by ψ_v is $(-1)^v$.

Example 3.3. Evaluate

$$\int_0^\infty e^{-ax^2}\,dx \qquad \text{and} \qquad \int_{-\infty}^\infty e^{-ax^2}\,dx.$$

The first integral may be taken along either the x or the y axis without affecting its value:

$$I = \int_0^\infty e^{-ax^2}\,dx = \int_0^\infty e^{-ay^2}\,dy.$$

Consequently, the square of the integral can be expressed as a double integral,

$$I^2 = \int_0^\infty e^{-ax^2}\,dx \int_0^\infty e^{-ay^2}\,dy$$

$$= \int_0^\infty \int_0^\infty e^{-a(x^2+y^2)}\,dx\,dy,$$

over the first quadrant of the xy plane.

This double integral is easily simplified by transforming to polar coordinates, in which

$$x^2 + y^2 = r^2$$

and the differential element of area is

$$dA = r\,d\varphi\,dr.$$

Then

$$I^2 = \int_0^\infty \int_0^{\pi/2} e^{-ar^2}\,r\,d\varphi\,dr = \int_0^\infty \varphi\Big|_0^{\pi/2} e^{-ar^2}\,r\,dr$$

$$= \frac{\pi}{2}\left(-\frac{e^{-ar^2}}{2a}\right)\Bigg|_0^\infty = \frac{\pi}{4a},$$

and

$$I = \frac{1}{2}\left(\frac{\pi}{a}\right)^{1/2}.$$

Because the given integrand is even, the integral from $-\infty$ to 0 equals that from 0 to ∞ and

$$\int_{-\infty}^{\infty} e^{-ax^2}\,dx = \left(\frac{\pi}{a}\right)^{1/2}.$$

Example 3.4. What fraction of $\int_0^\infty e^{-ax^2}\,dx$ is $\int_0^b e^{-ax^2}\,dx$?

The *error function* of u is

$$\operatorname{erf} u = \frac{2}{\sqrt{\pi}}\int_0^u e^{-w^2}\,dw$$

by definition. In the given integral, we let

$$x = \frac{w}{a^{1/2}},$$

then introduce the first result from Example 3.3, as well as the definition of the error function, to obtain

$$\int_0^b e^{-ax^2}\,dx = \frac{\sqrt{\pi}}{2\sqrt{a}}\,\frac{2}{\sqrt{\pi}}\int_0^{a^{1/2}b} e^{-w^2}\,dw$$

$$= \left(\int_0^\infty e^{-ax^2}\,dx\right)\operatorname{erf}(a^{1/2}b).$$

The desired fraction is $\operatorname{erf}(a^{1/2}b)$. Standard tables of error functions are available.

Example 3.5. Normalize the lowest state function for the harmonic oscillator.

When v is 0, formula (3.49) yields

$$\psi = N\,e^{-w^2/2}.$$

The corresponding particle density is

$$\rho = \psi^*\psi = N^2\,e^{-w^2} = N^2\,e^{-ax^2},$$

if N is assumed to be real. Setting the total probability of finding a particle in the state equal to 1 gives us

$$\int_{-\infty}^{\infty} \rho\,dx = \int_{-\infty}^{\infty} N^2\,e^{-ax^2}\,dx = N^2\int_{-\infty}^{\infty} e^{-ax^2}\,dx = 1.$$

Then introducing the value of the last integral from Example 3.3 leads to

$$N^2\left(\frac{\pi}{a}\right)^{1/2} = 1$$

or

$$N = \left(\frac{a}{\pi}\right)^{1/4}.$$

Note that formula (3.53) produces the same result.

Example 3.6. Show that $\psi = A\, e^{\pm w^2/2}$ satisfies the Schrödinger equation for the harmonic oscillator asymptotically as w^2 increases without limit.

Differentiate the given function twice:

$$\frac{d\psi}{dw} = \pm Aw\, e^{\pm w^2/2},$$

$$\frac{d^2\psi}{dw^2} = Aw^2\, e^{\pm w^2/2} \pm A\, e^{\pm w^2/2}.$$

As $w \to \pm\infty$, and w^2 increases without limit, the next to the last term overwhelms the last term and

$$\frac{d^2\psi}{dw^2} \simeq Aw^2\, e^{\pm w^2/2}.$$

Similarly, in (3.36) w^2 becomes very large with respect to b and b can be neglected:

$$\frac{d^2\psi}{dw^2} - w^2\psi \simeq 0.$$

The given ψ, together with the approximate derived $d^2\psi/dw^2$, meet this condition.

Example 3.7. Describe how and where an analytic solution of the Schrödinger equation for a pure state of the harmonic oscillator may go bad.

From Example 3.6, the Schrödinger equation for the oscillator is satisfied by

$$A\, e^{-w^2/2} \quad \text{at} \quad w = -\infty,$$

$$B\, e^{-w^2/2} \quad \text{at} \quad w = \infty,$$

and by

$$C\, e^{w^2/2} \quad \text{at} \quad w = -\infty,$$

$$D\, e^{w^2/2} \quad \text{at} \quad w = \infty,$$

where A, B, C, and D are constants. The last two forms cause $|\psi|$ to increase without limit as $|w|$ and the proportionate $|x|$ increase, as long as C and D are not zero. The probability density,

$$\rho = \psi^*\psi,$$

would also become infinite, and the integral of ρ over all x would be infinite, a meaningless result. We have to reject such solutions as being extraneous.

But when we choose the suitable form

$$A\,e^{-w^2/2} \qquad \text{at} \quad w = -\infty$$

and employ (3.36) to continue the solution through all w's, we generally find ψ ending in the unsuitable form

$$D\,e^{w^2/2} \qquad \text{at } w = \infty.$$

Only for b's satisfying (3.47) does the ψ join with the suitable

$$B\,e^{-w^2/2} \qquad \text{at} \quad w = \infty.$$

Independent of the $D\,e^{w^2/2}$ form at $w = \infty$ is the $B\,e^{-w^2/2}$ asymptotic form. This joins with the unsuitable

$$C\,e^{w^2/2} \qquad \text{at} \quad w = -\infty,$$

except for b's satisfying (3.47). Independent of each acceptable solution for

$$b = 2v + 1$$

there is also an unacceptable one, in which

$$C\,e^{w^2/2} \qquad \text{at} \quad w = -\infty$$

joins with

$$D\,e^{w^2/2} \qquad \text{at} \quad w = \infty.$$

The only suitable solutions describing pure states are those of (3.49).

3.7. Transitions between Vibrational Energy Levels

An internal mode of motion of a molecule may gain or lose energy (a) by interacting with other nontranslational modes in the same molecule, (b) by interacting with other modes in neighboring molecules, and (c) by interacting with the electromagnetic field. The processes resulting from the first two interactions are not governed by any simple selection rules, so they will not be considered here in any detail. However, let us note that a concentration of energy sufficient to cause a vibrational excitation may cause an accompanying rotational excitation or de-excitation because spacings between pertinent rotational levels are so much smaller than the spacing between the vibrational levels.

When a transition involves changes in dipole moment, or in polarizability, of a molecule, the corresponding mode interacts strongly with the electromagnetic field. Commonly, a single photon disappears or appears, or is altered, in each elementary step. Infrared spectra, in which absorption or emission is studied, will be considered first. Raman spectra, resulting when photons merely interact and exchange energy with modes of motion in a given material, will be considered later.

A symmetric molecule has a mode of vibration throughout which the system remains symmetric. Such a mode cannot exhibit a transition moment; hence it does not yield any infrared spectrum and the mode is said to be inactive in the infrared. On the other hand, an asymmetric mode of any molecule can exhibit an appreciable transition dipole moment. In the approximation that the mode is harmonic, the nonzero moment appears between states having ψ_v's with polynomial parts differing in degree by 1. The resulting selection rule, determining when absorption or emission of photons may occur, is

$$\Delta v = \pm 1. \tag{3.55}$$

From (3.52), the energy of the oscillator then changes by

$$\Delta E_{\text{vib}} = \left(v_0 \pm 1 + \frac{1}{2}\right)\hbar\omega_{\text{vib}} - \left(v_0 + \frac{1}{2}\right)\hbar\omega_{\text{vib}}$$

$$= \pm\hbar\omega_{\text{vib}} = \pm h\nu_{\text{vib}} \tag{3.56}$$

if v_0 is the initial v, ω_{vib} the classical angular frequency, and ν_{vib} the $\omega_{\text{vib}}/2\pi$ for the mode.

When *no* other transition accompanies the process, the energy of the photon absorbed or emitted equals the energy that the oscillator gains or loses:

$$h\nu_0 = \pm\Delta E_{\text{vib}}, \tag{3.57}$$

where ν_0 is the photon frequency. Also,

$$\bar{k}_0 = \frac{h\nu_0}{hc} = \frac{h\nu_{\text{vib}}}{hc} = \bar{k}_{\text{vib}}, \tag{3.58}$$

the wave number of the photon absorbed or emitted equals the wave number for the oscillator. The corresponding part of the spectrum is called the Q line or *branch*. Whether it is present depends on the nature of the molecule.

3.8. Vibrational Spectrum of a Diatomic Molecule

The simplest molecules to consider are diatomic, of type A–A, in which both atoms are the same, and of type A–B, in which the atoms are different. These are modeled by a single particle of mass μ in a parabolic potential, as we noted in Section 3.2.

An A–A molecule is generally symmetric in its lowest electronic state, and remains symmetric through the possible vibrational and rotational changes. Consequently, the transition moment between any two levels vanishes and the molecule does *not* exhibit a vibrational–rotational infrared spectrum. Hydrogen, H_2, is an example.

On the other hand, an A–B molecule is not symmetric and does yield finite transition moments between adjacent levels. Equation (3.55) applies in the approximation that anharmonicities can be neglected. However, each photon absorbed or emitted carries $\hbar$ angular momentum and possesses odd parity.

This angular momentum and parity can be supplied by a rearrangement of the electronic motion, without involving an excitation of an electron, when the molecule possesses an odd electron, with an angular momentum about the internuclear axis. A Q branch, at wave number

$$k = k_0, \tag{3.59}$$

then appears. The NO molecule behaves in this manner.

With any A—B, the rotational motion may change to absorb or supply the angular momentum and parity of the photon. Limitation (2.51) applies.

An absorbed (or emitted) photon excites (or de-excites) both vibration and rotation when the signs on Δv and ΔJ are the same:

$$\Delta v = \pm 1 \qquad \text{and} \qquad \Delta J = \pm 1. \tag{3.60}$$

Since the energy required to increase (or decrease) v combines additively with the energy required to increase (or decrease) J, and since the defined wave numbers are proportional to the corresponding energies, expressions (3.59) and (2.57) [or (2.58)] combine additively. We obtain

$$k = k_0 + 2B(J_0 + 1) \qquad \text{with} \quad J_0 = 0, 1, 2, \ldots \tag{3.61}$$

governing absorption and

$$k = k_0 + 2BJ_0 \qquad \text{with} \quad J_0 = 1, 2, 3, \ldots \tag{3.62}$$

governing emission. The resulting lines are said to form the *R branch* of the spectrum.

On the other hand, the energy needed to increase (or decrease) v is made up of energy from the photon and from a rotational change when

$$\Delta v = \pm 1 \qquad \text{and} \qquad \Delta J = \mp 1. \tag{3.63}$$

The proportionate wave numbers combine additively. Governing absorption is the sum

$$k_0 = k + 2BJ_0, \tag{3.64}$$

whence the photon wave number is

$$k = k_0 - 2BJ_0 \qquad \text{where} \quad J_0 = 1, 2, 3, \ldots. \tag{3.65}$$

Similarly, the photon wave number for emission is

$$k = k_0 - 2B(J_0 + 1) \qquad \text{where} \quad J_0 = 0, 1, 2, \ldots. \tag{3.66}$$

The resulting lines form the *P branch*.

If there were no variation in moment of inertia with J and with v, the energy levels and the resultant absorption spectrum would appear as in Figure 3.2. Actually, such variations are always present. Furthermore, each line spreads over a range of wave numbers, for the reasons noted in Section 2.9. The relative intensity of a 'line' depends on the extent to which the initial rotational level is populated at the temperature of the absorbing (or emitting) system.

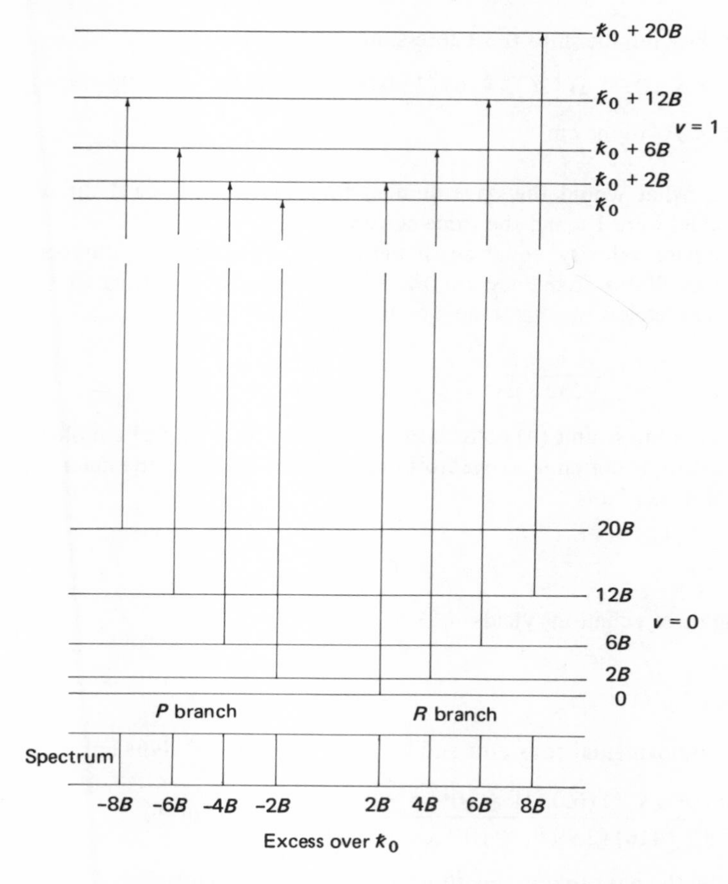

Fig. 3.2. Energy levels and permissible transitions of an ideal linear molecule, with the corresponding spectrum.

Example 3.8. If the wave number $ƙ_0$ about which the vibrational–rotational lines of HCl are arranged is 2990 cm^{-1}, what is the force constant for the molecule?

From (3.58), the frequency of the oscillator is related to the central wave number $ƙ_0$ by the equation

$$\nu_{vib} = cƙ_0.$$

Furthermore, the final equation in Example 3.1 can be solved for the force constant:

$$f = \mu(2\pi\nu_{\text{vib}})^2 = \mu(2\pi c \tilde{k}_0)^2.$$

Substitute the reduced mass from Example 2.5,

$$\mu = 1.628 \times 10^{-24} \text{ g},$$

and the given wave number into this expression:

$$f = (1.628 \times 10^{-24} \text{ g}) [2(3.1416)(2.9979 \times 10^{10} \text{ cm s}^{-1})(2990 \text{ cm}^{-1})]^2$$
$$= 5.16 \times 10^5 \text{ dyne cm}^{-1}.$$

Example 3.9. What would the wave number for a given mode be if the reduced mass for its model were 1 u and the force constant 10^5 dyne cm^{-1}?

Solve the phase velocity equation of light waves for the wave number; then substitute the oscillator frequency for the light frequency, following (3.58), and introduce the expression for this frequency from Example 3.1:

$$\tilde{k}_0 = \frac{\nu_0}{c} = \frac{\nu_{\text{vib}}}{c} = \frac{1}{2\pi c}\left(\frac{f}{\mu}\right)^{1/2}.$$

Since each atomic mass unit (u) contributes 1 gram to the weight of a mole of pure substance and 1 mole contains Avogadro's number N of particles, the mass in grams of a particle of mass μ' u is

$$\mu = \frac{\mu'}{N}.$$

Combining these equations yields

$$\tilde{k}_0 = \frac{1}{2\pi c}\left(\frac{f}{\mu'/N}\right)^{1/2}.$$

Introduce the fundamental constants and let μ' be 1 u, f be 10^5 dyne cm^{-1}:

$$\tilde{k}_0 = \frac{[(10^5 \text{ g s}^{-2})(6.0221 \times 10^{23} \text{ g}^{-1})]^{1/2}}{2(3.1416)(2.9979 \times 10^{10} \text{ cm s}^{-1})} = 1302.79 \text{ cm}^{-1}.$$

Now, solving the next-to-last equation for the force constant,

$$f = \frac{\mu'}{N}(2\pi c \tilde{k}_0)^2$$

and dividing by this relationship when f is 10^5 dyne cm^{-1}, μ' is 1 u gives us

$$\frac{f}{10^5 \text{ dyne cm}^{-1}} = \frac{\dfrac{\mu'}{N}(2\pi c \tilde{k}_0)^2}{\dfrac{1}{N}(2\pi c \times 1302.79 \text{ cm}^{-1})^2}$$

whence

$$f = \mu' \left(\frac{\tilde{k}_0}{1302.79 \text{ cm}^{-1}} \right)^2 \times 10^5 \text{ dyne cm}^{-1}.$$

The force constant is proportional to the reduced mass and to the square of the wave number.

3.9. Anharmonicities and their Net Effects

A chemical bond may be ionic, or covalent, or mixed in nature. A purely *ionic* bond arises from the complete transfer of one or more valence electrons from one participant atom to the other. Electrostatic attraction then pulls the atoms together until the effect of squeezing of inner shells in the atoms balances the attraction. An example is afforded by NaCl. An ideal *covalent* bond arises when the valence electrons from the participating atoms are used by each of these atoms equally, in a shared manner. In the formation, each bonding electron becomes less confined, losing both kinetic and potential energy. A simple example appears in H_2. A *mixed* bond arises when different atoms share electrons. In general, one atom has a stronger affinity for the valence electrons than the other and at equilibrium this atom becomes fractionally negative, the partner fractionally positive. An example occurs in HCl.

Compressing a bond tends to increase the confinement of the electrons and to increase the electrostatic repulsion between the positively charged nuclei. If nuclear forces did not intervene, the resulting potential would increase without limit as the internuclear distance r went to zero. On the other hand, stretching a bond takes the system toward the separated atom situation; thus it progressively lowers the bonding to zero, making the effective potential increase to a constant value.

Parabolic potential (3.12) does not allow for either of these effects. Instead, expression $\frac{1}{2} fx^2$ fails to increase fast enough as x decreases in its negative range and fails to level off as x increases in its positve range. A function that does behave properly in these ranges is sketched in Figure 3.3, together with the parabolic potential that approximates this function near the point of equilibrium.

This parabolic potential leads to state function (3.49) and energy (3.52). The vibrational quantum number v equals the number of nodes in ψ_v where

$$x = r - r_e \tag{3.67}$$

is finite. Now, distorting the parabolic potential to the realistic potential does not affect the curve in the immediate neighborhood of r_e. Motion largely confined to this neighborhood because energy E is low is not much affected. If $v + \frac{1}{2}$ in the equation

$$E_{vib} = \left(v + \frac{1}{2}\right) \hbar \omega_0 = \left(v + \frac{1}{2}\right) h\nu_0 \tag{3.68}$$

could approach zero, this equation would presumably not be altered appreciably near the limit.

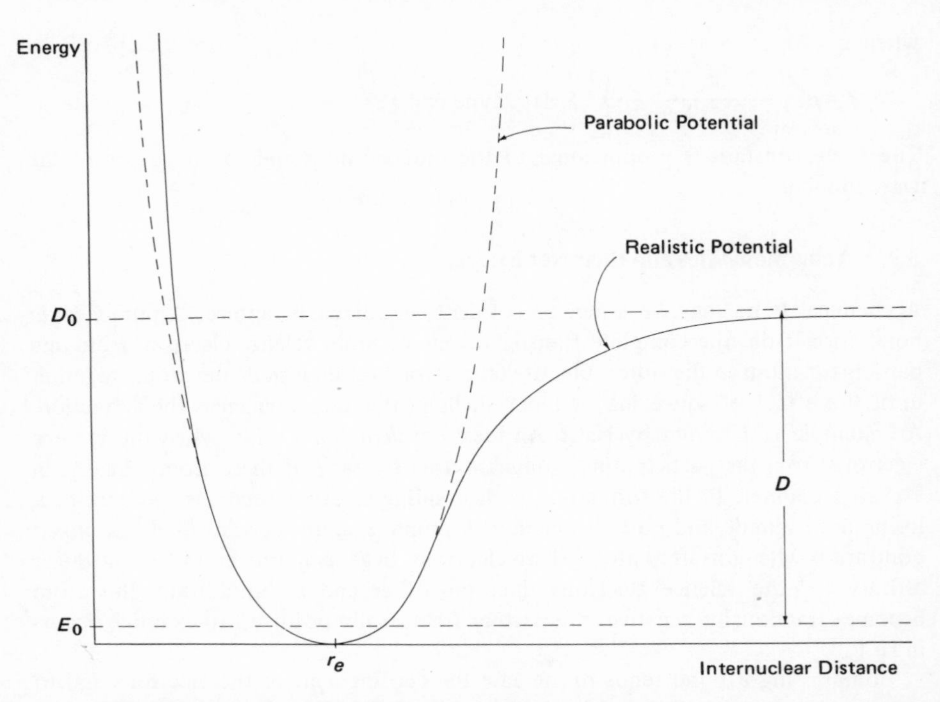

Fig. 3.3. The approximate parabolic potential and the actual potential binding two atoms together.

Correlating with each of the low-lying harmonic oscillator state functions is a solution for the realistic potential with the same number of nodes in the finite region about r_e and an asymptotic approach to zero on each side beyond this region. A *vibrational quantum number* v equal to this number of nodes is assigned to the solution. Now, increasing v from zero shortens the wavelength at each position, raises the energy, and causes the asymmetric nature of the potential to become more significant. The flattening of the potential on the right also has more effect than the steepening on the left; the model particle becomes much less confined than it would be in the parabolic potential. At energy D_0 it becomes essentially free. So we expect successive levels to come closer together until they meet when the D_0 level is reached.

An analytic function that reproduces this behavior is a power series in $v + \frac{1}{2}$, with the first term given by (3.68). In practice, we generally need employ only one additional term:

$$E_{\text{vib}} = \left(v + \frac{1}{2}\right)\hbar\omega_0 - \left(v + \frac{1}{2}\right)^2 b\hbar\omega_0$$

$$= \left(v + \frac{1}{2}\right)h\nu_0 - \left(v + \frac{1}{2}\right)^2 bh\nu_0 \quad \text{with } v = 0, 1, 2, \ldots. \tag{3.69}$$

Here h is Planck's constant, v_0 the harmonic oscillator frequency, and b a new parameter called the *anharmonicity constant*.

The steepening of potential V at low r and the flattening of V at high r lowers the level at which the state function for a given v appears, compresses the function on the left, and expands it on the right. Parity is no longer a good quantum number and additional changes in v are permitted. In addition to

$$\Delta v = \pm 1, \tag{3.70}$$

weaker lines corresponding to

$$\Delta v = \pm 2, \pm 3, \ldots \tag{3.71}$$

appear. Since the additional frequencies are approximate multiples of ν_0, they are called overtones or *harmonics*.

At energy D_0 successive levels come together, increasing v no longer increases E, and

$$\frac{\mathrm{d}E}{\mathrm{d}v} = 0. \tag{3.72}$$

But the derivative of (3.69) with respect to v is

$$\frac{\mathrm{d}E}{\mathrm{d}v} = \hbar\omega_0 - 2\left(v + \frac{1}{2}\right) b\hbar\omega_0. \tag{3.73}$$

Setting this equal to zero, solving for $v + \frac{1}{2}$,

$$v + \frac{1}{2} = \frac{1}{2b}, \tag{3.74}$$

and substituting back into (3.69) yields

$$D_0 = \frac{1}{2b}\hbar\omega_0 - \frac{1}{4b^2} b\hbar\omega_0 = \frac{\hbar\omega_0}{4b}. \tag{3.75}$$

Subtracting the approximate ground state energy

$$E = \frac{1}{2}\hbar\omega_0 \tag{3.76}$$

from D_0 gives us the dissociation energy

$$D = \frac{\hbar\omega_0}{4b} - \frac{1}{2}\hbar\omega_0. \tag{3.77}$$

An analytic function that behaves as the realistic potential is the closed form suggested by Phillip M. Morse:

$$V = D_0 [1 - \mathrm{e}^{-a(r - r_\mathrm{e})}]^2. \tag{3.78}$$

Example 3.10. Removing the rotational shifts from the vibrational spectrum of HCl leaves a fundamental at 2886 cm^{-1} and an overtone at 5668 cm^{-1}. Calculate the harmonic oscillator wave number $\bar{k}_0$ and the anharmonicity constant b for HCl.

Divide Equation (2.50) by hc and introduce the phase velocity relationship to obtain the corresponding wave number:

$$\frac{E}{hc} = \frac{h\nu}{hc} = \frac{\nu}{c} = \bar{k}.$$

Similarly reduce formula (3.69) to wave number units,

$$\frac{E}{hc} = \left[\left(v+\frac{1}{2}\right) - \left(v+\frac{1}{2}\right)^2 b\right]\frac{h\nu_0}{hc},$$

and rewrite the result in the form

$$\bar{k} = \left[\left(v+\frac{1}{2}\right) - \left(v+\frac{1}{2}\right)^2 b\right]\bar{k}_0.$$

Because the energy of an absorbed (or emitted) photon equals the change in energy of the molecule, its wave number equals the change in this $\bar{k}$. Consequently, the photon absorbed when v goes from 0 to 1 without a shift in J would have the wave number

$$\bar{k}_1 = \left[\frac{3}{2} - \left(\frac{3}{2}\right)^2 b - \frac{1}{2} + \left(\frac{1}{2}\right)^2 b\right]\bar{k}_0 = (1-2b)\bar{k}_0,$$

while the photon absorbed when v similarly goes from 0 to 2 would have the wave number

$$\bar{k}_2 = \left[\frac{5}{2} - \left(\frac{5}{2}\right)^2 b - \frac{1}{2} + \left(\frac{1}{2}\right)^2 b\right]\bar{k}_0 = (2-6b)\bar{k}_0.$$

Substitute the given data into these equations,

$$2886\ \text{cm}^{-1} = (1-2b)\bar{k}_0,$$

$$5668\ \text{cm}^{-1} = (2-6b)\bar{k}_0,$$

multiply the first equation by 2, and subtract the second:

$$104\ \text{cm}^{-1} = 2b\bar{k}_0.$$

Use this result to eliminate $2b\bar{k}_0$ from the first equation:

$$2886\ \text{cm}^{-1} = \bar{k}_0 - 104\ \text{cm}^{-1},$$

whence

$$\bar{k}_0 = 2990\ \text{cm}^{-1}$$

and

$$b = \frac{104\ \text{cm}^{-1}}{2(2990\ \text{cm}^{-1})} = 0.0174.$$

3.10. Spectra of Polyatomic Molecules

The position of the center of mass of each atom in a molecule is determined by three independent variables. If the molecule contains N atoms, the total number needed is $3N$. But three independent coordinates are needed to specify the position of the molecule's center of mass, two to specify the molecule's orientation in space when the system is linear, and three when it is nonlinear. A molecule, consequently, exhibits $3N - 5$ vibrational modes when it is linear and $3N - 6$ vibrational modes when it is not linear.

A typical vibrational mode is modeled by a single particle of appropriate mass μ in a potential $\frac{1}{2}fx^2$. The constants μ and f generally differ from mode to mode. The state of a mode is approximately described by a harmonic oscillator function (3.49); the complete vibrational motion is described by a product of such functions (with one from each mode).

A transition between two states of a given mode interacts with the electromagnetic field if a change in electric dipole moment accompanies the transition. Whenever the motion is symmetric in each direction from the center, the transition moment is zero and the mode is inactive in the infrared. When the motion is not symmetric in a direction, the transition moment differs from zero for neighboring ψ_v's, as noted in Section 3.7. The selection rule is

$$\Delta v = \pm 1, \tag{3.79}$$

where v is the vibrational quantum number for the mode.

(a) Symmetric Stretch

(b) Antisymmetric Stretch

(c) Bending Motion

Fig. 3.4. Normal vibrational modes of CO_2.

As an example, consider the linear CO_2 molecule. According to the appropriate expression in the first paragraph of this section, it has four vibrational modes. Phases in these are indicated in Figure 3.4. The symmetric stretching mode is indeed inactive with respect to absorption or emission of single protons. On the other hand, the asymmetric stretching mode and the two bending modes are active.

Associated with changes in the asymmetric stretching motion is an absorption band around 2349 cm^{-1}, while changes in either bending oscillation yield a band around 667 cm^{-1}. Because of anharmonicity, a v may shift by more than a unit. Because of interactions among the modes, combination levels, in which more than one vibrational quantum number changes, are also found.

Now, each photon possesses $\hbar$ angular momentum and an odd parity. On absorption or emission by a molecule of a given substance, these may alter the magnitude of rotation of the molecule, causing

$$\Delta J = \pm 1. \tag{3.80}$$

Consequently, P and R branches are observed. When the molecule is a rotating nonlinear system, or a linear system with net electronic angular momentum along the molecular axis, the absorption or emission may merely alter the direction of the angular momentum. Then a person observes the Q branch, for which

$$\Delta J = 0. \tag{3.81}$$

A Q branch also occurs associated with degenerate modes that couple to produce rotations. The bending modes of linear molecules are examples. Thus, the up-and-down bending of CO_2 in Figure 3.4 couples with the perpendicular, horizontal bending to produce angular momentum around the molecular axis, and a Q branch.

Under conditions generally employed, the detailed structure of a rotation–vibration spectrum is resolved only for the lightest molecules. Where it is not resolved into broadened lines, a person works with the contour, rather than with the individual rotational lines.

3.11. Raman Spectra

Throughout the infrared, visible, and ultraviolet ranges, the mass of a photon is small with respect to the mass of any given molecule. On striking a molecule, the photon may bounce off as from an infinitely massive body, with little change in its energy. Alternatively, it may transfer energy to or accept energy from the molecule, in the process first reported by Chandrasekhara V. Raman.

When close to a molecule, the photon acts through the electromagnetic field, polarizing the structure. If the polarizability of the molecule then undergoes a net change, a transfer of energy takes place. The energy lost or gained by the photon equals the energy gained or lost by the molecule.

A mode of motion is active in a Raman interaction if a possible change in the motion behaves under symmetry operations of the molecule as one or more components of the polarizability. A mode of vibration of an unsymmetric molecule exhibits such action when its

$$\Delta v = \pm 1. \tag{3.82}$$

But the change (3.82) in a mode of a molecule symmetric with respect to a center is active only if the mode is similarly symmetric.

The anharmonicity always present in a mode introduces a contribution from the change

$$\Delta v = \pm 2. \tag{3.83}$$

Alterations in two different modes at nearly the same wave numbers lead to interactions and to a mixing of the modes.

As an example, the CO_2 molecule again serves. When the pertinent v changes by 1, the symmetric stretching motion of Figure 3.4 is Raman active, while the antisymmetric stretching and bending motions are not. But a change of v by 2 for a bending mode mixes with the $\Delta v = 1$ alteration of the symmetric stretching mode to produce Raman shifts of 1286 cm^{-1} and 1388 cm^{-1}.

In very favorable circumstances, a fine structure due to rotational transitions can be observed. But the photon's angular momentum may or may not be reoriented in the Raman interaction. Since this momentum equals $\hbar$ about an axis, the rotation of the target molecule can change by as much as two units. But a change by one unit would not meet the symmetry requirements, so in the Raman spectrum, we have

$$\Delta J = 0 \qquad \text{or} \qquad \Delta J = \pm 2. \tag{3.84}$$

Raman lines produced by photons that have lost energy in their interactions with molecules are called *Stokes lines*; those of photons that have gained energy, *anti-Stokes lines*. Stokes lines for which $\Delta J = -2$ (anti-Stokes lines for which $\Delta J = 2$) are said to form the *O branch*, while Stokes lines for which $\Delta J = 2$ (anti-Stokes lines for which $\Delta J = -2$) form the *S branch*. As before, the lines for which $\Delta J = 0$ form the *Q branch*. (See Figures 3.5 and 3.6.)

3.12. The State Sum for a Vibrational Mode

The contribution of a vibrational mode of a typical molecule in a macroscopic system to the thermodynamic properties of the material can be calculated in the same way as the contribution of a translational mode.

From (3.52), the states of a harmonic oscillator have the energies

$$\frac{1}{2}\hbar\omega, \left(\frac{1}{2} + 1\right)\hbar\omega, \left(\frac{1}{2} + 2\right)\hbar\omega, \ldots . \tag{3.85}$$

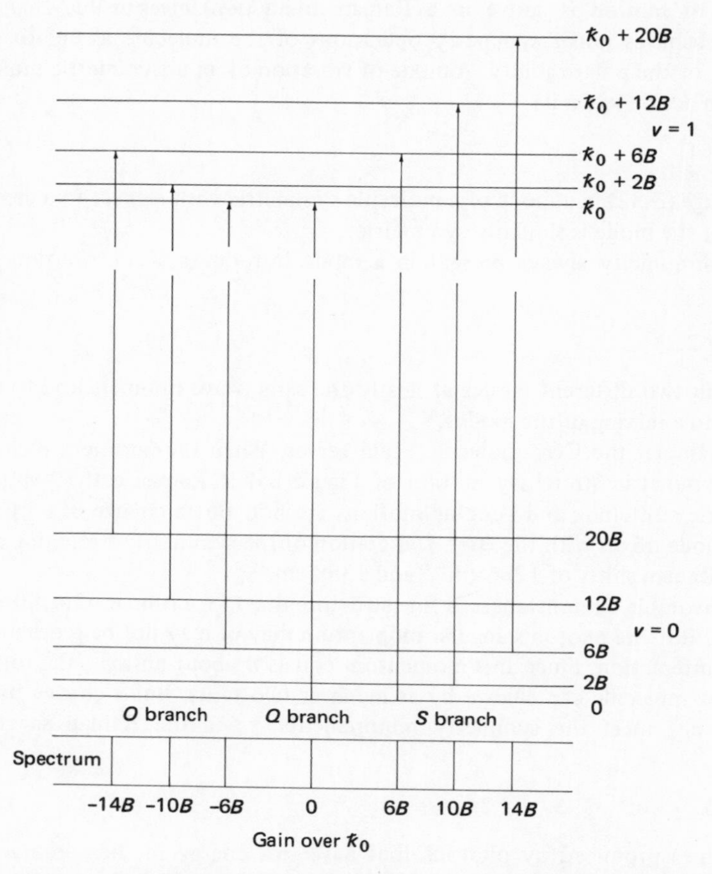

Fig. 3.5. Energy levels and permitted Raman displacements for a diatomic molecule, with the corresponding spectrum.

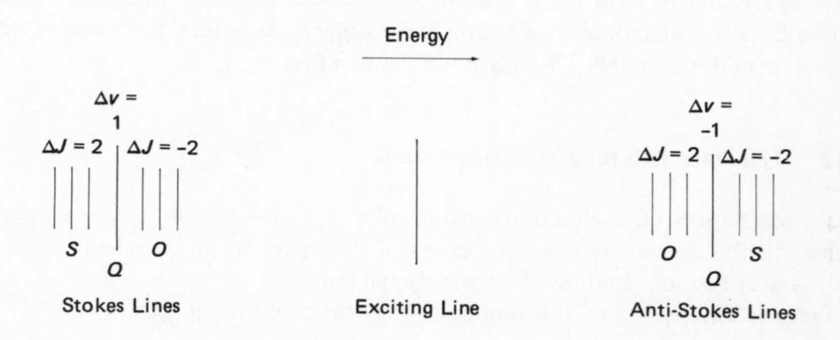

Fig. 3.6. Schematic picture of a Raman spectrum for a diatomic molecule.

Substituting these into the formula for the state sum, (1.89), gives us

$$Z_{\text{vib}} = e^{-\hbar\omega/2kT}\,(1 + e^{-\hbar\omega/kT} + e^{-2\hbar\omega/kT} + \ldots). \tag{3.86}$$

Now, the series

$$S = 1 + a + a^2 + \ldots \tag{3.87}$$

can be rearranged in the closed form

$$S = \frac{1}{1-a}. \tag{3.88}$$

Consequently, (3.86) can be rearranged to

$$Z_{\text{vib}} = \frac{e^{-\hbar\omega/2kT}}{1 - e^{-\hbar\omega/kT}}. \tag{3.89}$$

When the temperature is high enough so that

$$\hbar\omega \ll kT, \tag{3.90}$$

(3.89) reduces to

$$Z_{\text{vib}} \simeq \frac{1}{1-(1-\hbar\omega/kT)} = \frac{kT}{\hbar\omega}, \tag{3.91}$$

whence

$$\ln Z_{\text{vib}} \simeq \ln T + \ln\frac{k}{\hbar\omega}. \tag{3.92}$$

Substituting (3.92) into (1.93) gives us the energy

$$E_{\text{vib}} = NkT^2\left(\frac{1}{T}\right) = NkT. \tag{3.93}$$

When N is Avogadro's number, Nk equals the gas constant R and we obtain

$$E_{\text{vib}} = RT \tag{3.94}$$

for a vibrational mode.

3.13. Operator Formulations of the Harmonic Oscillator Equation

A differentiating operator has algebraic as well as transforming properties. Both aspects are investigated and exploited in operator techniques, as will be developed throughout Chapter 5. As a preliminary exercise, let us here recast the Schrödinger equation for the harmonic oscillator in useful operator forms.

The behavior of a harmonic oscillator is governed by (3.27), (3.35), and (3.36). If we let

$$\frac{\mathrm{d}}{\mathrm{d}w} = \mathrm{D} \tag{3.95}$$

and

$$\frac{\mathrm{d}^2}{\mathrm{d}w^2} = \mathrm{D}^2, \tag{3.96}$$

then (3.35) becomes

$$\mathrm{D}^2 \psi = (w^2 - b)\psi, \tag{3.97}$$

whence

$$(\mathrm{D}^2 - w^2)\psi = -b\psi. \tag{3.98}$$

Because D has transforming as well as algebraic properties the expression in parenthesis does not factor into $(\mathrm{D} - w)(\mathrm{D} + w)$ or into $(\mathrm{D} + w)(\mathrm{D} - w)$. Instead,

$$\begin{aligned}(\mathrm{D} - w)(\mathrm{D} + w)\psi &= (\mathrm{D} - w)(\mathrm{D}\psi + w\psi) \\ &= \mathrm{D}^2\psi + w\mathrm{D}\psi + \psi - w\mathrm{D}\psi - w^2\psi \\ &= (\mathrm{D}^2 - w^2 + 1)\psi \end{aligned} \tag{3.99}$$

and

$$\begin{aligned}(\mathrm{D} + w)(\mathrm{D} - w)\psi &= (\mathrm{D} + w)(\mathrm{D}\psi - w\psi) \\ &= \mathrm{D}^2\psi - w\mathrm{D}\psi - \psi + w\mathrm{D}\psi - w^2\psi \\ &= (\mathrm{D}^2 - w^2 - 1)\psi. \end{aligned} \tag{3.100}$$

We obtain the harmonic oscillator equation in factored form by substituting (3.98) into the right side of (3.99),

$$(\mathrm{D} - w)(\mathrm{D} + w)\psi = -(b - 1)\psi, \tag{3.101}$$

or into the right side of (3.100),

$$(\mathrm{D} + w)(\mathrm{D} - w)\psi = -(b + 1)\psi. \tag{3.102}$$

Now, we can formally introduce

$$b = 2v + 1 \tag{3.103}$$

without specifying v, obtaining

$$(\mathrm{D} - w)(\mathrm{D} + w)\psi = -2v\psi \tag{3.104}$$

and

$$(\mathrm{D} + w)(\mathrm{D} - w)\psi = -(2v + 2)\psi. \tag{3.105}$$

Remember that these equations limit the movement of the model particle of mass μ in the potential

$$V = \frac{1}{2} fx^2. \tag{3.106}$$

Variable w is related to coordinate x by (3.32),

$$w = a^{1/2} x, \tag{3.107}$$

for which

$$a^2 = \frac{f\mu}{\hbar^2} \tag{3.108}$$

and

$$ab = \frac{2\mu E}{\hbar^2}. \tag{3.109}$$

Consequently, energy E is related formally to v by the equation

$$E = \frac{\hbar^2}{2\mu} ab = \left(v + \frac{1}{2}\right) \hbar\omega \tag{3.110}$$

as in (3.52).

Discussion Questions

3.1. Into what independent and nearly independent motions can a general motion of the atoms in a molecule be resolved?

3.2. How many vibrational modes does an N-atom molecule exhibit? How does the motion of a single particle represent one of these modes?

3.3. Construct the one-particle system that possesses the kinetic and potential energies associated with the vibration of a diatomic molecule.

3.4. How is the classical frequency of oscillation related to the force constant for the mode?

3.5. Why cannot movement on a path along which the potential energy changes be governed by a single traveling de Broglie wave? What two traveling waves are needed?

3.6. How does one employ a symmetry argument in determining the variation of each of these traveling waves about a given point? How can a single differential equation governing the changes in the superposed wave be obtained?

3.7. How is the dependence on time eliminated from this differential equation? How is energy E introduced?

3.8. Construct and solve the Schrödinger equation for a homogeneous beam of noninteracting particles.

3.9. Construct the Schrödinger equation for the harmonic oscillator.

3.10. Why should the state function for the lowest level of a harmonic oscillator appear as a normal distribution curve?

3.11. How should the state functions for the successive excited levels appear?

3.12. Why do we look at relationships involving derivatives of e^{-w^2}?

3.13. How is a relationship identical with the Schrödinger equation for the harmonic oscillator obtained by differentiation?

3.14. How does the derivation in Question 3.13 enable us to construct the suitable state functions for the harmonic oscillator?

3.15. How does this derivation yield the quantization of the energy of a harmonic oscillator?

3.16. What roles are played by (a) the normalization constant, (b) the transcendental factor, (c) the polynomial factor, in the state function?

3.17. How is the ground state harmonic oscillator state function normalized?

3.18. Describe how the analytic solutions of the harmonic oscillator Schrödinger equation go bad when the energy is not properly chosen.

3.19. Describe the nature of the unsuitable ψ_v at a properly chosen energy level.

3.20. What is a Hermite polynomial?

3.21. What is parity? Why does ψ_v have a definite parity? Why does increasing v by 1 change the parity of ψ_v?

3.22. How may a mode of motion gain or lose energy?

3.23. What must a mode exhibit to be active in the infrared region?

3.24. What angular momentum and parity does a photon possess? How can a vibrational transition supply or remove these?

3.25. Relate the allowed vibrational–rotational transitions to the quantized energy levels for a diatomic molecule.

3.26. How is the force constant for a diatomic molecule obtained?

3.27. Discuss the nature of chemical bonding.

3.28. How should the potential energy of a diatomic molecule vary with the internuclear distance?

3.29. How does the anharmonicity affect the quantization of vibrational energy? What is the anharmonicity constant?

3.30. Define the vibrational quantum number v.

3.31. What is the dissociation energy? How may it be obtained from spectroscopic data?

3.32. How can a single particle model the behavior of a mode of an N-atom system? Why do the parameters for the various vibrational modes differ?

3.33. How is the complete vibrational wave function related to the functions for the individual modes? Why?

3.34. What is the Raman effect?

3.35. When is a mode inactive (a) in the infrared, (b) in the Raman spectrum?

3.36. How do harmonics and combination bands arise?

3.37. What are O, P, Q, R, S branches? How and where do they appear?

3.38. When is the contribution of a vibrational mode to the thermodynamic internal energy (a) NkT, (b) $\frac{1}{2}N\hbar\omega$?

3.39. What properties does a differentiating operator exhibit?

3.40. How we factor the operator $(D^2 - w^2 \pm 1)$?

3.41. Write the Schrödinger equation, for the harmonic oscillator, to state that the successive actions of two first-order differential operators produce the same effect as a constant operator acting on ψ.

Problems

3.1. The state function for a harmonic oscillator is not zero in the regions where $V > E$, but instead it drops asymptotically to zero as x rises (or falls) without limit. Calculate the probability that the model particle is in the region where $V \leqslant E$, within the classical range, when the oscillator is in its ground state.

3.2. By a specific integration, normalize the state function for the harmonic oscillator when the mode is in its first excited state, with $v = 1$.

3.3. What are the asymptotic solutions to the Schrödinger equation if

$$V = \frac{1}{2} fx^2 \qquad \text{when} \quad x < 0$$

and

$$V = \frac{1}{2} gx^2 \qquad \text{when} \quad x > 0?$$

3.4. Gaseous $^1H^{35}Cl$ exhibits maximum absorption of infrared radiation at the wave numbers (in cm^{-1}):

3014, 2998, 2981, 2963, 2945,
2926, 2906, 2865, 2844, 2821.

What are its rotational constant B and pure vibrational wave number κ_0?

3.5. Calculate the force constants for H_2 and D_2 from the observed vibrational wave numbers 4159.2 and 2990.3 cm^{-1}, respectively.

3.6. In H_2 the second and third vibrational levels are 4159.2 and 8082.4 cm^{-1} above the first, or ground, level. What is the corresponding harmonic oscillator wave number κ_0?

3.7. From the data in Problem 3.6, calculate the dissociation energy D of H_2.

3.8. Under irradiation, gaseous HCl shifts the wave numbers of photons by the following amounts (in cm^{-1}):

+143.8, +183.3, +222.2,
−101.1, −142.7, −187.5, −229.4, −271.0.

A Raman band also occurs centered 2886.0 cm^{-1} from the exciting line. Interpret these data and deduce (a) the rotational constant B, (b) the vibrational wave number κ_0.

3.9. Show that the v zeros of the vibrational eigenfunction ψ_v are all real and distinct.

3.10. Prove that $(d/dw)H_v(w) = 2vH_{v-1}(w)$.

3.11. If a harmonic oscillator is in its first excited state, what is the probability that $V > E$ at any given moment?

3.12. Derive the polynomial factor for the state function describing the second excited state of a harmonic oscillator. Then determine the location of its nodes.

3.13. If a particle moves in the potential $V = \frac{1}{2} fx^4$, how do its state functions behave where $|x|$ is very large?

3.14. Gaseous HBr absorbs strongly at the following wave numbers (cm^{-1}):

2671, 2658, 2645, 2631, 2617, 2602, 2587,
2571, 2539, 2523, 2506, 2488, 2470, 2451,

What are the rotational constant B and the pure vibrational wave number k_0 of the HBr molecule?

3.15. Calculate the force constants for CO and NO if the vibrational wave numbers are 2143.3 and 1876.0 cm^{-1}, respectively.

3.16. If the force constant for vibration of $^{23}Na^{35}Cl$ is 1.17×10^5 dyne cm^{-1}, what is its wave number k_0?

3.17. From the data in Example 3.10, calculate the vibrational energy of an HCl molecule in its ground and first excited states.

3.18. Calculate the dissociation energy of HCl from the wave numbers 8347 and 10 923 cm^{-1} of the second and third overtones in its vibrational spectrum.

3.19. Show that

$$H_{v+1}(w)\, e^{-w^2} = -\frac{d}{dw}\,[H_v(w)\, e^{-w^2}].$$

3.20. Prove that:

$$H_{v+1}(w) - 2wH_v(w) + 2vH_{v-1}(w) = 0.$$

References

Books

Borowitz, S.: 1967, *Fundamentals of Quantum Mechanics*, Benjamin, New York, pp. 203–226.

Colthup, N. B., Daly, L. H., and Wiberley, S. E.: 1964, *Introduction to Infrared and Raman Spectroscopy*, Academic Press, New York, pp. 1–97.

Horak, M., and Vitek, A.: 1979, *Interpretation and Processing of Vibrational Spectra*, Wiley, New York, pp. 1–413.

Potts, W. J., Jr: 1963, *Chemical Infrared Spectroscopy*, Wiley, New York, pp. 1–91.

White, R. L.: 1966, *Basic Quantum Mechanics*, McGraw-Hill, New York, pp. 67–93.

Articles

Ashby, R. A.: 1975, 'Flames: A Study in Molecular Spectroscopy', *J. Chem. Educ.* **52**, 632–637.

Brabson, G. D.: 1973, 'Calculation of Morse Wave Functions with Programmable Desktop Calculators', *J. Chem. Educ.* **50**, 397–399.

Buchdahl, H. A.: 1974, 'Remark on the Solutions of the Harmonic Oscillator Equation', *Am. J. Phys.* **42**, 47–50.

Gettys, W. E., and Graben, H. W.: 1975, 'Quantum Solution for the Biharmonic Oscillator', *Am. J. Phys.* **43**, 626–629.

Gibbs, R. L.: 1975, 'The Quantum Bouncer', *Am. J. Phys.* **43**, 25–28.

Jinks, K. M.: 1975, 'A Particle in a Chemical Box', *J. Chem. Educ.* **52**, 312–313.

Manka, C. K.: 1972, 'More on Numerical Solutions to Simple One-Dimensional Schrödinger Equations', *Am. J. Phys.* **40**, 1539–1542.

Mazur, P., and Barron, R. H.: 1974, 'On a Variation of a Derivation of the Schrödinger Equation', *Am. J. Phys.* **42**, 600–602.

Meyer-Vernet, N.: 1982, 'Strange Bound States in the Schrödinger Wave Equation: When Usual Tunneling Does Not Occur', *Am. J. Phys.* **50**, 354–356.

Mohammad, S. N.: 1979, 'Calculations of Vibrational–Rotational Coupling Constants in Diatomic Molecules', *Nuovo Cimento* **49B**, 124–134.
Noid, D. W., Koszykowski, M. L., and Marcus, R. A.: 1980, 'Molecular Vibration and the Normal Mode Approximation', *J. Chem. Educ.* **57**, 624–626.
Winn, J. S.: 1981, 'Analytic Potential Functions for Diatomic Molecules: Some Limitations', *J. Chem. Educ.* **58**, 37–38.

Chapter 4

Radial Motion in a Coulombic Field

4.1. A More Complicated Oscillation

Two submicroscopic particles bound together as a system are subject to (a) movement of their center of mass, (b) unidirectional and back-and-forth rotation about this center, and (c) oscillation in the interparticle distance, also about the center of mass. The first two movements have been considered in Chapters 1 and 2; but the third motion is generally different from and more complicated than the simple harmonic motion treated in Chapter 3.

Classically, if no external force appears at either particle, or if the net external forces acting on the two particles have the same direction and are proportional to the masses (as in a uniform gravitational field), then the system is represented by movements of single particles. Presumably, the result carries over into quantum mechanics. If m_1 is the mass of the first particle and m_2 the mass of the second, translation of the center of mass is modeled by movement of a hypothetical particle of mass $m_1 + m_2$ associated with the center. The rotation and oscillation are modeled by movements of a particle of mass $m_1 m_2/(m_1 + m_2)$ at the interparticle separation $\mathbf{r}$ from the center of the internal potential field $V(\mathbf{r})$, erected as an inertial frame.

Any increase in potential energy, associated with a shift in position of a particle, causes some reflection of its de Broglie wave. For the governing differential equation to allow this partial feedback, the equation must be second order. If the equation is to produce the appropriate wave behavior, as observed in other contexts, over each infinitesimal distance, it must be the Schrödinger equation, in the approximation that the relativistic effects of Einstein can be neglected. When the potential energy V is a function of r alone, the radial variable r can be separated from the angular variables θ and φ, and two separate second-order equations constructed.

For each energy E, two independent solutions of the radial equation exist. When V becomes constant (and can be considered negligible) at large r's, the solutions exhibit simple asymptotic forms. One of these is not useable when the particle is bound because it increases without limit as r increases. The other one is useable only if it extends to the center without yielding an infinite probability of finding the particle in a small region about this point. Only at *discrete* negative energies does the extension succeed. Then the intermediate region exhibits a smooth rise, or a waviness, also rising, with radial nodes.

The energy associated with a suitable state function depends on the de Broglie wavelength and the potential at each point. Thus it depends on the total number of nodal surfaces in either the real or the imaginary part of ψ, those introduced by the rotational factor as well as those introduced by the radial factor. Furthermore, this energy depends on the specific form of the potential.

In this chapter, we will assume that the potential varies inversely with the interparticle distance. So, the results will apply to the hydrogen atom and to any hydrogen-like ion, where V is Coulombic about the center.

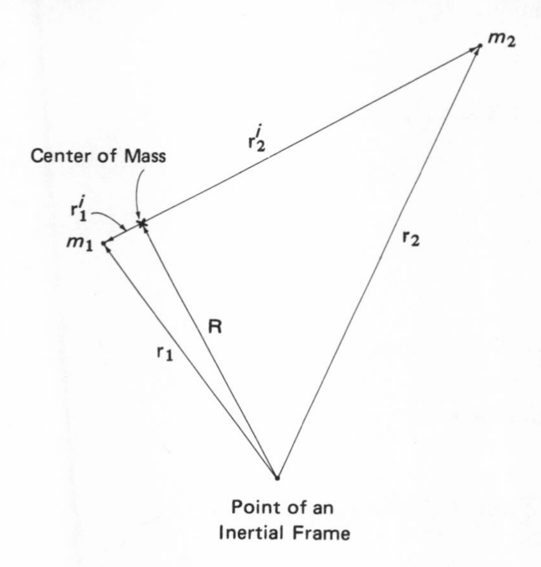

Fig. 4.1. Vectors locating masses m_1, m_2, and their center of mass in an inertial frame.

4.2. Representing the Axial and Angular Motions of Two Particles Bound Together

A fundamental assumption in our theory is that if each particle in a given system were at some point in the imposed space, moving at a certain velocity with respect to the imposed time scale, the unit would possess a kinetic energy and a potential energy calculated in the same way as the corresponding classical quantities. As a consequence, a simple system, such as a single particle in a specified potential field, can represent those aspects of a multiparticle unit for which it exhibits the same classical kinetic and potential energy functions.

Consider a particle of mass m_1 binding a particle of mass m_2 by a potential V.

Draw vectors $\mathbf{r}_1$ and $\mathbf{r}_2$ from a point in an inertial frame to these particles, as shown in Figure 4.1. Vector $\mathbf{R}$ locating the center of mass is defined by the equation

$$m_1\mathbf{r}_1 + m_2\mathbf{r}_2 = (m_1 + m_2)\mathbf{R}. \tag{4.1}$$

Add it to the figure. Also, draw $\mathbf{r}_1^i$ and $\mathbf{r}_2^i$ from the center of mass to the first and second particles. For these, (4.1) reduces to

$$m_1\mathbf{r}_1^i + m_2\mathbf{r}_2^i = 0. \tag{4.2}$$

Differentiate (4.2) with respect to the imposed time t, letting a dot over a symbol indicate the differentiation:

$$m_1\dot{\mathbf{r}}_1^i + m_2\dot{\mathbf{r}}_2^i = 0. \tag{4.3}$$

Since $\mathbf{r}_j$ equals the vector sum of $\mathbf{R}$ and $\mathbf{r}_j^i$, we have

$$\dot{\mathbf{r}}_j = \dot{\mathbf{R}} + \dot{\mathbf{r}}_j^i. \tag{4.4}$$

Also, we let

$$m_1 + m_2 = M, \tag{4.5}$$

the total mass.

Now, the kinetic energy of the two particles is

$$\begin{aligned} T &= \frac{1}{2}\Sigma m_j\dot{\mathbf{r}}_j \cdot \dot{\mathbf{r}}_j = \frac{1}{2}\Sigma m_j(\dot{\mathbf{R}} + \dot{\mathbf{r}}_j^i) \cdot (\dot{\mathbf{R}} + \dot{\mathbf{r}}_j^i) \\ &= \frac{1}{2}\Sigma m_j\dot{\mathbf{R}} \cdot \dot{\mathbf{R}} + \dot{\mathbf{R}} \cdot \Sigma m_j\dot{\mathbf{r}}_j^i + \frac{1}{2}\Sigma m_j\dot{\mathbf{r}}_j^i \cdot \dot{\mathbf{r}}_j^i \\ &= \frac{1}{2}M\dot{\mathbf{R}} \cdot \dot{\mathbf{R}} + \frac{1}{2}\Sigma m_j\dot{\mathbf{r}}_j^i \cdot \dot{\mathbf{r}}_j^i = T^e + T^i. \end{aligned} \tag{4.6}$$

In the second equality, (4.4) has been introduced; in the fourth equality, (4.5) and (4.3).

Since $\mathbf{r}_2^i$ and $\mathbf{r}_1^i$ are oppositely directed, we have

$$\mathbf{r}_2^i = r_2^i\mathbf{l}, \tag{4.7}$$

$$\mathbf{r}_1^i = -r_1^i\mathbf{l}, \tag{4.8}$$

where $\mathbf{l}$ is a unit vector. The time derivative of $\mathbf{l}$ equals the angular velocity of rotation ω times a unit vector $\mathbf{n}$ pointing perpendicular to $\mathbf{l}$ in the direction of rotation:

$$\dot{\mathbf{l}} = \omega\mathbf{n}. \tag{4.9}$$

So, differentiating (4.7) and (4.8) yields

$$\dot{\mathbf{r}}_2^i = \dot{r}_2^i \mathbf{l} + r_2^i \omega \mathbf{n}, \tag{4.10}$$

$$\dot{\mathbf{r}}_1^i = -\dot{r}_1^i \mathbf{l} - r_1^i \omega \mathbf{n}, \tag{4.11}$$

and the internal kinetic energy is

$$T^i = \frac{1}{2} \Sigma m_j \dot{\mathbf{r}}_j^i \cdot \dot{\mathbf{r}}_j^i = \frac{1}{2} \Sigma m_j (\dot{r}_j^i)^2 + \frac{1}{2} \Sigma m_j (r_j^i \omega)^2$$

$$= \frac{1}{2} \Sigma m_j (\dot{r}_j^i)^2 + \frac{1}{2} I \omega^2 = T^{\mathrm{rad}} + T^{\mathrm{rot}}. \tag{4.12}$$

The total kinetic energy of the system equals the translational kinetic energy T^e plus the rotational kinetic energy T^{rot} plus the radial kinetic energy T^{rad}.

When the system moves through space freely, with no fields opposing or aiding either its translation or rotation, the potential V is a function only of the magnitude of the vector

$$\mathbf{r} = \mathbf{r}_2 - \mathbf{r}_1 \tag{4.13}$$

drawn from the first particle to the second.

Now, we can place a particle of mass

$$\mu = \frac{m_1 m_2}{m_1 + m_2} \tag{4.14}$$

distance $\mathbf{r}$ from the origin of an inertial frame. The kinetic energy associated with this particle swinging around the origin at a given r equals the corresponding rotational energy of the physical system, according to Section 2.2. The kinetic energy associated with this particle changing its distance from the origin equals the kinetic energy of oscillation of the physical system, according to Section 3.2. If, in addition, the potential to which this particle is subjected is the same

$$V = V(r) \tag{4.15}$$

as in the actual system, this particle models the rotational and oscillatory behavior of the physical system faithfully. Quantity μ is called the reduced mass for the physical system, as before.

4.3. Differential Equations Governing Variations in a State Function

A submicroscopic particle cannot be tracked along a definite curve. When in a particular state, a system of particles has to be considered possessing each possible configuration in space with only a certain probability. The probability density is related to a wave function Ψ by Equations (1.1) and (1.46). Symmetry considerations determine how Ψ varies with coordinates and with time.

For simplicity, we will here consider only those motions that are represented, or modeled, by movements of a single particle. There are then only three coordinates on which potential V and wave function Ψ depend. Let us proceed as in Section 3.3, approximating the potential in an arbitrary small volume element by its average. From the principle of continuity, this action does not appreciably alter Ψ outside the element. Furthermore, the wave function and its first derivatives must match across the walls of the element, as in Section 1.15.

We consider the model particle, of mass μ, traveling in the element with a definite energy E. It then has the kinetic energy

$$\frac{p^2}{2\mu} = E - V. \tag{4.16}$$

Movement through the infinitesimal region in one direction is governed by a constituent wave function labeled Ψ_+; movement in the opposite direction, by a function labeled Ψ_-. Successive points along either path through the element are equivalent. Furthermore, each part of Ψ_+, or of Ψ_-, should produce a similar effect. Therefore, from symmetry, we have

$$\mathrm{d}\Psi_+ = ik\Psi_+ (\mathrm{d}x') - i\omega\Psi_+ \,\mathrm{d}t, \tag{4.17}$$

where the x' axis points in the direction of motion. And since the movement in the opposite direction involves the same kinetic and potential energies, the coefficient of $\Psi_- \,\mathrm{d}x'$ for it is merely the negative of the corresponding one in (4.17), while the coefficient for $\Psi_- \,\mathrm{d}t$ is the same:

$$\begin{aligned} \mathrm{d}\Psi_- &= ik\Psi_- (-\mathrm{d}x') - i\omega\Psi_- \,\mathrm{d}t \\ &= -ik\Psi_- \,\mathrm{d}x' - i\omega\Psi_- \,\mathrm{d}t. \end{aligned} \tag{4.18}$$

From (1.12) and (1.18) we obtain

$$k\,\mathrm{d}x' = \mathbf{k} \cdot \mathbf{dr} = k_1 \,\mathrm{d}x_1 + k_2 \,\mathrm{d}x_2 + k_3 \,\mathrm{d}x_3 \tag{4.19}$$

if we label the three Cartesian reference axes 1, 2, and 3. So (4.17) and (4.18) can be rewritten in the form

$$\begin{aligned} \mathrm{d}\Psi_\pm &= \pm i\Psi_\pm \mathbf{k} \cdot \mathbf{dr} - i\Psi_\pm \omega \,\mathrm{d}t \\ &= \pm i\Psi_\pm (k_1 \,\mathrm{d}x_1 + k_2 \,\mathrm{d}x_2 + k_3 \,\mathrm{d}x_3) - i\Psi_\pm \omega \,\mathrm{d}t. \end{aligned} \tag{4.20}$$

As before, we expect each constituent function to be analytic,

$$\mathrm{d}\Psi_\pm = \frac{\partial \Psi_\pm}{\partial x_1} \mathrm{d}x_1 + \frac{\partial \Psi_\pm}{\partial x_2} \mathrm{d}x_2 + \frac{\partial \Psi_\pm}{\partial x_3} \mathrm{d}x_3 + \frac{\partial \Psi_\pm}{\partial t} \mathrm{d}t. \tag{4.21}$$

Comparing (4.20) and (4.21) leads to

$$\frac{\partial \Psi_\pm}{\partial x_j} = \pm ik_j \Psi_\pm. \tag{4.22}$$

The propagation of submicroscopic particles in a homogeneous beam can be studied by subjecting the beam to diffraction. We find that (4.22) is satisfied with

$$k_j = \frac{p_j}{\hbar}. \tag{4.23}$$

We call k_j the jth component of wavevector $\mathbf{k}$.

Within the altered element, each k_j is constant, so differentiating (4.22) with respect to x_j yields simply

$$\frac{\partial^2 \Psi_\pm}{\partial x_j^2} = \pm ik_j \frac{\partial \Psi_\pm}{\partial x_j} = -k_j^2 \Psi_\pm. \tag{4.24}$$

The complete wave function equals the superposition of the two parts:

$$\Psi = \Psi_+ + \Psi_-. \tag{4.25}$$

But since each part satisfies (4.24), the complete function satisfies the same equation:

$$\frac{\partial^2 \Psi}{\partial x_j^2} = -k_j^2 \Psi. \tag{4.26}$$

Equation (4.26) gives the spatial rate of change of each first derivative, on moving across the replacement element in the pertinent direction. Since the wave function and its first derivative at the walls equal the corresponding expressions just outside, we argue that (4.26) must also apply to the original unaltered system.

The dependence on time of each constituent of the given energy state is the same. As a consequence, the temporal part can be factored from (4.26) to give

$$\frac{\partial^2 \psi}{\partial x_j^2} = -k_j^2 \psi. \tag{4.27}$$

Next, we sum both side of (4.27) over all three coordinates, obtaining

$$\Sigma \frac{\partial^2 \psi}{\partial x_j^2} = -\Sigma k_j^2 \psi. \tag{4.28}$$

The operator acting upon ψ on the left is the Laplacian operator,

$$\Sigma \frac{\partial^2}{\partial x_j^2} = \nabla^2, \tag{4.29}$$

while the sum Σk_j^2 on the right equals k^2, by the Pythagorean theorem.

Consequently, (4.28) can be rewritten in the form

$$\nabla^2 \psi = -k^2 \psi. \tag{4.30}$$

Combining (4.30) with (4.23) and (4.16) yields

$$\nabla^2 \psi = \frac{2\mu(V-E)}{\hbar^2} \psi, \tag{4.31}$$

a three-dimensional *Schrödinger equation*. Deductions from it agree very closely with experimental results wherever the refinements of Einsteinian relativity can be neglected.

In any given problem, one seeks the energies that allow physically suitable solutions to exist. With more than one spatial dimension, the search can be very difficult, but when the potential allows mutually perpendicular motions throughout a region to be independent, these motions are described by independent factors in the wave function, following Section 1.7. The coordinates on which the factors depend can be separated in (4.31) and ordinary differential equations, valid in the region, can be constructed. These can be solved in principle, obtaining the conditions sought.

4.4. An Orthogonal-Coordinate Representation of ∇^2

Before we can carry through the more difficult separations in the Schrödinger equation, we have to consider how ∇^2 may be expressed in terms of orthogonal coordinates. We will use the fact that the del-squared operator appears in the differential equation governing the movements of conserved effects.

Now, a transformation from Cartesian coordinates x_1, x_2, x_3 to a set of generalized coordinates q_1, q_2, q_3 is generated by equations of the form

$$x_1 = x_1(q_1, q_2, q_3), \tag{4.32}$$

$$x_2 = x_2(q_1, q_2, q_3), \tag{4.33}$$

$$x_3 = x_3(q_1, q_2, q_3). \tag{4.34}$$

Coefficients h_1, h_2, and h_3 are defined so the distance a physical point moves when q_1 increases by dq_1 with the other coordinates fixed is $h_1\ dq_1$, the distance traversed when only q_2 changes is $h_2\ dq_2$, and the distance traced out when only q_3 increases is $h_3\ dq_3$. Because we are interested in independent motions, we take these elements to be mutually perpendicular, as Figure 4.2 shows. The coordinates are then said to be *orthogonal*.

Let us consider a hypothetical conserved effect moving by the central point of the volume element, and through the various faces. The amount transported from one layer to the next varies directly with the cross-sectional area, with the time, and with the driving force. About a given concentration of the effect, this force is proportional to the negative of the gradient of the concentration, other things being the same.

Amounts will be measured in a certain unit and the results reported as number N, or when infinitesimal as $đN$. We let C be the concentration of the effect about the pertinent point, dS the appropriate infinitesimal area, dt the infinitesimal time.

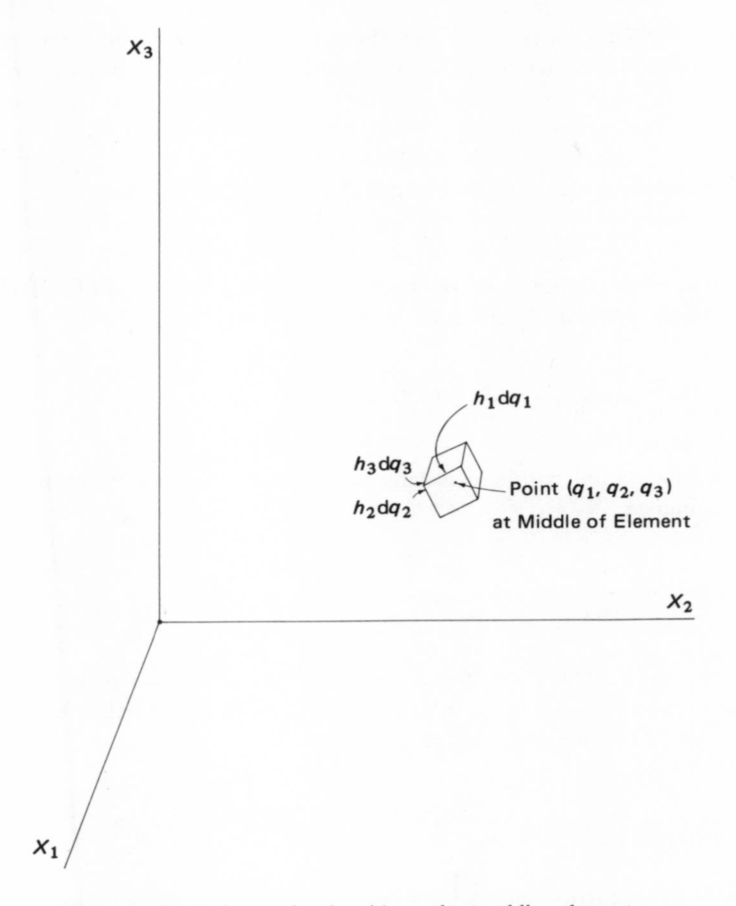

Fig. 4.2. Volume element bordered by orthogonal line elements.

For movement through the volume element by point (q_1, q_2, q_3) in the direction of increasing q_1, we have the area

$$dS_{q_1} = h_2 \, dq_2 \times h_3 \, dq_3 \tag{4.35}$$

and the concentration gradient $\partial C/h_1 \, \partial q_1$; with D the coefficient of proportionality, we have

$$d^3 N_{q_1} = -D \frac{\partial C}{h_1 \, \partial q_1} \, dS_{q_1} \, dt. \tag{4.36}$$

Hence

$$\frac{\partial d^2 N_{q_1}}{\partial t} = -D \frac{h_2 h_3}{h_1} \frac{\partial C}{\partial q_1} \, dq_2 \, dq_3 . \tag{4.37}$$

Equation (4.37) gives the rate at which the effect moves by the cross section of area (4.35) through the center of the volume element. The rate for movement into the element by way of the left face, where the first coordinate equals $q_1 - \mathrm{d}q_1/2$, is

$$-D \frac{h_2 h_3}{h_1} \frac{\partial C}{\partial q_1} \mathrm{d}q_2 \, \mathrm{d}q_3 - \frac{\partial}{\partial q_1} \left(-D \frac{h_2 h_3}{h_1} \frac{\partial C}{\partial q_1} \mathrm{d}q_2 \, \mathrm{d}q_3 \right) \frac{\mathrm{d}q_1}{2}, \tag{4.38}$$

while the rate at which it moves out of the element by way of the right face, where the first coordinate equals $q_1 + \mathrm{d}q_1/2$, is

$$-D \frac{h_2 h_3}{h_1} \frac{\partial C}{\partial q_1} \mathrm{d}q_2 \, \mathrm{d}q_3 + \frac{\partial}{\partial q_1} \left(-D \frac{h_2 h_3}{h_1} \frac{\partial C}{\partial q_1} \mathrm{d}q_2 \, \mathrm{d}q_3 \right) \frac{\mathrm{d}q_1}{2}. \tag{4.39}$$

The rate of accumulation within the element caused by movements across these faces is the difference

$$\frac{\partial}{\partial q_1} \left(D \frac{h_2 h_3}{h_1} \frac{\partial C}{\partial q_1} \right) \mathrm{d}q_1 \, \mathrm{d}q_2 \, \mathrm{d}q_3. \tag{4.40}$$

Similar rates of accumulation appear between the other two pairs of faces. Adding all of these and dividing by the volume of the element, which is $h_1 h_2 h_3 \times \mathrm{d}q_1 \, \mathrm{d}q_2 \, \mathrm{d}q_3$, yields the rate of change in concentration

$$\frac{\partial C}{\partial t} = \frac{1}{h_1 h_2 h_3} \left[\frac{\partial}{\partial q_1} \left(D \frac{h_2 h_3}{h_1} \frac{\partial C}{\partial q_1} \right) + \frac{\partial}{\partial q_2} \left(D \frac{h_3 h_1}{h_2} \frac{\partial C}{\partial q_2} \right) + \right.$$
$$\left. + \frac{\partial}{\partial q_3} \left(D \frac{h_1 h_2}{h_3} \frac{\partial C}{\partial q_3} \right) \right]. \tag{4.41}$$

When coefficient D is constant, Equation (4.41) reduces to

$$\frac{\partial C}{\partial t} = D \frac{1}{h_1 h_2 h_3} \left[\frac{\partial}{\partial q_1} \left(\frac{h_2 h_3}{h_1} \frac{\partial C}{\partial q_1} \right) + \frac{\partial}{\partial q_2} \left(\frac{h_3 h_1}{h_2} \frac{\partial C}{\partial q_2} \right) + \right.$$
$$\left. + \frac{\partial}{\partial q_3} \left(\frac{h_1 h_2}{h_3} \frac{\partial C}{\partial q_3} \right) \right]. \tag{4.42}$$

In Cartesian coordinates, (4.42) becomes

$$\frac{\partial C}{\partial t} = D \left(\frac{\partial^2 C}{\partial x_1^2} + \frac{\partial^2 C}{\partial x_2^2} + \frac{\partial^2 C}{\partial x_3^2} \right). \tag{4.43}$$

We recognize the expression multiplying D as the Cartesian form of $\nabla^2 C$. Consequently, Equation (4.42) tells us that

$$\nabla^2 C = \frac{1}{h_1 h_2 h_3}\left[\frac{\partial}{\partial q_1}\left(\frac{h_2 h_3}{h_1}\frac{\partial C}{\partial q_1}\right) + \frac{\partial}{\partial q_2}\left(\frac{h_3 h_1}{h_2}\frac{\partial C}{\partial q_2}\right) + \right.$$

$$\left. + \frac{\partial}{\partial q_3}\left(\frac{h_1 h_2}{h_3}\frac{\partial C}{\partial q_3}\right)\right]. \tag{4.44}$$

In this formula, q_1, q_2, q_3 are the generalized orthogonal coordinates and h_1, h_2, h_3 are the multipliers that change dq_1, dq_2, dq_3 to the perpendicular line elements.

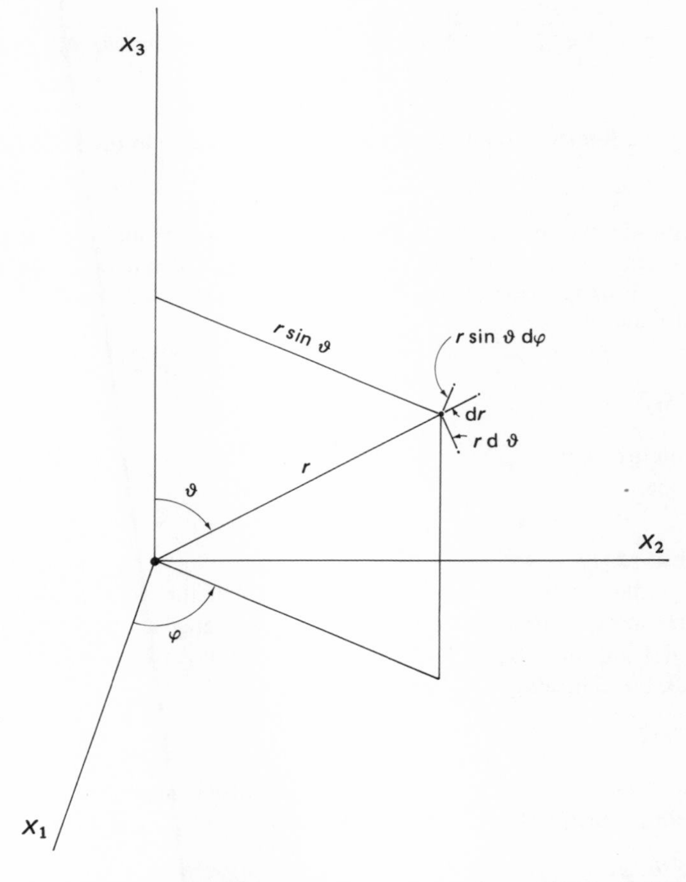

Fig. 4.3. The three line elements traced out when each spherical coordinate increases infinitesimally by itself.

Example 4.1. What is $\nabla^2 \psi$ in spherical coordinates?

The spherical coordinates of an arbitrary point are defined with respect to a center and Cartesian axes based on the center, as Figure 4.3 shows. The element traced out by an infinitesimal displacement with only r changing is dr, with only θ changing $r\, d\theta$, with only φ changing $r \sin\theta\, d\varphi$.

If we set $q_1 = r$, $q_2 = \theta$, and $q_3 = \varphi$, we have

$$h_1 = 1, \qquad h_2 = r, \qquad h_3 = r\sin\theta.$$

If we also replace C with ψ, Equation (4.44) becomes

$$\nabla^2 \psi = \frac{1}{r^2 \sin\theta}\left[\frac{\partial}{\partial r}\left(\frac{r^2 \sin\theta}{1}\frac{\partial \psi}{\partial r}\right) + \frac{\partial}{\partial \theta}\left(\frac{r\sin\theta}{r}\frac{\partial \psi}{\partial \theta}\right) + \frac{\partial}{\partial \varphi}\left(\frac{r}{r\sin\theta}\frac{\partial \psi}{\partial \varphi}\right)\right]$$

$$= \frac{1}{r^2}\frac{\partial}{\partial r}\left(r^2 \frac{\partial \psi}{\partial r}\right) + \frac{1}{r^2 \sin\theta}\frac{\partial}{\partial \theta}\left(\sin\theta \frac{\partial \psi}{\partial \theta}\right) + \frac{1}{r^2 \sin^2\theta}\frac{\partial^2 \psi}{\partial \varphi^2}.$$

4.5. Separating the Radial Variable from the Angular Variables in the Schrödinger Equation

We have been considering two particles, of masses m_1 and m_2, bound together by a potential $V(r)$, where $\mathbf{r}$ is the vector drawn from the center of m_1 to the center of m_2. We have seen how rotation and oscillation of this system are described by a single particle of reduced mass

$$\mu = \frac{m_1 m_2}{m_1 + m_2} \tag{4.45}$$

moving in the potential field

$$V = V(r) \tag{4.46}$$

considered to be at rest.

In this field, radius vector $\mathbf{r}$ extends from the center of the field to the model particle. Cartesian axes are drawn through the center and angles θ and φ are measured as Figure 4.3 indicates. Since V depends only on r, wavevector k varies only with r and (4.30), the Schrödinger equation, becomes

$$\nabla^2 \psi + k^2(r)\psi = 0. \tag{4.47}$$

We suppose that radial movement is independent of angular movement when the system is in a definite energy state. Consequently, we are led to try the form

$$\psi = R(r)Y(\theta, \varphi) \tag{4.48}$$

in (4.47).

First, operate on both sides of (4.48) with del,

$$\nabla\psi = \nabla(RY) = R\,\nabla Y + Y\,\nabla R, \tag{4.49}$$

then with del dot,

$$\nabla\cdot\nabla\psi = \nabla\cdot R\,\underline{\nabla Y} + \nabla\cdot\underline{R}\,\nabla Y + \nabla\cdot\underline{Y}\,\nabla R + \nabla\cdot Y\,\underline{\nabla R}. \tag{4.50}$$

The underline indicates the factor on which the differentiation of the secondly applied del acts. Rearranging the terms algebraically yields

$$\begin{aligned}\nabla^2\psi &= R\nabla\cdot\nabla Y + \nabla R\cdot\nabla Y + \nabla Y\cdot\nabla R + Y\nabla\cdot\nabla R\\ &= R\,\nabla^2 Y + Y\,\nabla^2 R.\end{aligned} \tag{4.51}$$

The two middle terms drop out because a gradient lies in the direction in which the space rate of change of the function is greatest; and R changes most rapidly in the direction in which only r changes, Y changes most rapidly in a perpendicular direction, in which only θ and φ vary.

Identity (4.51) converts (4.47) to

$$R\nabla^2 Y + Y\nabla^2 R + k^2(r)RY = 0. \tag{4.52}$$

Furthermore, the formula in Example 4.1 tells us that

$$\nabla^2 R = \frac{1}{r^2}\frac{\partial}{\partial r}\left(r^2\frac{\partial R}{\partial r}\right) = \frac{1}{r^2}\frac{\mathrm{d}}{\mathrm{d}r}\left(r^2\frac{\mathrm{d}R}{\mathrm{d}r}\right) \tag{4.53}$$

and

$$\nabla^2 Y = \frac{1}{r^2\sin\theta}\frac{\partial}{\partial\theta}\left(\sin\theta\frac{\partial Y}{\partial\theta}\right) + \frac{1}{r^2\sin^2\theta}\frac{\partial^2 Y}{\partial\varphi^2}, \tag{4.54}$$

when the vanishing terms are dropped. Therefore, multiplying (4.52) by $r^2(RY)^{-1}$ separates the variables. Rearranging the result then leads to

$$r^2\left[\frac{1}{R}\nabla^2 R + k^2(r)\right] = -\frac{1}{Y}r^2\,\nabla^2 Y. \tag{4.55}$$

Since the left side of (4.55) is a function only of r, no change in θ or φ can cause it to vary. Since the right side of (4.55) is a function only of θ and of φ, no change in r can cause it to vary. We thus have an expression that might vary only when r varies equal to one that does not vary when r varies. The expression must therefore equal a constant.

To determine what this constant is, we consider the rotational part of the motion by itself. Apply (4.30), the Schrödinger equation, to function Y alone, obtaining

$$\nabla^2 Y + k_Y^2(r)Y = 0. \tag{4.56}$$

Now, the wavevector is related to the corresponding linear momentum by de Broglie's equation (1.30),

$$p_Y = k_Y \hbar, \tag{4.57}$$

and to the corresponding kinetic energy in the classical manner; thus

$$T^{\mathrm{rot}} = \frac{p_Y^2}{2\mu} = \frac{k_Y^2 \hbar^2}{2\mu}. \tag{4.58}$$

Since (4.58) is the rotational energy of a two-particle system, it is quantized as (2.49) indicates. Replacing J with l and I with μr^2 gives us

$$\frac{k_Y^2 \hbar^2}{2\mu} = l(l+1)\frac{\hbar^2}{2\mu r^2}, \tag{4.59}$$

whence

$$k_Y^2 = l(l+1)\frac{1}{r^2}. \tag{4.60}$$

Relationship (4.60) converts (4.56) to

$$\nabla^2 Y + \frac{l(l+1)}{r^2} Y = 0. \tag{4.61}$$

Substituting the corresponding $-Y^{-1} r^2 \nabla^2 Y$ into (4.55) reduces it to

$$r^2 \left[\frac{1}{R}\nabla^2 R + k^2(r)\right] = l(l+1), \tag{4.62}$$

whence

$$\nabla^2 R + \left[k^2(r) - \frac{l(l+1)}{r^2}\right] R = 0. \tag{4.63}$$

Now, de Broglie's equation for total momentum

$$p = k\hbar \tag{4.64}$$

converts the energy equation

$$E = \frac{p^2}{2\mu} + V \tag{4.65}$$

to

$$E = \frac{k^2 \hbar^2}{2\mu} + V, \tag{4.66}$$

whence

$$k^2 = \frac{2\mu}{\hbar^2}(E - V). \tag{4.67}$$

Consequently, Equation (4.63) can be rewritten in the form

$$\nabla^2 R + \left\{ \frac{2\mu}{\hbar^2} [E - V(r)] - \frac{l(l+1)}{r^2} \right\} R = 0, \tag{4.68}$$

or, with (4.53), in the form

$$\frac{1}{r^2} \frac{\mathrm{d}}{\mathrm{d}r} \left(r^2 \frac{\mathrm{d}R}{\mathrm{d}r} \right) + \left\{ \frac{2\mu}{\hbar^2} [E - V(r)] - \frac{l(l+1)}{r^2} \right\} R = 0, \tag{4.69}$$

whence

$$\frac{\mathrm{d}^2 R}{\mathrm{d}r^2} + \frac{2}{r} \frac{\mathrm{d}R}{\mathrm{d}r} + \left\{ \frac{2\mu}{\hbar^2} [E - V(r)] - \frac{l(l+1)}{r^2} \right\} R = 0. \tag{4.70}$$

Note that $l(l+1)/r^2$ is k_Y^2, by (4.60). This equals $2\mu/\hbar^2$ times the kinetic energy associated with the angular motion, according to (4.67). Equation (4.70) could therefore be rewritten in the form

$$\frac{\mathrm{d}^2 R}{\mathrm{d}r^2} + \frac{2}{r} \frac{\mathrm{d}R}{\mathrm{d}r} + \frac{2\mu}{\hbar^2} [E - V(r) - T^{\mathrm{rot}}] R = 0. \tag{4.71}$$

When the square of the local wavevector depends only on distance of the particle from a point, variation in an angle about the point affects the radial motion, to or from the point, only through the corresponding kinetic energy.

Example 4.2. How does the wave function for a particle in a given energy state behave in a region where the potential is constant?

In a region where V does not vary, each energy state has a wavevector of fixed magnitude. Furthermore, Equations (4.17) and (4.18) apply, as in Chapter 1. Motion parallel to a line is described by (1.137) or (1.139).

When $E > V$, the wavevector k is real and the particle moves as a free translator within the region with

$$k = \frac{[2\mu(E - V)]^{1/2}}{\hbar}.$$

When $E < V$, *direct* movement into the region is limited by the attenuation constant

$$\kappa = \frac{k}{i} = \frac{[2\mu(V - E)]^{1/2}}{\hbar}.$$

Now, a positive kinetic energy T^{per} may be associated with movement perpendicular to the line of advance into the constant V region. In the approximation that this energy is fixed, movement into the region is limited by

$$\kappa = \frac{[2\mu(V + T^{\mathrm{per}} - E)]^{1/2}}{\hbar}.$$

Commonly, the interaction between two particles drops to zero as the distance between them increases. The resulting potential becomes constant. When the two particles are bound together, E is less than V in the far field. The movement of the model particle into the region is correspondingly attenuated. The movement perpendicular to the line of advance of the model particle appears as rotation of the system. At a given l, the kinetic energy associated with this becomes small as r increases.

Example 4.3. How does the radial factor for the wave function describing movement in a spherically symmetric field behave out where V becomes negligibly small?

At large r, where the second, fourth, and fifth terms in (4.70) are negligible, the equation reduces to the form

$$\frac{d^2R}{dr^2} + \frac{2\mu E}{\hbar^2} R = 0$$

that integrates to

$$R = A\,e^{-\kappa r} + B\,e^{\kappa r}$$

if

$$\kappa = \frac{(-2\mu E)^{1/2}}{\hbar}.$$

Constant κ is real as long as the particle is bound and E negative. But then, the probability density must go to zero as r increases without limit, and we must have

$$B = 0.$$

4.6. The Radial Equation for an Electron–Nucleus System

When an electron moves in the field of a nucleus, without other particles being present, the potential of the system varies inversely with the interparticle distance r. The differential equation for radial factor $R(r)$, which we have constructed, is valid and can be readily solved.

Consider a nucleus of mass m_1 and charge Ze attracting a particle of mass m_2 and charge $-e$ at distance r. The potential energy of the system is

$$V = -\frac{Ze^2}{4\pi\epsilon_0 r} \tag{4.72}$$

where ϵ_0 is the permittivity of space. Equation (4.70) for the radial factor in ψ then becomes

$$\frac{d^2R}{dr^2} + \frac{2}{r}\frac{dR}{dr} + \left[\frac{2\mu}{\hbar^2}\left(E + \frac{Ze^2}{4\pi\epsilon_0 r}\right) - \frac{l(l+1)}{r^2}\right]R = 0. \tag{4.73}$$

To simplify this equation, let

$$r = \frac{1}{2}anx \tag{4.74}$$

and

$$E = \mp\frac{b}{n^2}. \tag{4.75}$$

We obtain

$$\frac{d^2R}{dx^2} + \frac{2}{x}\frac{dR}{dx} + \left[\mp\frac{1}{4} + \frac{n}{x} - \frac{l(l+1)}{x^2}\right]R = 0 \tag{4.76}$$

if

$$a = \frac{4\pi\epsilon_0\hbar^2}{\mu Ze^2} \tag{4.77}$$

and

$$b = \frac{\mu Z^2 e^4}{32\pi^2\epsilon_0^2\hbar^2} \tag{4.78}$$

When the kinetic energy is less than the negative potential energy at each point, energy E is negative and the minus sign is employed. When the particles are not bound together, the kinetic energy is greater than $-V$ and the positive sign is employed.

The negative sign in (4.76) yields the asymptotic solution

$$R = A\,e^{-x/2} + B\,e^{x/2}, \tag{4.79}$$

while the positive sign in (4.76) leads to the distinct asymptotic solution

$$R = C\,e^{-ix/2} + D\,e^{ix/2}. \tag{4.80}$$

A radial factor that increases without limit as x increases does not describe a physical system; it implies that the probability of finding the particle at infinite distance from the origin is infinite. Consequently, formula (4.79) is not acceptable unless coefficient B is zero. Only one of the independent solutions may be suitable when m_2 is bound to m_1. But when the particles are not bound to each other, formula (4.80) holds and the asymptotic behavior does not eliminate a solution.

To determine how the radial factor behaves near the singular point at the origin, we substitute the series

$$R = a_1x^L + a_2x^{L+1} + a_3x^{L+2} + \ldots \tag{4.81}$$

into (4.76), obtaining

$$[L(L-1) + 2L - l(l+1)]\,a_1x^{L-2} + \ldots = 0. \tag{4.82}$$

Since this equation must not impose any condition on x, it is an identity; the coefficient of each power of x is zero. Thus, we have

$$L(L-1) + 2L - l(l+1) = 0 \tag{4.83}$$

or

$$L(L+1) = l(l+1), \tag{4.84}$$

whence

$$L = l \tag{4.85}$$

or

$$L = -(l+1). \tag{4.86}$$

Equation (4.85) leads to an R containing x^l as a factor. The limit of the function as x vanishes is a constant when $l = 0$, zero when $l = 1, 2, \ldots$. Either behavior is satisfactory. Equation (4.86) yields an R whose leading term is a_1/x^{l+1}. For all integral l's, this expression becomes infinite at $x = 0$ with such strength that the R cannot be normalized. These solutions have to be rejected. When l is zero, calculation of the next term in (4.82) leads to the condition

$$n = 0, \tag{4.87}$$

and by (4.74), to $r = 0$. Consequently, this solution is spurious.

When the two particles are bound together, the radial function behaves as Ex^l near $x = 0$ and as $A\,\mathrm{e}^{-x/2}$ where $x = \infty$. Only certain n's in (4.76) permit the acceptable small-x solution to join with the acceptable large-x solution as x increases. The resulting R appears as

$$R = u(x)x^l\,\mathrm{e}^{-x/2}. \tag{4.88}$$

Factor $u(x)$ serves to introduce the appropriate waviness and nodes into the state function at intermediate x's.

Formally substituting (4.88) into (4.76), with the negative sign chosen, gives us

$$x\frac{\mathrm{d}^2u}{\mathrm{d}x^2} + (2l + 2 - x)\frac{\mathrm{d}u}{\mathrm{d}x} + (n - l - 1)\,u = 0, \tag{4.89}$$

a differential equation with two independent solutions for each n and l. Thus, the transformation does not eliminate the unsuitable R's. On solving (4.89), one has to choose the forms that do not override the desired behavior of factors x^l and $\mathrm{e}^{-x/2}$, at $x = 0$ and $x = \infty$, in the radial function. An expression that becomes constant as $x \to 0$ and does not increase faster than x^j as $x \to \infty$ meets this condition.

Example 4.4. Show that a function which behaves as

$$R = \frac{N/\left(\frac{1}{2}an\right)^{l+1}}{x^{l+1}} = \frac{N}{r^{l+1}}$$

near $x = 0$ is not normalizable when l is a positive integer.

In normalization integral (1.47), the differential volume may be differential area $r^2\ \mathrm{d}\Omega$, on a shell distance r from the center, times the thickness of the shell $\mathrm{d}r$:

$$\mathrm{d}^3\mathbf{r} = \mathrm{d}r(r^2\ \mathrm{d}\Omega).$$

Here $\mathrm{d}\Omega$ is the solid angle subtended by the area, about the origin. Consequently, the factored function

$$\psi = R(r)Y(\theta, \varphi),$$

substituted into (1.47), yields

$$1 = \int_{\substack{\text{all}\\ \text{space}}} \psi^* \psi\ \mathrm{d}^3\mathbf{r} = \int_0^\infty R^2(r) r^2\ \mathrm{d}r \int_0^{4\pi} Y^* Y\ \mathrm{d}\Omega.$$

If $R = N/r^{l+1}$ and l is a positive integer, we have

$$\int_0^a R^2 r^2\ \mathrm{d}r = \int_0^a \frac{N^2}{r^{2l}}\ \mathrm{d}r = N^2 \left.\frac{r^{-2l+1}}{-2l+1}\right|_0^a = \infty$$

whenever N is not zero. No finite N can be found which makes this integral finite and allows ψ to be normalized.

4.7. Laguerre Polynomials

By differentiating a simple product, rearranging, and differentiating repeatedly, we can construct Equation (4.89). An explicit form for each suitable u and an expression for the corresponding energy can then be deduced.

We first multiply the exponential of $-x$ by x raised to an integral power j, thus forming a function that rises from zero at $x = 0$ to a maximum and subsequently falls back to zero asymptotically as x increases without limit:

$$y = x^j\ \mathrm{e}^{-x}. \tag{4.90}$$

Let us differentiate (4.90) once, rearrange, and simplify:

$$x \frac{\mathrm{d}y}{\mathrm{d}x} + (x - j)y = 0. \tag{4.91}$$

Then differentiate $j + 1$ times to obtain

$$x \frac{\mathrm{d}^2 z}{\mathrm{d}x^2} + (x + 1) \frac{\mathrm{d}z}{\mathrm{d}x} + (j + 1)z = 0, \tag{4.92}$$

where

$$z = \frac{\mathrm{d}^j y}{\mathrm{d}x^j}. \tag{4.93}$$

The expression obtained on multiplying z by e^x,

$$\mathrm{e}^x z = \mathrm{e}^x \frac{\mathrm{d}^j y}{\mathrm{d}x^j} = \mathrm{e}^x \frac{\mathrm{d}^j}{\mathrm{d}x^j}(x^j \mathrm{e}^{-x}) = L_j(x), \tag{4.94}$$

is called the *Laguerre polynomial* of degree j. Introducing this polynomial into (4.92),

$$x \frac{\mathrm{d}^2 L_j}{\mathrm{d}x^2} + (1 - x) \frac{\mathrm{d}L_j}{\mathrm{d}x} + jL_j = 0, \tag{4.95}$$

and differentiating k times leads finally to the equation

$$x \frac{\mathrm{d}^2 u}{\mathrm{d}x^2} + (k + 1 - x) \frac{\mathrm{d}u}{\mathrm{d}x} + (j - k)u = 0 \tag{4.96}$$

in which

$$u = N \frac{\mathrm{d}^k L_j}{\mathrm{d}x^k}. \tag{4.97}$$

We call

$$L_j^k(x) = \frac{\mathrm{d}^k}{\mathrm{d}x^k} L_j = \frac{\mathrm{d}^k}{\mathrm{d}x^k}\left[e^x \frac{\mathrm{d}^j}{\mathrm{d}x^j}(x^j \mathrm{e}^{-x})\right] \tag{4.98}$$

the *associated Laguerre polynomial* of degree $j - k$, indices j and k, while N is a normalization constant. Explicit expressions for the first seven Laguerre polynomials appear in Table 4.1.

TABLE 4.1.
Laguerre polynomials

Symbol	Formula
$L_0(x)$	1
$L_1(x)$	$-x+1$
$L_2(x)$	x^2-4x+2
$L_3(x)$	$-x^3+9x^2-18x+6$
$L_4(x)$	$x^4-16x^3+72x^2-96x+24$
$L_5(x)$	$-x^5+25x^4-200x^3+600x^2-600x+120$
$L_6(x)$	$x^6-36x^5+450x^4-2400x^3+5400x^2-4320x+720$

4.8. Quantization of the Radial Motion

Choosing j and k so the coefficients in (4.96) are the same as those in (4.89) determines n as an integer greater than azimuthal quantum number l. The corresponding steady-state energy varies inversely with the square of this n.

On comparison, we see that Equation (4.96) is the same as Equation (4.89) when

$$k = 2l + 1 \tag{4.99}$$

and

$$j = n + l. \tag{4.100}$$

From the way l was introduced with (4.59) and the known quantization of rotational energy, we have

$$l = 0, 1, 2, \ldots . \tag{4.101}$$

Since L_j is a polynomial of degree j, the kth derivative exists only when

$$j \geqslant k. \tag{4.102}$$

Formulas (4.99) and (4.100) convert this inequality to

$$n \geqslant l + 1. \tag{4.103}$$

Since j and l are positive integers or zero, Equation (4.100) implies that n is an integer. From (4.103), it is also positive:

$$n = 1, 2, 3, \ldots . \tag{4.104}$$

Equations (4.74), (4.77), (4.75), and (4.78) tell us that

$$r = \frac{1}{2} anx, \tag{4.105}$$

where

$$a = \frac{4\pi\epsilon_0 \hbar^2}{\mu Z e^2} = \frac{a_0}{Z}, \tag{4.106}$$

and

$$E = -\frac{b}{n^2}, \tag{4.107}$$

where

$$b = \frac{\mu Z^2 e^4}{32\pi^2 \epsilon_0^2 \hbar^2} = \frac{Z^2 e^2}{2(4\pi\epsilon_0 a_0)}. \tag{4.108}$$

In these formulas, $\hbar$ is Planck's constant divided by 2π, ϵ_0 the permittivity of space, μ the reduced mass of the system, Z the number of charges on the nucleus, e the charge on a proton, and a the *Bohr radius* of the atom in its ground state. Because it determines the energy, by (4.107), n is called the *principal quantum number*.

The rest energy of a particle of mass μ is μc^2. Factoring it and $Z^2/(2n^2)$ out of the expression for an energy level yields

$$E = -\frac{Z^2}{2n^2} \alpha^2 \mu c^2, \tag{4.109}$$

where

$$\alpha = \frac{e^2}{4\pi\epsilon_0 \hbar c}. \tag{4.110}$$

We can show that energy E of the particle in the Coulombic field is one-half the average potential energy. Therefore, Equation (4.109) indicates that $Z^2\alpha^2$ is the fraction of rest energy that can appear as electrostatic potential energy.

Introducing corrections for magnetic and relativistic effects would lead to a power series in $Z^2\alpha^2$ containing (4.109) as the first term. Therefore, expression $Z^2\alpha^2$ is also a measure of the relative magnitude of these corrections. As a consequence, α is called the *fine structure constant*. However, α is fundamentally involved in determining the course structure of the spectrum, as Equation (4.109) states.

Substituting (4.97), (4.98), (4.99), (4.100), (4.105) into (4.88) yields

$$R = N \left(\frac{2r}{na}\right)^l e^{-r/na} L_{n+l}^{2l+1}\left(\frac{2r}{na}\right). \tag{4.111}$$

Expression (4.111) constitutes the radial wave function for motion in a Coulombic field in closed form. Further calculations show that

$$N = -\left\{\left(\frac{2}{na}\right)^3 \frac{(n-l-1)!}{2n[(n+l)!]^3}\right\}^{1/2} \tag{4.112}$$

makes

$$\int_0^\infty R^2 r^2 \, dr = 1. \tag{4.113}$$

When the principal quantum number is 1, 2, or 3, the radial functions appear as in Table 4.2.

TABLE 4.2.
Hydrogen-like radial wave functions for normalization (4.113)

Quantum numbers		$R_{nl}(\rho)$ with $\rho = Zr/a_0$
n	l	
1	0	$(Z/a_0)^{3/2}\, 2\mathrm{e}^{-\rho}$
2	0	$\frac{(Z/a_0)^{3/2}}{2\sqrt{2}} (2-\rho)\, \mathrm{e}^{-\rho/2}$
2	1	$\frac{(Z/a_0)^{3/2}}{2\sqrt{6}} \rho\, \mathrm{e}^{-\rho/2}$
3	0	$\frac{2(Z/a_0)^{3/2}}{81\sqrt{3}} (27 - 18\rho + 2\rho^2)\, \mathrm{e}^{-\rho/3}$
3	1	$\frac{2\sqrt{2}\,(Z/a_0)^{3/2}}{81\sqrt{3}} (6-\rho)\rho\, \mathrm{e}^{-\rho/3}$
3	2	$\frac{4(Z/a_0)^{3/2}}{81\sqrt{30}} \rho^2\, \mathrm{e}^{-\rho/3}$

Example 4.5. Evaluate the Bohr radius, assuming the mass of the nucleus is infinite.

Into the formula for a_0 from (4.106), introduce accepted values of the fundamental constants and the mass of the electron:

$$a_0 = \frac{4\pi\epsilon_0\hbar^2}{\mu e^2} = \frac{\hbar^2}{c^2 \cdot 10^{-7}\, \mu e^2}$$

$$= \frac{(1.0546 \times 10^{-34})^2\, (\mathrm{J\,s})(\mathrm{kg\,m^2\,s^{-1}})}{(2.9979 \times 10^8)^2\, (10^{-7}\,\mathrm{C^{-2}\,J\,m})(9.1095 \times 10^{-31}\,\mathrm{kg})(1.6022 \times 10^{-19}\,\mathrm{C})^2}$$

$$= 0.52918 \times 10^{-10}\ \mathrm{m} = 0.52918\ \text{Å}.$$

Formula (4.106) then yields

$$a = \frac{0.52918}{Z}\ \text{Å}.$$

Example 4.6. What do the fine structure constant and its reciprocal equal?
We have

$$\alpha = \frac{e^2}{4\pi\epsilon_0 \hbar c} = \frac{c^2 \cdot 10^{-7} \cdot e^2}{\hbar c}$$

$$= \frac{(2.9979 \times 10^8)^2 (10^{-7} \text{ C}^{-2} \text{ J m})(1.6022 \times 10^{-19} \text{ C})^2}{(1.0546 \times 10^{-34} \text{ J s})(2.9979 \times 10^8 \text{ m s}^{-1})}$$

$$= 7.2973 \times 10^{-3}$$

and

$$\frac{1}{\alpha} = \frac{\hbar}{ce^2} \times 10^7 = 137.04.$$

Example 4.7. What is the rest energy of an electron?
Substituting accepted values of the fundamental constants into the pertinent formula yields

$$E = mc^2$$

$$= (9.1095 \times 10^{-31} \text{ kg})(2.9979 \times 10^8 \text{ m s}^{-1})^2$$

$$= 8.1872 \times 10^{-14} \text{ J}$$

or

$$E = \frac{8.1872 \times 10^{-14} \text{ J}}{1.6022 \times 10^{-19} \text{ J (eV)}^{-1}}$$

$$= 0.51100 \times 10^6 \text{ eV} = 0.51100 \text{ MeV}.$$

Example 4.8. How does the electronic energy in a hydrogen-like system vary with the pertinent quantum numbers?
Rewrite Equation (4.109) in the form

$$E = -\frac{Z^2}{n^2} \frac{\mu}{m_e} \frac{\alpha^2 m_e c^2}{2} = -\frac{Z^2}{n^2} \mu_r \frac{\alpha^2 m_e c^2}{2}$$

in which Z is the number of charges on the nucleus, n the principal quantum number, μ the reduced mass of the system, μ_r the ratio μ/m_e, and m_e the mass of an electron. From Examples 4.6 and 4.7, obtain

$$\frac{\alpha^2 m_e c^2}{2} = \frac{5.1100 \times 10^5 \text{ eV}}{2(137.04)^2} = 13.606 \text{ eV}.$$

and

$$E = -13.606 \frac{Z^2}{n^2} \mu_r \text{ eV}.$$

The energy of an electron in the field of a single particle, carrying positive charge, is independent of the azimuthal and magnetic quantum numbers.

Example 4.9. Since spectroscopic results are most accurately expressed in wave numbers, rewrite the formula for energy of a hydrogen-like atom in reciprocal centimeters.

The energy of a photon is related to its frequency by Equation (1.48). Let us divide this equation by hc, introduce the phase velocity relationship and the definition of wave number:

$$\frac{E}{hc} = \frac{h\nu}{hc} = \frac{\nu}{c} = \frac{1}{\lambda} = k.$$

Equality (4.109), for the energy of a hydrogen-like atom, was rewritten in Example 4.8. Let us divide the result by hc to obtain the energy in wave numbers. Then, if

$$R = \frac{\alpha^2 m_e c^2}{2hc}$$

$$= \frac{8.1872 \times 10^{-14} \text{ J}}{2(137.04)^2 (6.6262 \times 10^{-34} \text{ J s}) (2.9979 \times 10^{10} \text{ cm s}^{-1})}$$

$$= 1.0974 \times 10^5 \text{ cm}^{-1},$$

the formula is

$$k = -R \frac{Z^2}{n^2} \mu_r = -1.0974 \times 10^5 \frac{Z^2}{n^2} \mu_r \text{ cm}^{-1}.$$

Now, accurate spectroscopic measurements yield the coefficient

$$R = 109\,737.31 \text{ cm}^{-1},$$

which is called the *Rydberg constant.*

4.9. Electronic Spectrum of a Hydrogen-like Atom or Ion

The energy levels of a particle bound to another particle by a Coulomb potential are given by formula (4.109) in the approximation that relativistic effects are negligible. A transition from one level to another occurs when the system gains or loses energy as a result of (a) interacting with neighboring atoms, molecules, ions, or (b) interacting with the electromagnetic field.

The general conservation laws govern all transitions in a statistical manner. However, they do not lead to any simple limitation on the change in quantum numbers in interactions of the first type.

Interactions with the electromagnetic field, on the other hand, involve absorption or emission of photons. Each photon carries $\hbar$ angular momentum and possesses odd parity. It transfers this angular momentum and parity to or from the molecule, thus leading to the changes

$$\Delta l = \pm 1 \tag{4.114}$$

and

$$\Delta m = 0 \quad \text{or} \quad \Delta m = \pm 1. \tag{4.115}$$

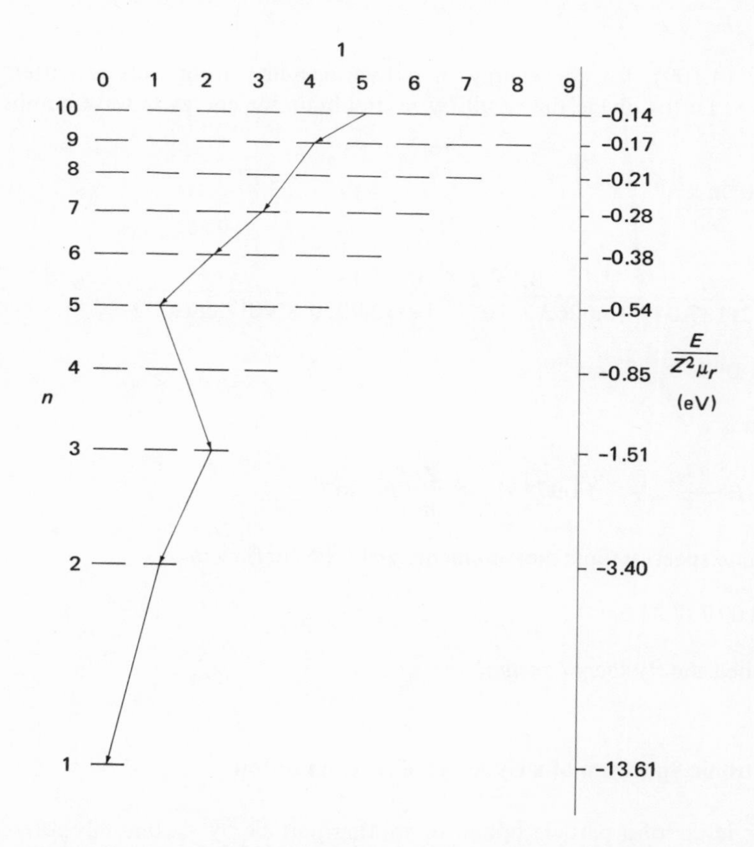

Fig. 4.4. Low-lying states of a hydrogen-like system and a cascade of transitions caused by the emission of successive photons.

The shift in n determines the energy of the photon on a statistical basis. (There is always a natural line width, as we noted in Section 2.9.) We have

$$\Delta n = \text{an integer}. \tag{4.116}$$

As before, Equations (4.114), (4.115), and (4.116) are called *selection rules*. Note Figure 4.4.

The states with different m's appear at different energy levels when an external magnetic field is applied. A splitting of the lines, determined by (4.115), then appears and is called the *Zeeman effect*.

Negative particles other than electrons can be added to atoms. Each known example is 200 or more times as massive as an electron. Consequently, at a given n it moves in a much smaller orbit. During the elementary capture process, the excess energy of such a particle tends to eject an electron from the atom, leaving the particle with a principal quantum number of the order of 100 or so.

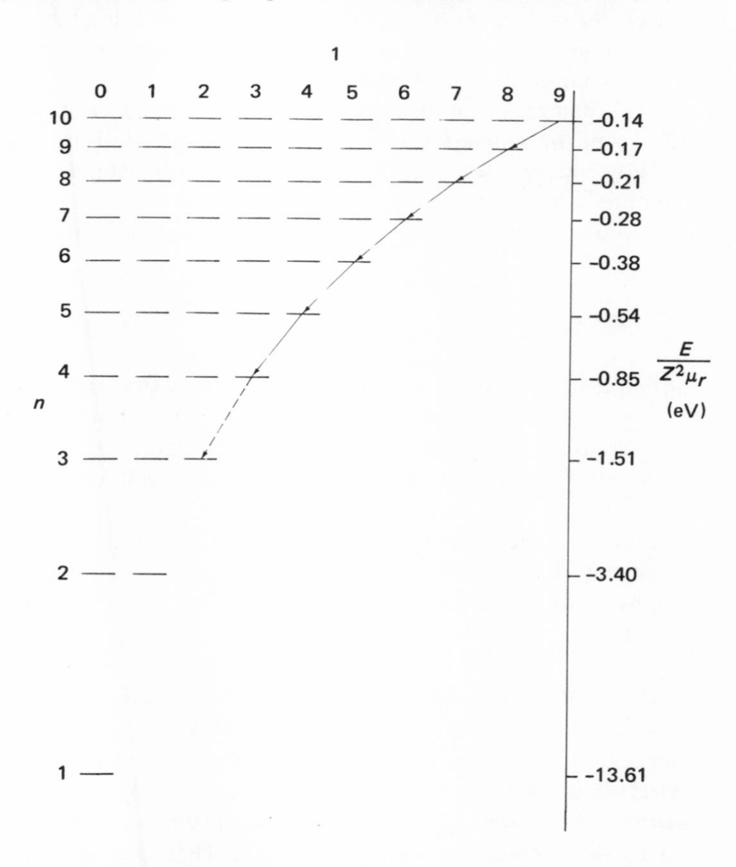

Fig. 4.5. Common sequence followed by a meson within an atom.

The entering particle continues to interact with the orbital electrons, losing energy to them and ejecting them one at a time. The particle can also lose energy by radiating photons. When the energy between successive n's becomes greater than the ionization energy of the most stable electron, photon emission predominates.

Eventually, the particle moves inside the bulk of the electron orbitals, the field becomes hydrogen-like, and Figure 4.5 applies. When a sufficiently low n (about 4 for a kaon) is reached, the particle spends a considerable portion of its time near the surface of the nucleus and it tends to react with a neutron or proton there. Such reaction terminates the independent existence of the particle.

4.10. Uninuclear Multi-electron Structures

Introducing additional electrons into a hydrogen-like system complicates the situation enormously. For then, a given electron moves in the field of a massive positively charged center *and* in the fields of each of the other orbital electrons. The atomic potential V depends not only on the distance of each electron from the center but also on every interelectronic distance.

As a result, the Schrödinger equation does not separate into one-electron equations. And for all except the simplest atoms, the complete equation is intractable. To make progress, however, we may consider each electron as subject merely to the average field of all other electrons and the nucleus.

In the crudest approximation, we consider the given electron to be moving in a Coulomb-type field

$$V = -\frac{\text{constant}}{r} \tag{4.117}$$

with the constant chosen to allow for the average shielding of the nucleus by the other electrons. The origin is located at the nucleus as before. For this V, the methods we have already developed are sufficient. The resultant states appear as in the hydrogen atom itself. Now, these states are filled in pairs, following the Pauli exclusion principle.

The mean distribution of each electron over r is obtained from the rough calculation, and this is employed to modify the potential seen by each electron. Thus, (4.117) is replaced by fields

$$V = V(r) \tag{4.118}$$

and the Schrödinger equation for each is solved. The same angular parts as before arise. The radial differential equations are solved numerically. The states are again filled as the Pauli exclusion principle requires.

A corrected electron distribution over r can now be obtained and made the basis for a new calculation. The procedure is repeated again and again until subsequent solutions agree. One thus obtains self-consistent results. These are known as the *Hartree* results for the given atom.

Each Hartree state is identified by the same angular factors and quantum numbers as is the solution in the zeroth-order approximation. The radial distribution, and the electron energies, are altered drastically, but the principal quantum number can still be related to the number of radial and angular nodes as in the hydrogen atom itself.

The primary determinant of energy is the principal quantum number n. So the electrons that have the same n in an atom are said to form a *shell*. A major secondary determinant of energy is the azimuthal quantum number l. Electrons that have the same n and l are said to form a *subshell*. A subshell is said to be filled when each of its conventional states is occupied as the Pauli exclusion principle allows.

Each subshell forms a different degenerate set, in the approximation that (4.118) holds. A single state in the subshell, therefore, is not spherically symmetric, but the complete set of states in the subshell is.

Since filled subshells are spherically symmetric, the Hartree approximation represents their effects quite well. The nonspherically symmetric part of the interaction in an incompletely filled subshell can be investigated by studying spectroscopic data. In fact, much of the work in explaining atomic spectra involves the investigation of such effects.

4.11. Useful Operator Formulations of the Radial Equation

As the harmonic oscillator equation was rewritten, the radial equation can be rearranged to contain the result of two first-order differentiating operators acting in sequence on R equaling a number times R.

We consider an electron bound, by itself, to a nucleus. The potential is given by (4.72). The radial factor in Ψ is governed by (4.76) with the minus sign taken. Let us rearrange this equation to

$$x^2 \frac{\mathrm{d}^2 R}{\mathrm{d}x^2} + 2x \frac{\mathrm{d}R}{\mathrm{d}x} - \frac{1}{4} x^2 R + nxR = l(l+1)R. \tag{4.119}$$

If we set

$$\frac{\mathrm{d}}{\mathrm{d}x} = \mathrm{D}, \tag{4.120}$$

we have

$$\left(x^2 \mathrm{D}^2 + 2x\mathrm{D} - \frac{1}{4} x^2 + nx\right) R = l(l+1)R. \tag{4.121}$$

As before, we can add a number times R to the left side of (4.121) and obtain a factorable expression.

Indeed since

$$[xD + f(x)]\,[xD - f(x) + 1] = x^2D^2 + xD - xf' - xfD + xD + fxD - f^2 + f$$
$$= x^2D^2 + 2xD - f^2 + f - xf', \qquad (4.122)$$

we need to set

$$f = \pm\frac{1}{2}x \mp n. \qquad (4.123)$$

For then

$$f' = \pm\frac{1}{2}, \qquad (4.124)$$

$$f^2 = \frac{1}{4}x^2 - xn + n^2, \qquad (4.125)$$

and

$$-f^2 + f - xf' = -\frac{1}{4}x^2 + nx - n^2 \mp n. \qquad (4.126)$$

Adding

$$(-n^2 \mp n)R \qquad (4.127)$$

to both sides of (4.121) leads to operator (4.122) being on the left. We thus obtain the equations

$$\left(xD + \frac{1}{2}x - n\right)\left(xD - \frac{1}{2}x + n + 1\right)R = [l(l+1) - n(n+1)]\,R \qquad (4.128)$$

and

$$\left(xD - \frac{1}{2}x + n\right)\left(xD + \frac{1}{2}x - n + 1\right)R = [l(l+1) - (n-1)n]R. \qquad (4.129)$$

The second operator in (4.128) can be made identical with the first in (4.129) by replacing $n + 1$ with n, n with $n - 1$:

$$\left(xD + \frac{1}{2}x - n + 1\right)\left(xD - \frac{1}{2}x + n\right)R_{n-1} = [l(l+1) - (n-1)n]R_{n-1}. \qquad (4.130)$$

The subscript on R indicates the principal quantum number for the function. Similarly, the second operator in (4.129) can be made identical with the first in (4.128) by replacing $n - 1$ with n, and n with $n + 1$:

$$\left(xD - \frac{1}{2}x + n + 1\right)\left(xD + \frac{1}{2}x - n\right)R_{n+1} = [l(l+1) - n(n+1)]R_{n+1}. \qquad (4.131)$$

Here the principal quantum number for function R is $n + 1$.

Discussion Questions

4.1. Why is the oscillation of the electron and nucleus in the hydrogen atom so different from that of the two nuclei in a diatomic molecule?

4.2. How can the internal movements, rotation and oscillation, of a hydrogen atom be modeled by motions of a single particle?

4.3. Why does the magnitude of the wavevector, rather than the wavevector itself, characterize the unmixed steady-state motion near each point in a field of varying potential?

4.4. How can a person introduce symmetry in a small region about an arbitrary point?

4.5. How does one employ this symmetry to obtain a spatial rate of change in the corresponding spatial rate of change in the wave function?

4.6. How is the result from Question 4.5 used to get a condition on the Laplacian of the state function? What is the equation summarizing this condition called?

4.7. Why are mutually independent motions associated with perpendicular directions about each spatial point?

4.8. When are mutually independent motions governed by separate factors in the complete wave function? Explain.

4.9. How can properties of the 'del-dot-del' operator be deduced from movements of a hypothetical effect?

4.10. Construct $\nabla^2 \psi$ in orthogonal generalized coordinates.

4.11. Express Schrödinger's equation involving the wavevector in spherical coordinates.

4.12. By substitution into this equation, show that, when the magnitude of the wavevector depends only on the radial coordinate, each pure energy state is governed by the product of a radial function and an angular function:

$$\psi = R(r)Y(\theta, \varphi).$$

4.13. Assume that the radial motion is independent of the angular motion as a result of V being a function of r alone and construct (4.71) directly from (4.31).

4.14. Explain how the wave function for a particle in a given energy state behaves in a region where its potential V is constant.

4.15. Construct the differential equation governing the radial function $R(r)$ for the hydrogen-like atom.

4.16. Determine how solutions for this equation behave (a) out, where r is very large, (b) in, where r is very small. Which of these are acceptable? Why?

4.17. What parameter determines which solution near $r = 0$ joins with each asymptotic solution at $r = \infty$? Why are only certain values of this parameter physically acceptable?

4.18. Are all solutions of the form

$$R = u(x)x^l \,\mathrm{e}^{-x/2}, \qquad \text{where} \quad r = \frac{1}{2}anx$$

acceptable? Why not?

4.19. Does $x^j\,\mathrm{e}^{-x}$ have any zeros in the range $0 < x < \infty$? How many zeros are introduced by each of the j differentiations? Does multiplication by e^x alter this result?

4.20. How many zeros does L_j possess? What does each of the k differentiations do to these zeros?

4.21. Describe the waviness that L_j^k introduces into the radial function R.

4.22. Why is n in the energy equation an integer greater than l?

4.23. Why is there both an upper and a lower bound to the energy of states for which the sign in (4.76) is negative?

4.24. Show that $R(r)$ has $n - l - 1$ zeros in the finite range of r.

4.25. What is the fine structure constant α? How does the energy of a hydrogen-like atom depend on $Z\alpha$?

4.26. What rules govern the transitions that occur during collisions?

4.27. What rules govern the emission and absorption of photons by hydrogen-like atoms?

4.28. Why do states with differing m's appear at different energies in an external magnetic field? What is the Zeeman effect?

4.29. Why does a meson in a multi-electron atom behave as a single electron in the field of the nucleus when its principal quantum number is low?

4.30. Identify the shells and subshells of an atom. How may these be occupied? Explain.

4.31. What is the Hartree approximation? Why does it represent a filled subshell satisfactorily? Why does it not represent a partially filled subshell accurately?

4.32. Why does a composite differentiating operator have transforming as well as combinatorial properties?

4.33. How do we obtain a sum and a difference plus one such that the product of the two contains only those single differentiating operators in (4.121)?

4.34. Rewrite the radial equation so it states that the successive action of two first order differentiating operators on R produce the same effect as a multiplying number.

Problems

4.1. Construct the Schrödinger equation for a hydrogen-like ion in rectangular coordinates.

4.2. Substitute the function $\psi = A\,\mathrm{e}^{-ar}$ into the equation from Problem 4.1 and through differentiation in rectangular coordinates obtain a and E for the $1s$ orbital.

4.3. A point may be located with respect to a Cartesian coordinate system by dropping a perpendicular from the point to the z axis, letting r be the length of this perpendicular, φ be the angle between this perpendicular and the zx plane, and z be the distance of the point above the xy plane. Express $\nabla^2\psi$ in terms of these cylindrical coordinates.

4.4. How does the Schrödinger equation for a single particle appear in the cylindrical coordinates r, φ, z? When may ψ have the structure

$$\psi = R(r)\,\Phi(\varphi)\,Z(z)?$$

4.5. Calculate the next two terms in (4.82). How do $l = 0, L = -1$ lead to the result $n = 0$?

4.6. Write down ψ for a 1s orbital of a hydrogen-like ion and evaluate the normalization constant by integration.

4.7. If a hydrogen-like system is in its ground state, what is the probability that an electron is out where $V > E$ at any given moment?

4.8. Determine the most probable distance of the electron from the nucleus in the 1s state.

4.9. What transition in H is associated with the emission of (a) a 97 492 cm^{-1} photon, (b) a 20 565 cm^{-1} photon?

4.10. How much energy is released when a pion, 273 times as heavy as an electron, moves from the $n = 7$ to the $n = 6$ level of sulfur?

4.11. Substitute the function $\psi = Ax\,\mathrm{e}^{-ar}$ into the equation from Problem 4.1 and, by differentiation with respect to rectangular coordinates, obtain a and E for the 2p orbital.

4.12. Substitute the function $\psi = B(1 + br)\,\mathrm{e}^{-cr}$ into the Schrödinger equation from Problem 4.1 and through differentiation with respect to rectangular coordinates determine b, c, and E for the 2s state.

4.13. Take the ψ for a 2s orbital of a hydrogen-like system and obtain the normalization constant by integration.

4.14. If the electron in a hydrogen-like atom is in its 2s orbital, what is the probability that it is out where $V > E$ at a given time t?

4.15. Determine the most probable distance of the electron from the nucleus in the 2s state.

4.16. Before reacting with a nucleus having a Z of 17, a captured kaon emits an 85 000 eV x-ray photon. If the ratio of the kaon mass to the electron mass is 966, what unit change in quantum number n produced this radiation?

4.17. How many reciprocal centimeters is the 2p level above the 1s level in (a) H^1, (b) $(Li^7)^{++}$? In atomic mass units, the mass of an electron is 0.000548580, the mass of a proton is 1.0072765, and the mass of $(Li^7)^{+++}$ is 7.0143595.

4.18. If ξ, η, φ are related to spherical coordinates r, θ, φ by the equations

$$\xi = r(1 - \cos\theta),$$

$$\eta = r(1 + \cos\theta),$$

$$\varphi = \varphi,$$

how are they related to conventional rectangular coordinates? Express

$$\mathrm{d}s^2 = \mathrm{d}x^2 + \mathrm{d}y^2 + \mathrm{d}z^2$$

in terms of ξ, η, φ are obtain the coefficients h_1, h_2, h_3.

4.19. Formulate Schrödinger's equation in the parabolic ξ, η, φ coordinate system of Problem 4.18.

4.20. Show that when V is central, variable φ separates from variables ξ and η in this Schrödinger equation. Show that when V is Coulombic, the separation can be completed.

References

Books

Eggers, D. F., Jr, Gregory, N. W., Halsey, G. D., Jr, and Rabinovitch, B. S.: 1964, *Physical Chemistry*, Wiley, New York, pp. 1–77.

Harris, L. and Loeb, A. L.: 1963, *Introduction to Wave Mechanics*, McGraw-Hill, New York, pp. 185–202.

Kuhn, H. G.: 1961, *Atomic Spectra*, Academic Press, New York, pp. 1–13, 84–246.

Sherwin, C. W.: 1959, *Introduction to Quantum Mechanics*, Holt, Rinehart and Winston, New York, pp. 62–107.

Straughan, B. P., and Walker, S.: 1976, *Spectroscopy*, Vol. 1, Chapman and Hall, London, pp. 1–107.

Articles

Armstrong, B. H., and Power, E. A.: 1963, 'Acceptable Solutions and Boundary Conditions for the Schrödinger and Dirac Equations', *Am. J. Phys.* **31**, 262–268.

Goodisman, J.: 1973, 'On the Negative-Energy Wave Functions for the Coulomb Field', *Am. J. Phys.* **41**, 1255–1257.

Holladay, W. G., Thomas, J. B., Jr, and Smith, C. R.: 1963, 'Number of Energy Levels for a Debye-Hückel or Yukawa Potential', *Am. J. Phys.* **31**, 16–19.

Lain, L., Toree, A., and Alvarino, J. M.: 1981, 'Radial Probability Density and Normalization in Hydrogenic Atoms', *Am. J. Phys.* **58**, 617.

Mak, T. C. W., and Li, W.-K.: 1975, 'Relative Sizes of Hydrogenic Orbitals and the Probability Criterion', *J. Chem. Educ.* **52**, 90–91.

Peterson, C.: 1975, 'The Radial Equation for Hydrogen-Like Atoms', *J. Chem. Educ.* **52**, 92–94.

Prato, D. P.: 1977, 'Direct Derivation of Differential Operators in Curvilinear Coordinates', *Am. J. Phys.* **45**, 1003–1004.

Scaife, D. B.: 1978, 'Atomic Orbital Contours – A New Approach to an Old Problem', *J. Chem. Educ.* **55**, 442–445.

Seki, R.: 1971, 'On Boundary Conditions for an Infinite Square-Well Potential in Quantum Mechanics', *Am. J. Phys.* **39**, 929–931.

Srinivasan, K.: 1977, 'Simple Derivation of ∇^2 in Cylindrical and Spherical Polar Coordinates', *Am. J. Phys.* **45**, 767–768.

Wiegand, C. E.: 1972, 'Exotic Atoms', *Sci. American* **227** (5), 102–110.

Yano, A. F., and Yano, F. B.: 1972, 'Hydrogenic Wave Functions for an Extended, Uniformly Charged Nucleus', *Am. J. Phys.* **40**, 969–971.

Zablotney, J.: 1975, 'Energy Levels of a Charged Particle in the Field of a Spherically Symmetric Uniform Charge Distribution', *Am. J. Phys.* **43**, 168–172.

Chapter 5

Quantum Mechanical Operators

5.1. Dependence of Observables on the State Function

Insofar as one can tell, a submicroscopic particle does not move along a definite path through the imposed Euclidean space. Instead, there exists a potentiality in each volume element; at a given time, the particle may appear in an infinitesimal region $d^3\mathbf{r}$ with the probability

$$\rho\, d^3\mathbf{r} = \Psi^* \Psi\, d^3\mathbf{r}. \tag{5.1}$$

The magnitude of state function Ψ determines the probability density ρ; the phase angle of Ψ describes what can be learned about movements of the particle (in the nonrelativistic theory).

But the probability distribution of the particle in space together with its propagation properties produce all observable quantities. Consequently, the state function itself determines these quantities, as we first noted in Section 1.2. A series of operations, some or all carried out on Ψ, must lead to each physically significant number.

Certain basic properties involve only one operation. Indeed (4.31), the Schrödinger equation, can be rearranged so that it consists of an operator acting on ψ equaling E times ψ. Similarly, each component of the wavevector squared at a point can be obtained from (4.26).

The homogeneous linear equations yielding E, k^2, k_x^2, k_y^2, and k_z^2 have the eigenvalue form to be discussed in Section 5.2. Each of these equations is satisfied by an infinite number of solutions. Now, a transformation converts one acceptable solution into another. For the harmonic oscillator, and for the hydrogen-like atom, the pertinent operator is easily formulated.

An excited system may spontaneously lose energy in one or more forms and thereby gain stability. This possibility leads to a broadening of the higher level.

5.2. Eigenvalue Equations

An operator A that acts in the space of the independent variables in general converts a function Ψ of those variables into another function Φ:

$$A\Psi = \Phi, \tag{5.2}$$

However, with many operators, the initial function may be constructed so that the resulting function is merely a multiple of the initial one.

With certain Ψ's meeting specified conditions, we can find an A such that

$$\begin{aligned} A\Psi_1 &= a_1\Psi_1, \\ A\Psi_2 &= a_2\Psi_2, \\ &\;\;\vdots \\ A\Psi_N &= a_N\Psi_N. \end{aligned} \tag{5.3}$$

The particular functions

$$\Psi_1, \Psi_2, \ldots, \Psi_N \tag{5.4}$$

would then yield a set of numbers

$$a_1, a_2, \ldots, a_N. \tag{5.5}$$

The states of a system that are pure with respect to a given measurement correspond to functions (5.4). Results of the measurement may consequently appear as numbers (5.5) when A is the appropriate operator. Indeed, the Schrödinger equation can be put in form (5.3), with the observable being energy. Operators for other observables can also be constructed and the general theory being developed here applied.

The Ψ_j's satisfying (5.3), and meeting the specified conditions, are called *eigenfunctions*; the operator A is called an *eigenoperator*; the numbers $a_1, a_2, \ldots, a_N$ are called *eigenvalues*. Formula

$$A\Psi_j = a_j\Psi_j \tag{5.6}$$

is called an *eigenvalue equation*. Note that end index N may be any integer or infinity, depending on the system.

We have employed the rule that each equivalent part of a pure state function produces the same effect. Thus, both Ψ_j and $\lambda\Psi_j$, where λ is a number, real or complex, should yield the same eigenvalue a_j. We have

$$A(\lambda\Psi_j) = a_j(\lambda\Psi_j) = \lambda a_j\Psi_j = \lambda A\Psi_j \tag{5.7}$$

and

$$A\lambda = \lambda A. \tag{5.8}$$

Furthermore, pure states mix according to the superposition principle

$$\Psi = \Sigma c_j\Psi_j. \tag{5.9}$$

When all states contributing appreciably to Ψ have the same eigenvalue a_k, we have

$$A(\Sigma c_j \Psi_j) = a_k(\Sigma c_j \Psi_j) = \Sigma c_j a_k \Psi_j = \Sigma c_j A \Psi_j. \tag{5.10}$$

For our applications, let us generalize this result, assuming

$$A \,\Sigma c_j \Psi_j = \Sigma c_j A \Psi_j \tag{5.11}$$

regardless of which c_j's are different from zero. Now, any operator obeying (5.8) and (5.11) is said to be *linear*.

Let us next consider any two of the eigenvalue equations for a given operator

$$A \Psi_j = a_j \Psi_j, \tag{5.12}$$

$$A \Psi_k = a_k \Psi_k, \tag{5.13}$$

and the complex conjugate of each,

$$A^* \Psi_j^* = a_j^* \Psi_j^*, \tag{5.14}$$

$$A^* \Psi_k^* = a_k^* \Psi_k^*. \tag{5.15}$$

Multiply (5.12) by Ψ_j^* and integrate over (a) all space or (b) a typical region if the particles under consideration (a) are confined or (b) satisfy periodic boundary conditions, respectively. Let R represent the region over which the integration proceeds. Then

$$\int_R \Psi_j^* A \Psi_j \,\mathrm{d}^3\mathbf{r} = a_j \int_R \Psi_j^* \Psi_j \,\mathrm{d}^3\mathbf{r}. \tag{5.16}$$

Similarly, multiply (5.14) by Ψ_j and integrate:

$$\int_R \Psi_j A^* \Psi_j^* \,\mathrm{d}^3\mathbf{r} = a_j^* \int_R \Psi_j^* \Psi_j \,\mathrm{d}^3\mathbf{r}. \tag{5.17}$$

Again, R is the pertinent region. Whenever the eigenvalue is real, we have

$$a_j = a_j^* \tag{5.18}$$

and the right sides of (5.16), (5.17) are equal. So the left sides have to be equal:

$$\int_R \Psi_j^* A \Psi_j \,\mathrm{d}^3\mathbf{r} = \int_R \Psi_j A^* \Psi_j^* \,\mathrm{d}^3\mathbf{r}$$

$$= \int_R (A \Psi_j)^* \Psi_j \,\mathrm{d}^3\mathbf{r}. \tag{5.19}$$

Next, multiply (5.13) by Ψ_j^* and integrate over R:

$$\int_R \Psi_j^* A \Psi_k \,\mathrm{d}^3\mathbf{r} = a_k \int_R \Psi_j^* \Psi_k \,\mathrm{d}^3\mathbf{r}. \tag{5.20}$$

Also, multiply (5.14) by Ψ_k and similarly integrate, assuming a_j is real:

$$\int_R (A\,\Psi_j)^*\Psi_k\,\mathrm{d}^3\mathbf{r} = a_j \int_R \Psi_j^*\,\Psi_k\,\mathrm{d}^3\mathbf{r}. \tag{5.21}$$

Whenever the eigenvalues are equal, the right sides of (5.20), (5.21) are equal, and the left sides must be equal:

$$\int_R \Psi_j^* A\,\Psi_k\,\mathrm{d}^3\mathbf{r} = \int_R (A\,\Psi_j)^*\,\Psi_k\,\mathrm{d}^3\mathbf{r}. \tag{5.22}$$

Functions Ψ_j and Ψ_k describe pure states with eigenvalues a_j and a_k. When these values differ, the states must be completely distinct – containing nothing in common. A mixture of the states is described by the superposition

$$\Psi = c_j\Psi_j + c_k\Psi_k. \tag{5.23}$$

The probability density integrated over all pertinent space is then

$$\int_R \Psi^*\Psi\,\mathrm{d}^3\mathbf{r} = c_j^*c_j \int_R \Psi_j^*\Psi_j\,\mathrm{d}^3\mathbf{r} + c_k^*c_k \int_R \Psi_k^*\Psi_k\,\mathrm{d}^3\mathbf{r}$$

$$+ c_j^*c_k \int_R \Psi_j^*\Psi_k\,\mathrm{d}^3\mathbf{r} + c_jc_k^* \int_R \Psi_j\Psi_k^*\,\mathrm{d}^3\mathbf{r}. \tag{5.24}$$

With the Ψ's normalized as (5.1) indicates, Equation (5.24) states that the total probability equals, presumably, the probability that the system is in the jth state plus the probability that it is in the kth state. The probability it is in the jth state should be independent of c_k and Ψ_k; therefore, this probability is identified with the first term on the right. Similarly, the probability the system is in the kth state is given by the second term and the last two terms must vanish. These can vanish in general only if

$$\int_R \Psi_j^*\Psi_k\,\mathrm{d}^3\mathbf{r} = 0. \tag{5.25}$$

When (5.25) holds, Ψ_j and Ψ_k are said to be *orthogonal*.

With (5.25), Equations (5.20) and (5.21) lead to

$$\int_R \Psi_j^* A\,\Psi_k\,\mathrm{d}^3\mathbf{r} = \int_R (A\,\Psi_j)^*\,\Psi_k\,\mathrm{d}^3\mathbf{r}. \tag{5.26}$$

This result applies however j and k are chosen.

But, the most general state functions are arbitrary mixtures of eigenfunctions. So let us multiply (5.26) by $c_j^*d_k$, employ the linearity of A to introduce c_j as the coefficient of Ψ_j, d_k as the coefficient of Ψ_k, and sum over both j and k:

$$\int_R \Sigma c_j^*\Psi_j^* A\,\Sigma d_k\Psi_k\,\mathrm{d}^3\mathbf{r} = \int_R (A\,\Sigma c_j\Psi_j)^*\,\Sigma d_k\Psi_k\,\mathrm{d}^3\mathbf{r}. \tag{5.27}$$

If we let

$$\Sigma c_j \Psi_j = \Psi_l, \qquad \Sigma d_k \Psi_k = \Psi_m, \tag{5.28}$$

then we have

$$\int_R \Psi_l^* A \Psi_m \, d^3\mathbf{r} = \int_R (A \Psi_l)^* \Psi_m \, d^3\mathbf{r}, \tag{5.29}$$

where Ψ_l and Ψ_m are any two possible state functions, pure or mixed.

An operator A for which (5.26) and (5.29) hold is said to be *Hermitian*. We will employ such operators in our eigenvalue equations for the reasons given.

When Ψ_j does not equal a constant times Ψ_k yet they both have the same eigenvalue, a person can construct

$$\Psi_l = c_1 \Psi_j + c_2 \Psi_k \tag{5.30}$$

such that

$$\int_R \Psi_j^* \Psi_l \, d^3\mathbf{r} = 0. \tag{5.31}$$

Then Ψ_j and Ψ_l are orthogonal. Similarly, if n linearly independent eigenfunctions have the same eigenvalue, n mutually orthogonal functions corresponding to that eigenvalue can be constructed. The maximum number of orthogonal normalized functions that can be formed at one time for a given eigenvalue is called the *degeneracy* for that level.

Example 5.1. Show that where A and B are separately Hermitian, the sum $A + B$ is Hermitian.

Let Ψ_j and Ψ_k be any well-behaved functions (possible mixed-state functions) of the coordinates and time in the region R over which A and B are Hermitian. From (5.29), we then have

$$\int_R \Psi_j^* A \Psi_k \, d^3\mathbf{r} = \int_R A^* \Psi_j^* \Psi_k \, d^3\mathbf{r}$$

and

$$\int_R \Psi_j^* B \Psi_k \, d^3\mathbf{r} = \int_R B^* \Psi_j^* \Psi_k \, d^3\mathbf{r}.$$

Add these equations and combine the integrals to obtain

$$\int_R \Psi_j^* (A + B) \Psi_k \, d^3\mathbf{r} = \int_R (A + B)^* \Psi_j^* \Psi_k \, d^3\mathbf{r}.$$

Since $A + B$ enters this equation as A enters (5.29), $A + B$ is Hermitian.

Example 5.2. Show that if A is Hermitian, a real number α times A is also Hermitian.

Use the constancy of α, the Hermiticity of A, and the realness of α to establish the overall equality:

$$\int_R \Psi_j^* \alpha A \Psi_k \, d^3\mathbf{r} = \alpha \int_R \Psi_j^* A \Psi_k \, d^3\mathbf{r}$$

$$= \alpha \int_R A^* \Psi_j^* \Psi_k \, d^3\mathbf{r}$$

$$= \int_R (\alpha A)^* \Psi_j^* \Psi_k \, d^3\mathbf{r}.$$

Example 5.3. Prove that if A and B are two Hermitian operators, then product AB is Hermitian whenever A and B commute.

First apply Hermitian switch (5.29) to operator A, then to operator B. Finally commute B^* and A^*. (These commute since B and A do.)

$$\int_R \Psi_j^* A (B\Psi_k) \, d^3\mathbf{r} = \int_R (A^* \Psi_j^*) B \Psi_k \, d^3\mathbf{r}$$

$$= \int_R B^* (A^* \Psi_j^*) \Psi_k \, d^3\mathbf{r}$$

$$= \int_R A^* B^* \Psi_j^* \Psi_k \, d^3\mathbf{r}.$$

The overall equation shows that AB is Hermitian.

Example 5.4. Determine when

$$\frac{\hbar}{i} \frac{\partial}{\partial x}$$

is Hermitian.

When Ψ_j and Ψ_k are any two well-behaved functions, we have

$$\iint \left(\int_a^b \Psi_j^* \frac{\hbar}{i} \frac{\partial}{\partial x} \Psi_k \, dx \right) dy \, dz = \iint \left(\int_a^b \Psi_j^* \frac{\hbar}{i} \, d\Psi_k \right) dy \, dz$$

$$= \iint \Psi_j^* \frac{\hbar}{i} \Psi_k \Big|_a^b dy \, dz - \iint \left(\int_a^b \frac{\hbar}{i} \, d\Psi_j^* \Psi_k \right) dy \, dz$$

$$= \iint \Psi_j^* \frac{\hbar}{i} \Psi_k \Big|_a^b dy \, dz + \iiint_a^b \left(\frac{\hbar}{i} \frac{\partial}{\partial x} \right)^* \Psi_j^* \Psi_k \, dx \, dy \, dz.$$

The integrated part is zero and the given operator is Hermitian (a) when Ψ_j and Ψ_k have the same values at $x = a$ and $x = b$ and (b) when Ψ_j and/or Ψ_k are zero at the limits. Situation (a) prevails when periodic boundary conditions are employed; situation (b) prevails when the particle is confined.

5.3. Formula for the Expectation Value

When a system is in a pure state with respect to a certain property, the property exhibits one value only. The relationship between the state function and the value is like that in the eigenvalue equation. For basic properties, eigenoperators can be found and the pertinent

$$A\Psi_j = a_j\Psi_j \tag{5.32}$$

constructed.

In (5.32), A is said to be the operator for observable a_j, while Ψ_j is the corresponding pure state function. For convenience, the Ψ_j's may all be taken orthogonal to each other and normalized to 1:

$$\int_R \Psi_j^* \Psi_k \, \mathrm{d}^3\mathbf{r} = \delta_{jk}. \tag{5.33}$$

Here δ_{jk} is *Kronecker's delta*, defined as 0 when $j \neq k$ and 1 when $j = k$.

Now, any mixed state may be described as a superposition of possible pure states:

$$\Psi = \Sigma\, c_j\Psi_j. \tag{5.34}$$

Since the total probability for the system to be in any of these states is 1, we have

$$\int_R \Psi^* \Psi \, \mathrm{d}^3\mathbf{r} = \int_R (\Sigma\, c_j\Psi_j)^* (\Sigma\, c_k\Psi_k) \, \mathrm{d}^3\mathbf{r}$$

$$= \Sigma\Sigma\, c_j^* c_k \delta_{jk} = \Sigma\, c_j^* c_j = 1. \tag{5.35}$$

Because the only term involving the ith state is

$$c_i^* c_i \equiv w_i, \tag{5.36}$$

this term is interpreted as the probability that the system is in the ith state. It is called the *weight* of the ith state.

Operating on (5.34) with A and introducing condition (5.32) yields the sum

$$A\Psi = \Sigma\, c_k A\Psi_k = \Sigma\, c_k a_k \Psi_k. \tag{5.37}$$

Multiplying overall (5.37) by the complex conjugate of (5.34) and integrating over the pertinent region leads to

$$\int_R \Psi^* A \Psi \, d^3\mathbf{r} = \int_R (\Sigma\, c_j \Psi_j)^* (\Sigma\, c_k a_k \Psi_k) \, d^3\mathbf{r}$$

$$= \Sigma\Sigma\, c_j^* c_k a_k \delta_{jk} = \Sigma\, c_j^* c_j a_j$$

$$= \Sigma\, w_j a_j \equiv \langle A \rangle. \tag{5.38}$$

Term $w_i a_i$ with i a free index is the weight of the ith state times the value of the observable in that state. Therefore, the quantity

$$\langle A \rangle = \int_R \Psi^* A \Psi \, d^3\mathbf{r} \tag{5.39}$$

is the average value of the observable when the wave function for the system is Ψ. We call integral (5.39) the *expectation value* of the observable corresponding to operator A.

Example 5.5. Evaluate the integral

$$\int_{-\infty}^{\infty} x^2 \, e^{-ax^2} \, dx.$$

Integrate, by parts, the exponential function from the integrand,

$$\int_a^b e^{-ax^2} \, dx = e^{-ax^2} x \Big|_a^b + 2 \int_a^b ax^2 \, e^{-ax^2} \, dx,$$

and rearrange,

$$\int_a^b x^2 \, e^{-ax^2} \, dx = \frac{1}{2a} \left[\int_a^b e^{-ax^2} \, dx - e^{-ax^2} x \Big|_a^b \right].$$

Let limits a and b become $-\infty$, ∞ and introduce the result from Example 3.3:

$$\int_{-\infty}^{\infty} x^2 \, e^{-ax^2} \, dx = \frac{\pi^{1/2}}{2a^{3/2}}.$$

Example 5.6. What is the expectation value for x^2 in the harmonic oscillator state

$$\Psi = \left(\frac{a}{\pi} \right)^{\frac{1}{4}} e^{-ax^2/2} \, e^{-i\omega t}?$$

Substitute

$$A = x^2$$

and the given Ψ into (5.39). Employ the result from Example 5.5 to evaluate:

$$\langle x^2 \rangle = \left(\frac{a}{\pi}\right)^{\frac{1}{2}} \int_{-\infty}^{\infty} x^2 \, e^{-ax^2} \, dx = \left(\frac{a}{\pi}\right)^{\frac{1}{2}} \frac{\pi^{\frac{1}{2}}}{2a^{3/2}}$$

$$= \frac{1}{2a}.$$

5.4. The Dirac Delta

Describing a particle at a definite position requires a probability density ρ which is zero everywhere except at the position, where the density becomes infinite in such a way that the spatial integral of ρ is 1. The resulting expression is not an analytic function. However, we need to consider its nature because the corresponding Ψ is an eigenfunction of the operator for position. A form for this operator can be deduced from the nature of the ρ.

First, let us define a generalized function $\delta(v)$ such that

$$\delta(x - a) = 0 \qquad \text{when} \quad x \neq a, \tag{5.40}$$

$$\delta(x - a) = \infty \qquad \text{when} \quad x = a, \tag{5.41}$$

and

$$\int_{-\infty}^{\infty} \delta(x - a) \, dx = 1. \tag{5.42}$$

We call $\delta(v)$ the one-dimensional *Dirac delta*. It can be portrayed as the limit of the function sketched in Figure 5.1, as the width decreases to zero, the height increases without limit.

Because the Dirac delta is symmetric about $x = a$, we have

$$\delta(x - a) = \delta(a - x). \tag{5.43}$$

Of considerable interest is the integral

$$\int_{-\infty}^{\infty} \delta(x - a) f(x) \, dx. \tag{5.44}$$

in which $f(x)$ is a well-behaved function. The properties of the Dirac delta make the integrand zero except at $x = a$, where $f(x) = f(a)$. As a result, function $f(x)$ is constant where the integrand differs from zero. It can be factored out and the integral evaluated:

$$\int_{-\infty}^{\infty} \delta(x - a) f(x) \, dx = f(a) \int_{-\infty}^{\infty} \delta(x - a) \, dx$$

$$= f(a). \tag{5.45}$$

In the last step, value (5.42) has been introduced.

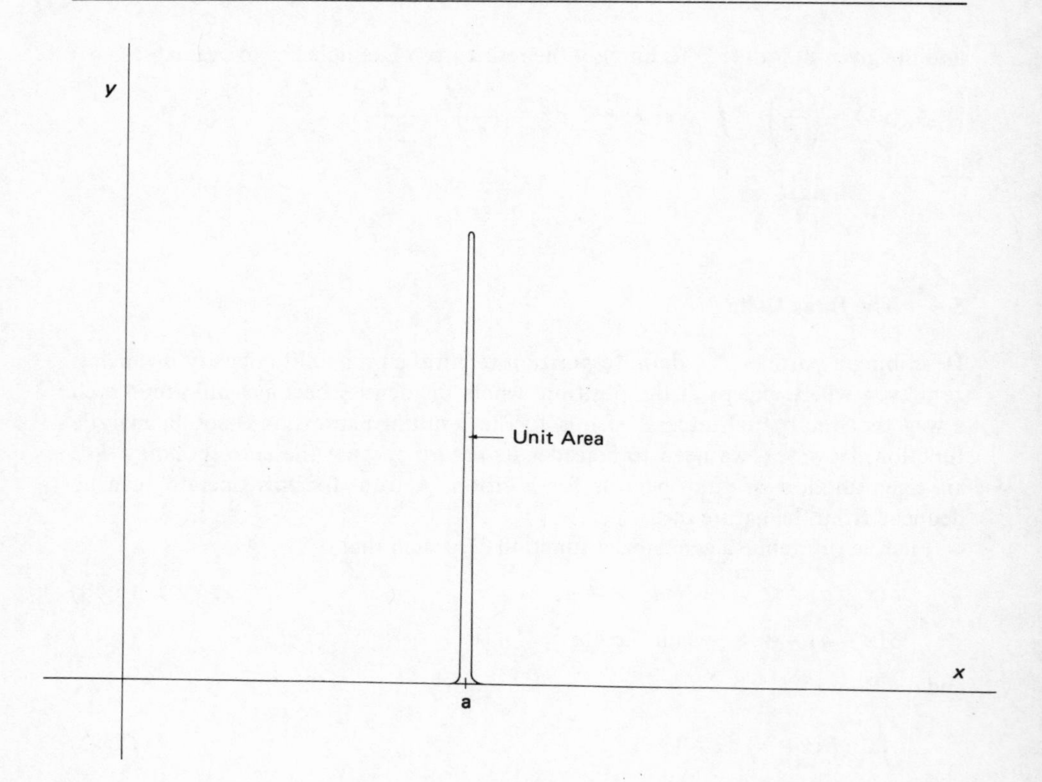

Fig. 5.1. Function $y = f(x)$ that turns into $\delta(x - a)$ when its height becomes infinite, its width infinitesimal.

The three-dimensional *Dirac delta,* $\delta(\mathbf{v})$, is defined by the conditions

$$\delta(\mathbf{r}-\mathbf{a}) = 0 \quad \text{when} \quad \mathbf{r} \neq \mathbf{a}, \tag{5.46}$$

$$\delta(\mathbf{r}-\mathbf{a}) = \infty \quad \text{when} \quad \mathbf{r} = \mathbf{a}, \tag{5.47}$$

with

$$\int_{\substack{\text{all}\\ \text{space}}} \delta(\mathbf{r}-\mathbf{a})\, d^3\mathbf{r} = 1. \tag{5.48}$$

As before, this Dirac delta is symmetric with respect to reflection through its center:

$$\delta(\mathbf{r}-\mathbf{a}) = \delta(\mathbf{a}-\mathbf{r}). \tag{5.49}$$

Also, when function $f(\mathbf{r})$ is well-behaved, we obtain

$$\int_{\substack{\text{all}\\\text{space}}} \delta(\mathbf{r}-\mathbf{a}) f(\mathbf{r})\, \mathrm{d}^3\mathbf{r} = f(\mathbf{a}) \int_{\substack{\text{all}\\\text{space}}} \delta(\mathbf{r}-\mathbf{a})\, \mathrm{d}^3\mathbf{r}$$

$$= f(\mathbf{a}). \tag{5.50}$$

The probability density for a particle at position $\mathbf{a}$ has the properties of $\delta(\mathbf{r}-\mathbf{a})$:

$$\rho = \delta(\mathbf{r}-\mathbf{a}). \tag{5.51}$$

But then, we also have

$$\int_R \mathbf{r}\rho\, \mathrm{d}^3\mathbf{r} = \int_R \mathbf{r}\,\delta(\mathbf{r}-\mathbf{a})\, \mathrm{d}^3\mathbf{r}$$

$$= \mathbf{a} \tag{5.52}$$

and

$$\int_R \mathbf{r}\rho\, \mathrm{d}^3\mathbf{r} = \int_R \mathbf{r}\Psi^*\Psi\, \mathrm{d}^3\mathbf{r}$$

$$= \int_R \Psi^*\mathbf{r}\Psi\, \mathrm{d}^3\mathbf{r}. \tag{5.53}$$

The final expression in (5.53) has the form of (5.39) with $\mathbf{r}$ playing the role of A. So we interpret $\mathbf{r}$ as the quantum mechanical operator for position. Similarly, x is the operator for the x coordinate, y for the y coordinate, z for the z coordinate. Furthermore, the operator for a real $f(\mathbf{r})$ is simply $f(\mathbf{r})$. Each of these operators is Hermitian when region R is properly chosen.

Example 5.7. How is the generalized function described by Figure 5.2 related to the Dirac delta?

The Dirac delta vanishes everywhere except where its argument is zero. As a result, we have

$$\eta(x) = \int_{-\infty}^{x} \delta(\xi)\, \mathrm{d}\xi = 0 \qquad \text{when} \quad x < 0$$

and

$$\eta(x) = \int_{-\infty}^{x} \delta(\xi)\, \mathrm{d}\xi = 1 \qquad \text{when} \quad x > 0.$$

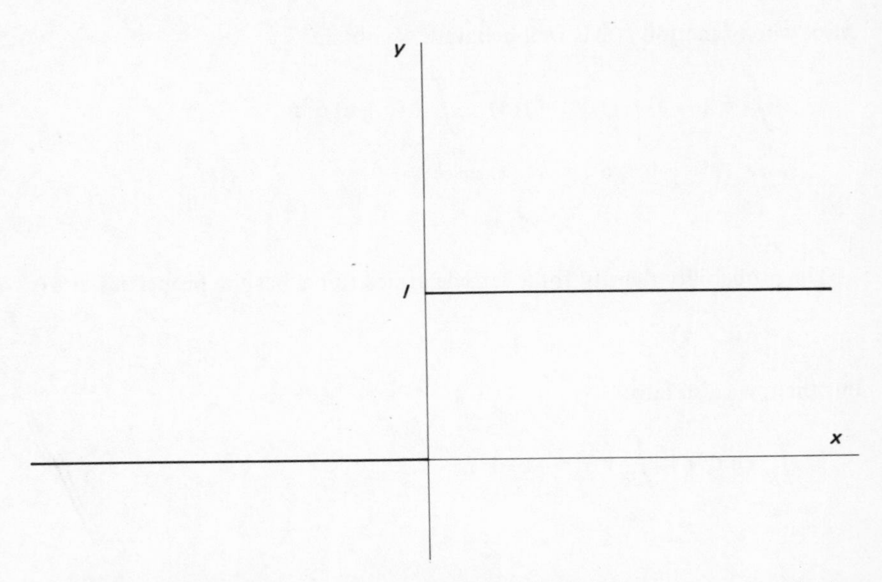

Fig. 5.2. Generalized function $y = \eta(x)$ which steps from 0, when $x < 0$, to 1 when $x > 0$.

The generalized function described as the integral from $-\infty$ to x of the one-dimensional Dirac delta equals zero when x is negative, one when x is positive. Differentiating the formula for η gives us

$$\frac{d\eta}{dx} = \delta(x).$$

Expression $\eta(x)$ is often called the *unit step function.*

5.5. Operators for Momenta and Energy

Hermitian operators for linear and angular momenta, and for energy, can also be readily constructed.

A particle possesses a definite linear momentum in a homogeneous beam. But the wave function for the beam is governed by (1.20):

$$d\Psi = ik_x \Psi\, dx + ik_y \Psi\, dy + ik_z \Psi\, dz - i\omega\Psi\, dt. \tag{5.54}$$

On comparing this with (1.2), we see that

$$\frac{\partial\Psi}{\partial x} = ik_x \Psi. \tag{5.55}$$

Let us multiply (5.55) by $\hbar/i$

$$\frac{\hbar}{i}\frac{\partial}{\partial x}\Psi = \hbar k_x \Psi \tag{5.56}$$

and combine with de Broglie's equation to obtain

$$\frac{\hbar}{i}\frac{\partial}{\partial x}\Psi = p_x \Psi. \tag{5.57}$$

Relationship (5.57) exhibits the same form as (5.6); the operator for linear momentum p_x is the differentiating operator on the left. Formally

$$p_x = \frac{\hbar}{i}\frac{\partial}{\partial x}. \tag{5.58}$$

Similarly, the operators for the y and z components of linear momentum are

$$p_y = \frac{\hbar}{i}\frac{\partial}{\partial y}; \tag{5.59}$$

$$p_z = \frac{\hbar}{i}\frac{\partial}{\partial z}. \tag{5.60}$$

A particle possessing a definite angular momentum around the z axis is governed by (2.23),

$$\mathrm{d}\Psi = ikr \sin\theta\, \Psi\, \mathrm{d}\varphi. \tag{5.61}$$

On comparing with (2.12), we see that

$$\frac{\partial \Psi}{\partial \varphi} = ikr \sin\theta\, \Psi. \tag{5.62}$$

Multiply by $\hbar/i$ and reduce the right side:

$$\begin{aligned}\frac{\hbar}{i}\frac{\partial}{\partial \varphi}\Psi &= \hbar kr \sin\theta\, \Psi = pr \sin\theta\, \Psi \\ &= \mathscr{M}_z \Psi.\end{aligned} \tag{5.63}$$

In the second equality, de Broglie's relationship has been introduced; in the third equality, the definition of angular momentum around the z axis.

Equation (5.63) is of the eigenvalue-equation form, with the operator relationship

$$\mathscr{M}_z = \frac{\hbar}{i}\frac{\partial}{\partial \varphi}. \tag{5.64}$$

In a given energy state, Equation (1.44) holds. On multiplying both sides by $\hbar/i$, we obtain

$$\frac{\hbar}{i}\frac{\partial}{\partial t}\Psi = -\hbar\omega\Psi, \tag{5.65}$$

an eigenvalue equation. Since $\hbar\omega$ equals the energy, by (1.48), we have the operator formula

$$E = -\frac{\hbar}{i}\frac{\partial}{\partial t}. \tag{5.66}$$

An integration by parts, as presented in Example 5.4, shows that each of the operators obtained is Hermitian when Ψ meets the appropriate conditions.

Example 5.8. What is the expectation value for p_x^2 in the harmonic oscillator state

$$\Psi = \left(\frac{a}{\pi}\right)^{\frac{1}{4}} e^{-ax^2/2}\, e^{-i\omega t}?$$

Applying (5.58) twice leads to

$$p_x^2 = \frac{\hbar}{i}\frac{\partial}{\partial x}\frac{\hbar}{i}\frac{\partial}{\partial x} = -\hbar^2\frac{\partial^2}{\partial x^2}.$$

Let this act on the spatial factor in Ψ:

$$p_x^2\psi = -\hbar^2\frac{\partial^2}{\partial x^2} N\,e^{-ax^2/2} = \hbar^2 N\frac{\partial}{\partial x}\,e^{-ax^2/2}\,(ax)$$

$$= \hbar^2 N\,e^{-ax^2/2}\,(a - a^2x^2).$$

Then construct the expectation integral:

$$\langle p_x^2\rangle = \int_{-\infty}^{\infty}\Psi^* p_x^2\Psi\,dx = \hbar^2\left(\frac{a}{\pi}\right)^{\frac{1}{2}}\int_{-\infty}^{\infty} e^{-ax^2}\,(a - a^2x^2)\,dx$$

$$= \hbar^2\left(\frac{a}{\pi}\right)^{\frac{1}{2}}\left[a\left(\frac{\pi}{a}\right)^{\frac{1}{2}} - a^2\left(\frac{\pi}{a}\right)^{\frac{1}{2}}\frac{1}{2a}\right] = \hbar^2\left(a - \frac{1}{2}a\right)$$

$$= \frac{1}{2}\hbar^2 a = \frac{1}{2}\hbar(f\mu)^{\frac{1}{2}}.$$

In the integration, results from Example 3.3 and 5.5 have been employed. The expression for a comes from (3.28).

The corresponding kinetic energy is

$$\langle T\rangle = \frac{1}{2\mu}\langle p_x^2\rangle = \frac{1}{2}\left[\frac{1}{2}\hbar\left(\frac{f}{\mu}\right)^{\frac{1}{2}}\right] = \frac{1}{2}E_0,$$

where according to (3.52), E_0 is the energy of the state.

5.6. The Hamiltonian Operator

Both the kinetic and potential energy parts of E are represented in the Schrödinger equation. Each part appears as a Hermitian operator, the sum as the Hamiltonian operator.

Consider a particle subject to a scalar potential $V(\mathbf{r})$. Since $V(\mathbf{r})$ is real and since it commutes with Ψ_j^*, it acts as a Hermitian operator. Also, suppose that the boundary conditions are chosen so operators (5.58), (5.59), and (5.60) are Hermitian. (Recall Example 5.4.)

In Examples 5.1 and 5.3, we showed that the sum of any Hermitian operators is a Hermitian operator and that the product of commuting Hermitian operators is Hermitian. Now p_x, p_y, and p_z are Hermitian and each commutes with itself. So p_x^2, p_y^2, p_z^2 are individually Hermitian and the kinetic energy operator

$$T = \frac{p_x^2}{2\mu} + \frac{p_y^2}{2\mu} + \frac{p_z^2}{2\mu} \tag{5.67}$$

is Hermitian. Here μ is the mass of the particle.

Furthermore, the total energy

$$H = T + V = \frac{1}{2\mu}(p_x^2 + p_y^2 + p_z^2) + V(\mathbf{r}), \tag{5.68}$$

in Hamiltonian form, is Hermitian. Substituting (5.58), (5.59), (5.60) into (5.68) leads to

$$\begin{aligned} H &= -\frac{\hbar^2}{2\mu}\left(\frac{\partial^2}{\partial x^2} + \frac{\partial^2}{\partial y^2} + \frac{\partial^2}{\partial z^2}\right) + V(\mathbf{r}) \\ &= -\frac{\hbar^2}{2\mu}\nabla^2 + V. \end{aligned} \tag{5.69}$$

Expression (5.69) is called the *Hamiltonian operator*.

Since the value of H is the energy E,

$$H = E, \tag{5.70}$$

an eigenvalue equation for E is

$$\left(-\frac{\hbar^2}{2\mu}\nabla^2 + V\right)\psi = E\psi. \tag{5.71}$$

This is a rearrangement of Schrödinger equation (4.31).

5.7. Shift Operators for the Harmonic Oscillator

Besides eigenoperators, operators that transform one eigenfunction into another can be constructed for simple quantum mechanical systems. The appropriate one enables us to calculate each solution in turn from a ground state function.

For the harmonic oscillator, the factored equations in Section 3.13 furnish a convenient starting point. Let us suppose that the parameter in (3.104) is v_2, that this number in (3.105) is v_1, and label the eigenfunctions accordingly:

$$(\mathrm{D} - w)(\mathrm{D} + w)\psi_{v_2} = -2v_2\,\psi_{v_2}, \tag{5.72}$$

$$(\mathrm{D} + w)(\mathrm{D} - w)\psi_{v_1} = -(2v_1 + 2)\psi_{v_1}. \tag{5.73}$$

Applying $(\mathrm{D} - w)$ to (5.73) yields

$$(\mathrm{D} - w)(\mathrm{D} + w)\,[(\mathrm{D} - w)\psi_{v_1}] = -(2v_1 + 2)\,[(\mathrm{D} - w)\psi_{v_1}], \tag{5.74}$$

while applying $(\mathrm{D} + w)$ to (5.72) gives us

$$(\mathrm{D} + w)(\mathrm{D} - w)\,[(\mathrm{D} + w)\psi_{v_2}] = -2v_2\,[(\mathrm{D} + w)\psi_{v_2}]. \tag{5.75}$$

On comparing (5.72) with (5.74), we see that

$$(\mathrm{D} - w)\psi_{v_1} \tag{5.76}$$

is an eigenfunction of $(\mathrm{D} - w)(\mathrm{D} + w)$ with the eigenvalue

$$-(2v_1 + 2) = -2v_2, \tag{5.77}$$

whence

$$v_2 = v_1 + 1. \tag{5.78}$$

Furthermore, we have

$$(\mathrm{D} - w)\psi_{v_1} = c_{v_1}^{+}\,\psi_{v_1+1}, \tag{5.79}$$

since no eigenfunctions of the harmonic oscillator are degenerate.

In (5.75),

$$(\mathrm{D} + w)\psi_{v_2} \tag{5.80}$$

is the eigenfunction of $(\mathrm{D} + w)(\mathrm{D} - w)$ with $-2v_2$ the eigenvalue. From (5.73), the eigenvalue equals $-(2v_1 + 2)$; so

$$-2v_2 = -(2v_1 + 2) \tag{5.81}$$

and

$$v_1 = v_2 - 1. \tag{5.82}$$

Furthermore, comparing (5.75) and (5.73) yields

$$(\mathrm{D} + w)\psi_{v_2} = c_{v_2}^{-}\,\psi_{v_2-1}. \tag{5.83}$$

Operator (D − w) transforms an eigenfunction for number v_1 into a constant times an eigenfunction for number $v_1 + 1$. On the other hand, operator (D + w) changes ψ for number v_2 into a constant times ψ for number $v_2 - 1$. If the acceptable solutions of the Schrödinger equation for the harmonic oscillator are arranged in order of increasing v, (D − w) takes the system from one such solution to the next higher solution; it is a *step-up* operator. Similarly, (D + w) takes the system from one solution to the next lower solution; it is a *step-down* operator. Both (D − w) and (D + w) are called *shift operators*.

If the harmonic oscillator is initially in its lowest state, action of the step-down operator on its ψ presumably yields zero:

$$(\mathrm{D} + w)\psi = 0. \tag{5.84}$$

Substituting (5.84) into (3.104), or into (5.72), then tells us that

$$v = 0 \tag{5.85}$$

in the state. Furthermore, each step-up operation increases v by 1. And, this operation serves to introduce one node into ψ. Consequently, parameter v is identified with the vibrational quantum number v with which we are already familiar.

5.8. Nodeless Solutions

The two simplest solutions of the harmonic oscillator Schrödinger equation arise when the shift operators, acting on ψ, yield zero.

Of particular interest is relationship (5.84),

$$(\mathrm{D} + w)\psi = 0, \tag{5.86}$$

that makes

$$v = 0. \tag{5.87}$$

For then

$$\frac{\mathrm{d}\psi}{\mathrm{d}w} + w\psi = 0 \tag{5.88}$$

or

$$\frac{\mathrm{d}\psi}{\psi} = -w\,\mathrm{d}w, \tag{5.89}$$

whence

$$\ln \psi = \ln N - \frac{w^2}{2} \tag{5.90}$$

and

$$\psi = N\,e^{-w^2/2}. \tag{5.91}$$

This is the ground-state eigenfunction obtained in Chapter 3. According to Example 3.5, it is normalized when

$$\psi = \left(\frac{a}{\pi}\right)^{\frac{1}{4}} e^{-w^2/2}. \tag{5.92}$$

In a similar way, we can set

$$(\mathrm{D} - w)\psi = 0. \tag{5.93}$$

Then (3.105), or (5.73), is satisfied with

$$v = -1. \tag{5.94}$$

In explicit form, (5.93) is

$$\frac{\mathrm{d}\psi}{\mathrm{d}w} - w\psi = 0 \tag{5.95}$$

whence

$$\frac{\mathrm{d}\psi}{\psi} = w\,\mathrm{d}w \tag{5.96}$$

and

$$\ln\psi = \ln B + \frac{w^2}{2} \tag{5.97}$$

or

$$\psi = B\,e^{w^2/2}. \tag{5.98}$$

This solution is not acceptable because it makes ρ increase without limit as $|w|$, and $|x|$, increase without limit.

5.9. Acceptable Wavy Solutions

By applying step-up operator $(\mathrm{D} - w)$ repeatedly to the acceptable nodeless solution, each successive excited eigenfunction for a harmonic oscillator can be generated in turn. Coefficient c_v^+ in (5.79) is determined by a normalization argument.

The model for a harmonic oscillator is the particle of mass μ in the potential

$$V = \frac{1}{2}fx^2 = \frac{1}{2}\frac{f}{a}w^2. \tag{5.99}$$

This field confines the particle, causing Ψ to be a standing wave. Thus, the spatial part, ψ, may be real. For quantum number v, used as an index on ψ, we then have

$$\int_{-\infty}^{\infty} \psi_v \psi_v \, dx = 1, \tag{5.100}$$

whence

$$\int_{-\infty}^{\infty} \psi_v \psi_v \, dw = \int_{-\infty}^{\infty} \psi_v \psi_v \, a^{\frac{1}{2}} \, dx = a^{\frac{1}{2}}. \tag{5.101}$$

Note that this result is independent of the index on ψ.

Multiplying each side of (5.79) by itself and integrating over all w's leads to

$$\int_{-\infty}^{\infty} [(D - w)\psi_v] \, (D - w)\psi_v \, dw$$

$$= \int_{-\infty}^{\infty} c_v^+ \psi_{v+1} \, c_v^+ \psi_{v+1} \, dw = c_v^{+2} \, a^{\frac{1}{2}}. \tag{5.102}$$

However, replacing $(D\psi_v)\,dw$ by its equivalent $d\psi_v$, integrating by parts, expanding the differential of $(D - w)\psi_v$ to the derivative with respect to w times dw, introducing D for d/dw, and bringing in (5.73) gives us

$$\int_{-\infty}^{\infty} [(D - w)\psi_v] \, (D - w)\psi_v \, dw = \int_{-\infty}^{\infty} d\psi_v \, (D - w)\psi_v -$$

$$- \int_{-\infty}^{\infty} w\psi_v \, (D - w)\psi_v \, dw = \psi_v (D - w)\psi_v \, \Big|_{-\infty}^{\infty} - \int_{-\infty}^{\infty} \psi_v \, d(D - w)\psi_v -$$

$$- \int_{-\infty}^{\infty} \psi_v w(D - w)\psi_v \, dw = 0 - \int_{-\infty}^{\infty} \psi_v \frac{d}{dw}(D - w)\psi_v \, dw -$$

$$- \int_{-\infty}^{\infty} \psi_v w(D - w)\psi_v \, dw = - \int_{-\infty}^{\infty} \psi_v (D + w)\,(D - w)\psi_v \, dw$$

$$= \int_{-\infty}^{\infty} \psi_v (2v + 2)\psi_v \, dw = (2v + 2)a^{\frac{1}{2}}. \tag{5.103}$$

Since the initial forms in (5.102) and (5.103) are the same, the final forms are equal:

$$c_v^{+2} a^{\frac{1}{2}} = 2(v + 1)a^{\frac{1}{2}}. \tag{5.104}$$

Canceling $a^{\frac{1}{2}}$ and taking the negative square root yields

$$c_v^+ = -[2(v + 1)]^{\frac{1}{2}}. \tag{5.105}$$

Now, rewrite (5.79) as

$$\psi_v = \frac{D - w}{c^+_{v-1}} \psi_{v-1} = (-1) \frac{D - w}{(2v)^{\frac{1}{2}}} \psi_{v-1} \tag{5.106}$$

and apply the result to build up ψ_v from (5.92):

$$\psi_v = (-1)^v \frac{(D - w)^v}{(2^v v!)^{\frac{1}{2}}} \left(\frac{a}{\pi}\right)^{\frac{1}{4}} e^{-w^2/2}. \tag{5.107}$$

Since operator $(D - w)$ acts to cause

$$(D - w)\, e^{-w^2/2} = -w\, e^{-w^2/2} - w\, e^{-w^2/2} = -2w\, e^{-w^2/2}$$
$$= e^{w^2/2}\, D\, e^{-w^2} \tag{5.108}$$

and

$$(D - w)\, e^{w^2/2}\, D^{v-1}\, e^{-w^2}$$
$$= w\, e^{w^2/2}\, D^{v-1}\, e^{-w^2} - w\, e^{w^2/2}\, D^{v-1}\, e^{-w^2} + e^{w^2/2}\, D^v\, e^{-w^2}$$
$$= e^{w^2/2}\, D^v\, e^{-w^2}, \tag{5.109}$$

we know that

$$(D - w)^v\, e^{-w^2/2} = e^{w^2/2}\, D^v\, e^{-w^2}$$
$$= e^{-w^2/2}\, e^{w^2} \frac{d^v}{dw^v} e^{-w^2} \tag{5.110}$$

by induction. Consequently, Equation (5.107) can be rewritten in the form

$$\psi_v = \frac{1}{(2^v v!)^{\frac{1}{2}}} \left(\frac{a}{\pi}\right)^{\frac{1}{4}} e^{-w^2/2} (-1)^v e^{w^2} \frac{d^v}{dw^v} e^{-w^2}$$
$$= N\, e^{-w^2/2} H_v(w). \tag{5.111}$$

This result agrees with formula (3.49), with N chosen as in (3.53). Thus, use of (3.53) is justified.

Example 5.9. From c^+_v, determine coefficient c^-_v and forms for normalized step-up and step-down operators.

Introduce substitution (5.83), and then (5.79), into the left side of (5.72) and equate the result to the right side:

$$(D - w)(D + w)\psi_v = (D - w)c^-_v\, \psi_{v-1} = c^+_{v-1}\, c^-_v\, \psi_v$$
$$= -2v\psi_v.$$

Cancel ψ_v from the last equality, substitute in value (5.105),

$$-(2v)^{\frac{1}{2}} c_v^- = -2v,$$

and solve for the desired coefficient:

$$c_v^- = (2v)^{\frac{1}{2}}.$$

From (5.79) and (5.83), we now obtain

$$\psi_{v+1} = \frac{\mathrm{D} - w}{c_v^+} \psi_v = -\frac{\mathrm{D} - w}{[2(v+1)]^{\frac{1}{2}}} \psi_v$$

and

$$\psi_{v-1} = \frac{\mathrm{D} + w}{c_v^-} \psi_v = \frac{\mathrm{D} + w}{(2v)^{\frac{1}{2}}} \psi_v.$$

5.10. The Inverse Operator as the Adjoint of an Operator

An operator that transforms one pure state into another is not, in general, Hermitian. Instead, its adjoint is its inverse, and its inverse, its adjoint.

Consider a quantum mechanical system with the normalized orthogonal eigenfunctions $\Psi_1, \Psi_2, \ldots, \Psi_l, \ldots, \Psi_m, \ldots$. Suppose that B converts Ψ_m to Ψ_l and that B^{-1} converts Ψ_l back to Ψ_m. Then

$$\int_R \Psi_l^* B \Psi_m \, \mathrm{d}^3\mathbf{r} = \int_R \Psi_l^* \Psi_l \, \mathrm{d}^3\mathbf{r} \tag{5.112}$$

and

$$\int_R (B^{-1} \Psi_l)^* \Psi_m \, \mathrm{d}^3\mathbf{r} = \int_R \Psi_m^* \Psi_m \, \mathrm{d}^3\mathbf{r}. \tag{5.113}$$

Since the eigenfunctions are normalized, the right sides of (5.112) and (5.113) are equal. Therefore,

$$\int_R \Psi_l^* B \Psi_m \, \mathrm{d}^3\mathbf{r} = \int_R (B^{-1} \Psi_l)^* \Psi_m \, \mathrm{d}^3\mathbf{r}. \tag{5.114}$$

But by definition, an operator $B^\dagger$ related to B by the formula

$$\int_R \Psi_l^* B \Psi_m \, \mathrm{d}^3\mathbf{r} = \int_R (B^\dagger \Psi_l)^* \Psi_m \, \mathrm{d}^3\mathbf{r} \tag{5.115}$$

is called the *adjoint* of B. On comparing (5.114) with (5.115), we find that the inverse of B is its adjoint:

$$B^{-1} = B^{\dagger} . \tag{5.116}$$

From Example 5.9, the normalized step-down operator for the harmonic oscillator, acting on the state in which $v = j$, is

$$a_j = \frac{\mathrm{D} + w}{(2j)^{\frac{1}{2}}}, \tag{5.117}$$

while the normalized step-up operator to produce the same state is

$$b_j = -\frac{\mathrm{D} - w}{(2j)^{\frac{1}{2}}}. \tag{5.118}$$

Neither of these is Hermitian.

But since b_j is the inverse of a_j, relationship (5.116) tells us that

$$b_j = a_j^{\dagger} . \tag{5.119}$$

The step-up operator is the adjoint of the step-down operator. Similarly,

$$a_j = b_j^{\dagger} . \tag{5.120}$$

Because of (5.119) and (5.120), either binary product

$$a_j^{\dagger} a_j \qquad \text{or} \qquad b_j^{\dagger} b_j \tag{5.121}$$

returns the system to its original state and so is Hermitian.

5.11. Shift Operators for a Hydrogen-like Atom

The Schrödinger equation for a harmonic oscillator yields shift operators because the equation can be rearranged to form a pair of eigenvalue equations with eigenoperators containing the same first-order factors in the two possible sequences. The Schrödinger equation for the radial equation is similarly presented in Section 4.11; so shift operators also exist for it.

Thus, radial motion within a hydrogen-like atom is governed by Equations (4.128) and (4.131). If we place subscripts on state function R indicating the principal quantum number and the azimuthal quantum number, (4.128) becomes

$$\left(x\mathrm{D} + \frac{1}{2}x - n\right)\left(x\mathrm{D} - \frac{1}{2}x + n + 1\right)R_{n,l} = [l(l+1) - n(n+1)]R_{n,l}. \tag{5.122}$$

The radial motion is also governed by (4.129) and (4.130). Adding subscripts to the state function to indicate the pertinent principal and azimuthal quantum numbers causes (4.129) to become

$$\left(x\mathrm{D} - \frac{1}{2}x + n\right)\left(x\mathrm{D} + \frac{1}{2}x - n + 1\right)R_{n,l} = [l(l+1) - (n-1)n]R_{n,l}. \quad (5.123)$$

Applying $(x\mathrm{D} - \frac{1}{2}x + n)$ to (4.130) yields

$$\left(x\mathrm{D} - \frac{1}{2}x + n\right)\left(x\mathrm{D} + \frac{1}{2}x - n + 1\right)\left[\left(x\mathrm{D} - \frac{1}{2}x + n\right)R_{n-1,l}\right]$$

$$= [l(l+1) - (n-1)n]\left[\left(x\mathrm{D} - \frac{1}{2}x + n\right)R_{n-1,l}\right]. \quad (5.124)$$

The eigenoperator in (5.123) is the operator preceding the bracketed expression in (5.124); the eigenvalues of these on the right sides are also the same. Since there is only one radial state for each n and l, we must therefore have

$$\left(x\mathrm{D} - \frac{1}{2}x + n\right)R_{n-1,l} = aR_{n,l} \quad (5.125)$$

where a is a constant.

Similarly, applying $(x\mathrm{D} + \frac{1}{2}x - n)$ to (4.131) yields

$$\left(x\mathrm{D} + \frac{1}{2}x - n\right)\left(x\mathrm{D} - \frac{1}{2}x + n + 1\right)\left[\left(x\mathrm{D} + \frac{1}{2}x - n\right)R_{n+1,l}\right]$$

$$= [l(l+1) - n(n+1)]\left[\left(x\mathrm{D} + \frac{1}{2}x - n\right)R_{n+1,l}\right]. \quad (5.126)$$

Now, (5.126) can be considered a form of (5.122); so we obtain

$$\left(x\mathrm{D} + \frac{1}{2}x - n\right)R_{n+1,l} = bR_{n,l} \quad (5.127)$$

where b is a constant. Replacing $n + 1$ with n, n with $n - 1$, then yields

$$\left(x\mathrm{D} + \frac{1}{2}x - n + 1\right)R_{n,l} = bR_{n-1,l}. \quad (5.128)$$

The rotational quantum number l may be zero or any positive integer:

$$l = 0, 1, 2, \ldots . \quad (5.129)$$

For energy E to have a lower limit with a given l, the sequence described by (5.128) must terminate, yielding zero for some n. According to (5.123), the first such zero arises when

$$n - 1 = l. \quad (5.130)$$

We then have

$$\left(xD + \frac{1}{2}x - n + 1\right) R_{n,l} = \left(xD + \frac{1}{2}x - l\right) R_{n,l} = 0, \tag{5.131}$$

whence

$$x\frac{dR}{dx} + \left(\frac{1}{2}x - l\right) R = 0 \tag{5.132}$$

or

$$\frac{dR}{R} = \frac{l}{x}\,dx - \frac{dx}{2}. \tag{5.133}$$

Integrating (5.133) leads to

$$\ln R = \ln A + l \ln x - \frac{x}{2} \tag{5.134}$$

or

$$R = Ax^l\, e^{-x/2} \qquad \text{for} \quad n = l + 1, \tag{5.135}$$

a suitable solution. Operating repeatedly on expression (5.135) as

$$\left(xD - \frac{1}{2}x + n + 1\right) R_{n,l} = aR_{n+1,l}, \tag{5.136}$$

derived from (5.125), indicates generates the other eigenfunctions for the same l.

Each step creates an expression of the form

$$R = u(x)x^l\, e^{-x/2} \tag{5.137}$$

in which the degree of the polynomial $u(x)$ is increased by 1. When (5.130) holds, this degree is zero. Each application of the operator increases n by 1; so the final degree is

$$n - l - 1. \tag{5.138}$$

Note that (5.137) agrees with (4.88); (5.138) agrees with the statements in Section 4.8.

5.12. Spontaneous Decay of an Unstable State

Implicit in the derivation of the time-independent Schrödinger equation, in Section 4.3, is the assumption that the state described is stable. Excited states are, however, generally unstable. A rotating polar molecule produces an electromagnetic wave, one or more photons, and thereby loses rotational energy. A vibrating polar molecule behaves similarly. An excited electron in an atom radiates an electromagnetic wave, one or more photons, and also loses energy. An unstable nucleus emits one or more particles and thereby gains stability.

Just as interactions with surrounding fields and particles lead to excitation, such processes also cause de-excitation. In addition, decay occurs spontaneously. Here we are concerned with the latter process.

Consider a set of identical systems undergoing a spontaneous shift to a lower level or state. Let the probability that a given particle in the set thus decays during interval dt be

$$\frac{1}{\tau}\,dt. \tag{5.139}$$

Since the probability per unit time of the decay we are considering is not affected by the presence of other particles, time τ is constant.

If we disregard the particle nature of the material, treating it instead as a continuum, we consider the number of identical unstable particles N to decrease continuously. Then the probability of decay equals $-dN/N$. Equating this infinitesimal fraction to (5.139) leads to the formula

$$-\frac{dN}{N} = \frac{dt}{\tau}. \tag{5.140}$$

Approximate law (5.140) integrates to yield

$$N = N_0\,e^{-t/\tau}. \tag{5.141}$$

The *mean life* of a state equals the sum of the ages at decay divided by the initial number of particles. When the initial number of particles is large enough, the continuum approximation holds and the sum may be replaced by an integral:

$$\begin{aligned}\bar{t} &= \frac{1}{N_0}\int_{N_0}^{0} t(-dN) = \frac{1}{N_0}\int_0^\infty t\,\frac{N\,dt}{\tau} = \frac{1}{\tau}\int_0^\infty t\,e^{-t/\tau}\,dt \\ &= \frac{1}{\tau}\,\frac{t\,e^{-t/\tau}}{-1/\tau}\bigg|_0^\infty + \int_0^\infty e^{-t/\tau}\,dt \\ &= -(t+\tau)\,e^{-t/\tau}\bigg|_0^\infty = \tau.\end{aligned} \tag{5.142}$$

Decay law (5.140) has been introduced in the second step, integrated form (5.141) in the third step, and an integration by parts in the fourth step. Note that the mean life equals parameter τ.

The state of a particle decaying by a single path is represented by a wave function Ψ with a probability density ρ that decreases according to law (5.141). Since

$$\rho = \Psi^* \Psi, \tag{5.143}$$

we must have

$$\Psi = \psi \, e^{-i\omega_R t} \, e^{-t/(2\tau)}. \tag{5.144}$$

Parameter ω_R is the angular frequency the state would exhibit if the state were stable.

The temporal factor in (5.144) can be expressed as a superposition of simple exponentials,

$$e^{-i\omega_R t} \, e^{-t/(2\tau)} = \frac{1}{(2\pi)^{\frac{1}{2}}} \int_{-\infty}^{\infty} \chi(\omega) \, e^{-i\omega t} \, d\omega, \tag{5.145}$$

each with amplitude

$$\frac{1}{(2\pi)^{\frac{1}{2}}} \chi(\omega) \, d\omega. \tag{5.146}$$

After developing Fourier integral theory, in the next chapter, we will determine the dependence of $\chi(\omega)$ on ω_R and τ.

Example 5.10. How much time is needed for half of a set of identical unstable particles to decay?

The time required for half of the given unstable particles to decay in a certain manner is called the *half-life* $t_{1/2}$ for the process. At this time,

$$N = \frac{N_0}{2}$$

and Equation (5.141) becomes

$$\frac{N_0}{2} = N_0 \, e^{-t_{1/2}/\tau},$$

whence

$$t_{1/2} = \tau(\ln 2) = (0.69315)\tau.$$

Discussion Questions

5.1. Explain what space-time aspects produce the observable properties of a particle in nature.

5.2. Why are these properties related to the state function for the particle?

5.3. What may an operator do to Ψ?
5.4. To what mathematical expressions do the pure states with respect to a given measurement correspond? Explain.
5.5. Why do we take the operator for an observable to be (a) linear, (b) Hermitian?
5.6. Why are the Ψ's for different values of an observable orthogonal?
5.7. What is degeneracy?
5.8. Why is the sum of two Hermitian operators Hermitian? When is a scalar times a Hermitian operator Hermitian? When is the product of two Hermitian operators Hermitian?
5.9. What is the statistical weight w of a pure state in a mixed state? How is this weight related to the expansion coefficients?
5.10. How is the formula for expectation value established?
5.11. How does a person describe (a) ρ, (b) ψ, for a particle at a definite position?
5.12. Define the Dirac delta in one, two, and three dimensions.
5.13. Describe the unit step function.
5.14. Why can we consider $\mathbf{r}$ to be the operator for position?
5.15. Determine whether

$$\text{(a)}\ \frac{\partial}{\partial x}, \quad \text{(b)}\ \hbar\frac{\partial}{\partial x}, \quad \text{(c)}\ i\frac{\partial}{\partial x}, \quad \text{(d)}\ \frac{\hbar}{i}\frac{\partial}{\partial x}$$

is Hermitian.
5.16. Explain how the operators for (a) linear momenta p_x, p_y, p_z, (b) angular momenta $\mathcal{M}_z$, $\mathcal{M}_x$, $\mathcal{M}_y$, (c) energy E can be obtained.
5.17. What is (a) the Hamiltonian expression for energy, (b) the Hamiltonian operator for energy?
5.18. Justify the form for the Hamiltonian operator.
5.19. Show that if two distinct operators can be combined in either order to form an operator whose eigenfunction is a state function, each of the original operators acts to convert one state function into a constant times another.
5.20. Explain why $(\mathrm{D} - w)$ and $(\mathrm{D} + w)$ are step-up and step-down operators for harmonic oscillator state functions.
5.21. How can the operators in Question 5.20 be normalized?
5.22. How are the nodeless solutions for the harmonic oscillator obtained? How are all acceptable solutions found from one of these?
5.23. What is the adjoint of an operator?
5.24. Why is the inverse of an operator that transforms one pure state into another equal to the adjoint?
5.25. Identify and describe the step-up and step-down operators for radial eigenfunctions of a hydrogen-like atom.
5.26. Derive the form of each operator in Question 5.25.
5.27. Why is there a lower limit on the energy for each l? What is the relationship between n and l at this energy?
5.28. By a simple integration, obtain $R(r)$ at the limit described in Question 5.27.
5.29. Why is the probability that a given excited state decay proportional to $\mathrm{d}t$?

5.30. Why is this probability equal to $-dN/N$?
5.31. Derive the decay law: $N = N_0\ e^{-t/\tau}$.
5.32. Show how an expression for the mean life of a decaying state is obtained.
5.33. Why is the ω for a decaying state spread over a band of angular frequencies? Why is the energy of a decaying state similarly spread?

Problems

5.1. Show that $\frac{1}{2}(AB + BA)$ is Hermitian when operators A and B are Hermitian.

5.2. Show when a scalar function $V(\mathbf{r})$ is Hermitian.

5.3. In what circumstances is $\frac{h}{i}\frac{\partial}{\partial \varphi}$ Hermitian?

5.4. Prove that an eigenfunction of $\frac{h}{i}\frac{\partial}{\partial x}$ is an eigenfunction of $-\frac{h^3}{i}\frac{\partial^3}{\partial x^3}$. Determine how the eigenvalues of the two operators are related.

5.5. What is the expectation value for r in the hydrogen atom state for which

$$\psi = \frac{1}{\pi^{1/2}a^{3/2}}\ e^{-r/a}?$$

5.6. What is the expectation value for potential energy in the ground hydrogen atom state?

5.7. Determine what $\frac{d^2}{dx^2}\frac{|x|}{2}$ represents.

5.8. Show that

(a) $x\,\delta(x) = 0$,

(b) $x\frac{d}{dx}\delta(x) = -\delta(x)$,

(c) $\delta(ax) = a^{-1}\,\delta(x)$.

5.9. Show that the adjoint of a product of operators is the product of the adjoints in the opposite order:

$$(AB)^{\dagger} = B^{\dagger}A^{\dagger}.$$

5.10. If the half-life of tritium is 12.26 years, what fraction of an initial amount of ^3H is left after one year?

5.11. Prove that $\frac{1}{2i}(AB - BA)$ is Hermitian if operators A and B are Hermitian.

5.12. When is $-\frac{h}{i}\frac{\partial}{\partial p_x}$ Hermitian in momentum space?

5.13. If operator A is a constant, what are its eigenvalues? What is the degeneracy of each of these?

5.14. Show that if Ψ_j is an eigenfunction of A, then Ψ_j is also an eigenfunction of $P(A)$, where $P(A)$ is a polynomial in A. Determine how the eigenvalues of A and $P(A)$ are related.

5.15. Compute the expectation value for r in the first excited state of hydrogen for which

$$\psi = \frac{1}{4(2\pi)^{1/2}a^{3/2}}\ \frac{r}{a}\ e^{-r/2a}\cos\theta.$$

5.16, What is the expectation value for potential energy in the first excited hydrogen atom state of Problem 5.15?

5.17. Determine what is represented by

$$\lim_{\epsilon \to 0} \frac{1}{\pi} \frac{\epsilon}{x^2 + \epsilon^2}.$$

5.18. Show that

$$\text{(a)}\ \delta(x^2 - a^2) = \frac{\delta(x - a) + \delta(x + a)}{2a},$$

$$\text{(b)}\ \int_{-\infty}^{\infty} \delta(a - x)\, \delta(x - b)\, \mathrm{d}x = \delta(a - b).$$

5.19. Show that the adjoint of a complex number is the complex conjugate of the number.

5.20. Each gram of carbon in living material emits 15 beta particles per minute, on the average. Assume that the source for these is ^{14}C and calculate the fraction of carbon atoms that are ^{14}C. The reciprocal mean life for ^{14}C is $2.28 \times 10^{-10}\ \text{min}^{-1}$.

References

Books

Fong, P.: 1962, *Elementary Quantum Mechanics*, Addison-Wesley, Reading, Mass., pp. 297–360.

Matthews, P. T.: 1963, *Introduction to Quantum Mechanics*, McGraw-Hill, New York, pp. 12–35.

Rojansky, V.: 1942, *Introductory Quantum Mechanics*, Prentice-Hall, New York, pp. 1–42.

Articles

Borneas, M.: 1972, 'A Quantum Equation of Motion with Higher Derivatives', *Am. J. Phys.* **40**, 248–251.

Das, R., and Sannigrahi, A. B.: 1981, 'The Factorization Method and Its Applications in Quantum Chemistry', *J. Chem. Educ.* **58**, 383–388.

David, C. W.: 1966, 'Ladder Operator Solution for the Hydrogen Atom Electronic Energy Levels', *Am. J. Phys.* **34**, 984–985.

Dean, C. E., and Fulling, S. A.: 1982, 'Continuum Eigenfunction Expansions and Resonances: A Simple Model', *Am. J. Phys.* **50**, 540–544.

Edwards, I. K.: 1979, 'Quantization of Inequivalent Classical Hamiltonians', *Am. J. Phys.* **47**, 153–155.

Flores, J., Henestroza, E., Mello, P. A., and Moshinsky, M.: 1981, 'Decay of a Compound Particle and the Einstein-Podolsky-Rosen Argument', *Am. J. Phys.* **49**, 59–63.

Gruber, G. R.: 1976, 'On the Transition from Classical to Quantum Mechanics in Generalized Coordinates', *Found. Phys.* **6**, 111–113.

Jordan, T. F.: 1975, 'Why $-i\nabla$ is the Momentum', *Am. J. Phys.* **43**, 1089–1093.

Jordan, T. F.: 1976, 'Conditions on Wave Functions Derived from Operator Domains', *Am. J. Phys.* **44**, 567–570.

Leaf, B.: 1979, 'Momentum Operators for Curvilinear Coordinate Systems', *Am. J. Phys.* **47**, 811–813.

Liboff, R. L., Nebenzahl, I., and Fleischmann, H. H.: 1973, 'On the Radial Momentum Operator', *Am. J. Phys.* **41**, 976–980.

Mucci, J. F., and Haskins, P. J.: 1973, 'Error Limits of Expectation Values in Quantum Mechanical Calculations', *Am. J. Phys.* **41**, 987–989.

Newmarch, J. D., and Golding, R. M.: 1978, 'Ladder Operators for Some Spherically Symmetric Potentials in Quantum Mechanics', *Am. J. Phys.* **46**, 658–660.

Peña, de la, L., and Montemayor, R.: 1980, 'Raising and Lowering Operators and Spectral Structure: A Concise Algebraic Technique', *Am. J. Phys.* **48**, 855–860.

Salsburg, Z. W.: 1965, 'Factorization of the Radial Equation for the Hydrogen Atom', *Am. J. Phys.* **33**, 36–39.

Shoemaker, D. P.: 1972, 'The Dirac Delta Function and the Density of States of Several Systems', *J. Chem. Educ.* **49**, 607–610.

Swenson, R. J., and Hermanson, J. C.: 1972, 'Energy Quantization and the Simple Harmonic Oscillator', *Am. J. Phys.* **40**, 1258–1260.

Villasenor-Gonzales, P., and Cisneros-Parra, J.: 1981, 'Quantum Operators in Generalized Coordinates', *Am. J. Phys.* **49**, 754–756.

Winter, R. G.: 1977, 'Construction of Some Soluble Quantal Problems', *Am. J. Phys.* **45**, 569–571.

Chapter 6

Wave Packets, Potentials, and Forces

6.1. Mixing Wave Functions

We have seen how a state function having a definite wavevector and angular frequency represents conditions in a uniform beam of particles. In like manner, a single function represents the free rotation of a system about an axis. Furthermore, a particle confined in a parabolic or Coulombic potential appears in states possessing a definite angular frequency and energy.

However, allowing a confined particle, or a rotating system, to radiate energy, as one or more photons, destroys the definiteness of the angular frequency and energy, as we saw in Section 5.12. Also, reducing a beam to a single particle, confined to a small moving region, destroys the definiteness of the wavevector and angular frequency. Just as a sum or integral of the simple functions is required to represent $T(t)$ in the decaying system, so also is a sum or integral required to represent the moving particle. This superposition forms what is called a *wave packet*.

By studying the movements of such a packet, we can deduce the connection between the angular frequency and energy. The specific time-dependent Schrödinger equation can be constructed. An expression for the rate of change of an expectation value can be deduced. Also, the rate of change of the expectation value for momentum can be related to the expectation value for force. Thus, we obtain a quantum mechanical form for Newton's second law.

Initially, we saw how each equivalent part of a state function for movement with a given momentum by a chosen point in space and time produces the same effect on any change in the function. Also, when different wavelets corresponding to different ways of behaving contribute to a motion, the wavelets simply add to each other. By induction, functions describing the various independent pure states of a system superpose linearly to form all possible state functions. Thus, a general Ψ can be expressed as a linear combination of independent eigenfunctions.

If the eigenfunctions are enumerable, each can be labeled by an index j. Then, if c_j is the multiplying constant determining the relative phasing of Ψ_j and the extent of its contribution to the overall state, we have

$$\Psi = \Sigma c_j \Psi_j(\mathbf{r}, t). \tag{6.1}$$

If the eigenfunctions are not enumerable, they form a continuum in which successive functions differ only infinitesimally from each other over a large region of space and time. The possible superpositions are then expressible by the integral

$$\Psi = \int c(k)\,\Psi(\mathbf{r}, t, k)\,\mathrm{d}k, \tag{6.2}$$

as we first noted in Section 1.8.

We call Ψ_j, or $\Psi(k)$, a *basis function*, while c_j, or $c(k)\,\mathrm{d}k$, is called the corresponding *component*. Expression (6.1) is like the formula expressing an arbitrary vector as a linear combination of independent base vectors. Function Ψ is analogous to the arbitrary vector, while Ψ_j is analogous to the jth base vector.

Each coefficient c_i is readily obtained from Ψ and Ψ_i when

$$\int_R \Psi_i^* \Psi_j\,\mathrm{d}^3\mathbf{r} = \delta_{ij}; \tag{6.3}$$

that is, when the basis functions are normalized and mutually orthogonal. For then, multiplying (6.1) by Ψ_i^* and integrating over the pertinent volume yields

$$\begin{aligned}\int_R \Psi_i^* \Psi\,\mathrm{d}^3\mathbf{r} &= \int_R \Psi_i^* \Sigma c_j \Psi_j\,\mathrm{d}^3\mathbf{r}\\ &= \Sigma c_j \int_R \Psi_i^* \Psi_j\,\mathrm{d}^3\mathbf{r}\\ &= \Sigma c_j\,\delta_{ij} = c_i.\end{aligned} \tag{6.4}$$

When the basis functions are merely orthogonal, we have

$$\int_R \Psi_i^* \Psi_j\,\mathrm{d}^3\mathbf{r} = N_i\,\delta_{ij}. \tag{6.5}$$

Multiplying (6.1) by Ψ_i^* and integrating now leads to

$$\int_R \Psi_i^* \Psi\,\mathrm{d}^3\mathbf{r} = \Sigma\, c_j N_i\,\delta_{ij} = N_i c_i, \tag{6.6}$$

where i is a free index, taking on the assigned value only, while j is a dummy index, running over all integers that may appear in Ψ_j.

Eigenfunctions combining as Equation (6.1) indicates include (a) those for given periodic boundary conditions and (b) those for confinement in a given box. If the length of a block edge, or the length of a wall, is allowed to increase without limit, the separation between the levels decreases to zero and successive eigenfunctions approach each other. In the limit, sum (6.1) then becomes integral (6.2).

6.2. Fourier Series and Fourier Integrals

The standing-wave eigenfunctions for each dimensional motion of a particle in a box superpose at a given time to form a Fourier series. Imposing the limiting process just described causes the series to be replaced by a Fourier integral. Furthermore, such an integral characterizes spatial variations in the Ψ of a one-dimensional freely moving wave packet at a given time, and temporal variations for a decaying state at a given place.

A *Fourier series* is a linear combination of all sine and cosine functions having the prescribed periodicity. When the period is 2π along the x axis and the interval within which an arbitrary smooth variation is allowed extends from $x = -\pi$ to $x = \pi$, we have

$$f(x) = \frac{1}{2}a_0 + \Sigma a_j \cos n_j x + \Sigma b_j \sin n_j x \tag{6.7}$$

where $f(x)$ is the resulting function, a_j and b_j are the jth coefficients, n_j equals integer j, while j is $1, 2, 3, \ldots$.

Let us change the independent variable in (6.7) from x to ξ and the dummy index from j to k:

$$f(\xi) = \frac{1}{2}a_0 + \Sigma a_k \cos n_k\xi + \Sigma b_k \sin n_k\xi. \tag{6.8}$$

As in deriving (6.6), let us multiply both sides of (6.8) (a) by $\frac{1}{2}$, (b) by $\cos n_j\xi$, (c) by $\sin n_j\xi$, and integrate over the interval from $\xi = -\pi$ to $\xi = \pi$. Since the constant and the different trigonometric functions in the series are mutually orthogonal, and since

$$\int_{-\pi}^{\pi} \left(\frac{1}{2}\right)\left(\frac{1}{2}\right) \mathrm{d}\xi = \frac{\xi}{4}\bigg|_{-\pi}^{\pi} = \frac{1}{2}\pi, \tag{6.9}$$

$$\int_{-\pi}^{\pi} (\cos n\xi)(\cos n\xi)\,\mathrm{d}\xi = \int_{-\pi}^{\pi}\left(\frac{1}{2} + \frac{1}{2}\cos 2n\xi\right)\mathrm{d}\xi$$

$$= \frac{\xi}{2}\bigg|_{-\pi}^{\pi} + 0 = \pi, \tag{6.10}$$

$$\int_{-\pi}^{\pi} (\sin n\xi)(\sin n\xi)\,\mathrm{d}\xi = \pi, \tag{6.11}$$

we find that

$$a_0 = \frac{1}{\frac{1}{2}\pi}\int_{-\pi}^{\pi} \frac{1}{2} f(\xi)\,\mathrm{d}\xi = \frac{1}{\pi}\int_{-\pi}^{\pi} f(\xi)\,\mathrm{d}\xi, \tag{6.12}$$

$$a_j = \frac{1}{\pi} \int_{-\pi}^{\pi} f(\xi) \cos n_j \xi \, d\xi, \tag{6.13}$$

$$b_j = \frac{1}{\pi} \int_{-\pi}^{\pi} f(\xi) \sin n_j \xi \, d\xi. \tag{6.14}$$

To extend the period of the series from 2π to $2\pi p$, we replace (6.7) by

$$f(x) = \frac{a_0}{2p} + \Sigma \frac{a_j}{p} \cos \left(n_j \frac{x}{p} \right) + \Sigma \frac{b_j}{p} \sin \left(n_j \frac{x}{p} \right). \tag{6.15}$$

Equations (6.13) and (6.14) then become

$$\frac{a_j}{p} = \frac{1}{\pi} \int_{\xi/p = -\pi}^{\pi} f(\xi) \cos \left(n_j \frac{\xi}{p} \right) d \frac{\xi}{p}, \tag{6.16}$$

or

$$a_j = \frac{1}{\pi} \int_{-\pi p}^{\pi p} f(\xi) \cos \left(\frac{n_j}{p} \xi \right) d\xi \tag{6.17}$$

and

$$b_j = \frac{1}{\pi} \int_{-\pi p}^{\pi p} f(\xi) \sin \left(\frac{n_j}{p} \xi \right) d\xi. \tag{6.18}$$

Letting

$$\frac{n_j}{p} = k_j \tag{6.19}$$

and noting that

$$\frac{1}{p} = \frac{n_j + 1}{p} - \frac{n_j}{p} = \Delta k, \tag{6.20}$$

we rewrite series (6.15) in the form

$$f(x) = \frac{a_0}{2p} + \Sigma a_j (\cos k_j x) \, \Delta k + \Sigma b_j (\sin k_j x) \, \Delta k. \tag{6.21}$$

When p increases without limit, the constant term drops out and the series becomes an *integral*

$$f(x) = \int_0^{\infty} [a(k) \cos kx + b(k) \sin kx] \, dk \tag{6.22}$$

with

$$a(k) = \frac{1}{\pi}\int_{-\infty}^{\infty} f(\xi)\cos k\xi\, d\xi, \tag{6.23}$$

$$b(k) = \frac{1}{\pi}\int_{-\infty}^{\infty} f(\xi)\sin k\xi\, d\xi. \tag{6.24}$$

Introducing the formulas for the coefficients, (6.23) and (6.24), into (6.22) yields

$$f(x) = \frac{1}{\pi}\int_{0}^{\infty} dk \int_{-\infty}^{\infty} f(\xi)\,(\cos k\xi \cos kx + \sin k\xi \sin kx)\, d\xi$$

$$= \frac{1}{\pi}\int_{0}^{\infty} dk \int_{-\infty}^{\infty} f(\xi)\cos\,[k(x-\xi)]\, d\xi. \tag{6.25}$$

Finally, the substitution

$$\cos u = \frac{e^{iu} + e^{-iu}}{2} \tag{6.26}$$

allows us to make the limits on both integrations the same:

$$f(x) = \frac{1}{2\pi}\int_{0}^{\infty} dk \int_{-\infty}^{\infty} [f(\xi)\, e^{ik(x-\xi)} + f(\xi)\, e^{-ik(x-\xi)}]\, d\xi$$

$$= \frac{1}{2\pi}\int_{-\infty}^{\infty} dk \int_{-\infty}^{\infty} f(\xi)\, e^{ik(x-\xi)}\, d\xi$$

$$= \frac{1}{(2\pi)^{\frac{1}{2}}}\int_{-\infty}^{\infty} e^{ikx}\, dk \left[\frac{1}{(2\pi)^{\frac{1}{2}}}\int_{-\infty}^{\infty} f(\xi)\, e^{-ik\xi}\, d\xi\right]. \tag{6.27}$$

The final expression in brackets is called the *Fourier transform* of function f. Let us label this transform $g(k)$:

$$g(k) = \frac{1}{(2\pi)^{\frac{1}{2}}}\int_{-\infty}^{\infty} f(\xi)\, e^{-ik\xi}\, d\xi. \tag{6.28}$$

From (6.27), the inverse Fourier transform of g is

$$f(x) = \frac{1}{(2\pi)^{\frac{1}{2}}}\int_{-\infty}^{\infty} g(k)\, e^{ixk}\, dk. \tag{6.29}$$

Example 6.1. What Fourier series represents the function

$$f(x) = 0 \qquad \text{when} \qquad -\pi < x < 0$$

and

$$f(x) = \pi \qquad \text{when} \quad 0 < x < \pi?$$

Introduce the given function into (6.12), (6.13), and (6.14):

$$a_0 = \frac{1}{\pi}\int_{-\pi}^{0} (0)\,d\xi + \frac{1}{\pi}\int_{0}^{\pi} \pi\,d\xi = \xi\Big|_0^{\pi} = \pi,$$

$$a_j = \frac{1}{\pi}\int_{-\pi}^{0} (0)\cos n_j\xi\,d\xi + \frac{1}{\pi}\int_{0}^{\pi} \pi\cos n_j\xi\,d\xi$$

$$= \frac{1}{n_j}\sin n_j\xi\Big|_0^{\pi} = 0,$$

$$b_j = \frac{1}{\pi}\int_{-\pi}^{0} (0)\sin n_j\xi\,d\xi + \frac{1}{\pi}\int_{0}^{\pi} \pi\sin n_j\xi\,d\xi$$

$$= -\frac{1}{n_j}\cos n_j\xi\Big|_0^{\pi} = \frac{1}{n_j}(1 - \cos n_j\pi).$$

Then construct series (6.7):

$$f(x) = \frac{\pi}{2} + 2\left(\frac{\sin x}{1} + \frac{\sin 3x}{3} + \frac{\sin 5x}{5} + \ldots\right).$$

Example 6.2. Show how a function $f(x)$ defined over the interval $0 < x < \pi$ may be represented by a series of (a) sine terms, (b) cosine terms.

The given function can be extended into the interval $-\pi < x < 0$ either as an odd function or as an even function. Thus, we may write

$$f(-x) = (-1)^p f(x)$$

with p equal to 1 and 2. Each sine term is odd; each cosine term is even. So a series of sine terms is odd and can represent the situation when $p = 1$. Similarly, a series of cosine terms is even and represents the function when we choose $p = 2$.

Indeed, with choice $p = 1$, we obtain

$$a_j = 0$$

and

$$b_j = \frac{2}{\pi}\int_0^{\pi} f(\xi)\sin n_j\xi\,d\xi.$$

Similarly, when $p = 2$, we have

$$b_j = 0$$

and

$$a_0 = \frac{2}{\pi} \int_0^\pi f(\xi)\, \mathrm{d}\xi,$$

$$a_j = \frac{2}{\pi} \int_0^\pi f(\xi) \cos n_j \xi\, \mathrm{d}\xi.$$

6.3. Broadening of a Level Caused by Instability

The temporal factor of the wave function for identical particles in an unstable state can be expressed as a weighted sum of one-frequency temporal factors. The spread over frequency, and the implied uncertainty in energy, vary inversely with the mean life of the state.

In Section 5.12, we saw that the Ψ for a spontaneously decaying state of a given system factors as (5.144) indicates. Presumably, we can construct the temporal part of the function by superposing single-frequency exponentials. Thus, expansion (5.145),

$$\mathrm{e}^{-i\omega_R t}\, \mathrm{e}^{-t/(2\tau)} = \frac{1}{(2\pi)^{\frac{1}{2}}} \int_{-\infty}^{\infty} \chi(\omega)\, \mathrm{e}^{-i\omega t}\, \mathrm{d}\omega, \tag{6.30}$$

was introduced.

This integral is indeed suitable since it has the same structure as the integral in (6.28), with t corresponding to k and ω corresponding to ξ. We can assume that the unstable state is prepared at the time $t = 0$. The temporal factor is then 0 when $t < 0$ and the accompanying equation, (6.29), becomes

$$\chi(\omega) = \frac{1}{(2\pi)^{\frac{1}{2}}} \int_0^\infty \exp\left[-it\left(\omega_R - \omega - \frac{i}{2\tau}\right)\right] \mathrm{d}t. \tag{6.31}$$

Carrying out the integration yields

$$\chi(\omega) = \frac{A}{\omega_R - \omega - i/(2\tau)} \tag{6.32}$$

with A a constant.

The density of the contribution of $\mathrm{e}^{-i\omega t}$ over frequency ω is measured by

$$\chi^*(\omega)\chi(\omega). \tag{6.33}$$

This is largest at the frequency at which the denominator of (6.32) has the smallest magnitude; that is, where

$$\omega = \omega_R \tag{6.34}$$

and

$$\chi^* \chi = \frac{A^*}{i/2\tau}\left(\frac{A}{-i/2\tau}\right) = 4\tau^2 A^* A. \tag{6.35}$$

The density falls to one-half its maximum at the point where

$$\chi^* \chi = 2\tau^2 A^* A. \tag{6.36}$$

This sitution arises where in (6.32) we have

$$\omega_R - \omega = \frac{1}{2\tau}. \tag{6.37}$$

Let us multiply (6.37) by $\hbar$,

$$\hbar\omega_R - \hbar\omega = \frac{\hbar}{2\tau}, \tag{6.38}$$

and identify $\hbar\omega_R$ as E_R, the energy at the maximum, and $\hbar\omega$ as E, here the energy at one-half the maximum:

$$E_R - E = \frac{\hbar}{2\tau}. \tag{6.39}$$

The energy difference Γ from the first half-maximum point to the second one, on the other side of the peak, is double quantity (6.39):

$$\Gamma = \frac{\hbar}{\tau}. \tag{6.40}$$

In words, Γ is a measure of the uncertainty in energy of the unstable state, τ a measure of the uncertainty in its time of decay. Representing the former by ΔE, the latter by Δt, we have

$$(\Delta E)(\Delta t) = \hbar. \tag{6.41}$$

Statement (6.41) is called the energy-time *uncertainty principle.*

Example 6.3. Calculate the mean life of a resonant state whose width Γ is 10.0 MeV.

From (1.34), the energy per electron volt equals the magnitude of e. So the accepted fundamental constants and the given Γ inserted into (6.40) yield

$$\tau = \frac{\hbar}{\Gamma} = \frac{1.0546 \times 10^{-34}\ \text{J s}}{(1.00 \times 10^{7}\ \text{eV})(1.6022 \times 10^{-19}\ \text{J eV}^{-1})}$$

$$= 6.58 \times 10^{-23}\ \text{s}.$$

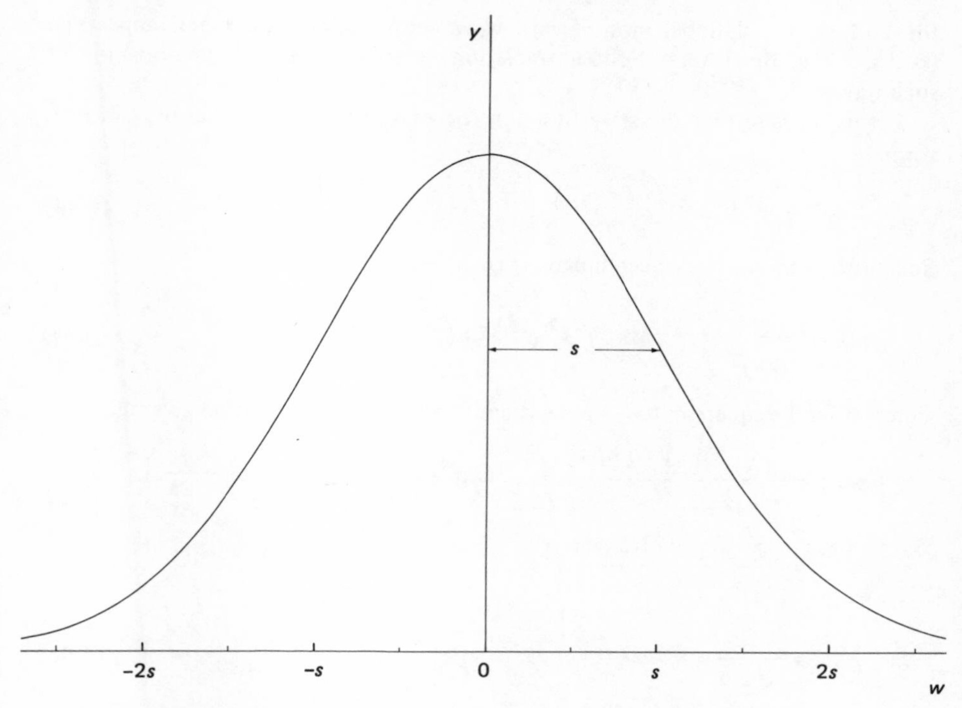

Fig. 6.1. A Gaussian wave.

6.4. A Gaussian Wave Packet and Its Wavevector Representation

A particle may move in a certain direction through free space as in a homogeneous beam. Its wave function then has structure (1.7). Or, it may be confined to a small region as if by a potential. A particularly simple form is the normal distribution curve as observed in the ground state of an oscillator.

A normal distribution curve is defined by the *Gaussian* function,

$$Y = A \exp\left(-\frac{w^2}{2s^2}\right), \tag{6.42}$$

of Figure 6.1. In this, w is the independent variable and parameter s^2 is the *variance*, s itself being the *standard deviation*. Interestingly, a coordinate representation of this shape, for a particle in free space, yields a wavevector representation of the same shape. Furthermore, the relationship between the standard deviation in the coordinate and the standard deviation in the wavevector is like (6.41).

First, we note that the exponential in the function

$$\psi = A\, e^{ikx} \tag{6.43}$$

for uniform translational motion with wavevector k along the x axis appears in (6.29). Thus, the inverse Fourier transform describes a general superposition of such waves.

Let us suppose that the state function for a given free particle has the Gaussian form

$$\psi = N\,\mathrm{e}^{-\alpha^2 x^2}. \tag{6.44}$$

Substituting this into the accompanying (6.28) leads to

$$\phi(k) = \frac{1}{(2\pi)^{\frac{1}{2}}} \int_{-\infty}^{\infty} N\,\mathrm{e}^{-\alpha^2 \xi^2}\,\mathrm{e}^{-ik\xi}\,\mathrm{d}\xi. \tag{6.45}$$

Completing the square in the exponent and executing the integration gives us

$$\begin{aligned}\phi(k) &= \frac{N}{(2\pi)^{\frac{1}{2}}}\,\frac{\mathrm{e}^{-k^2/(2\alpha)^2}}{\alpha} \int_{-\infty}^{\infty} \exp - \left(\alpha\xi + \frac{ik}{2\alpha}\right)^2 \mathrm{d}\left(\alpha\xi + \frac{ik}{2\alpha}\right) \\ &= \frac{N}{(2\pi)^{\frac{1}{2}}}\,\frac{\mathrm{e}^{-k^2/(2\alpha)^2}}{\alpha}\,\pi^{\frac{1}{2}} \\ &= \frac{N}{\sqrt{2}\,\alpha}\,\mathrm{e}^{-k^2/(2\alpha)^2}.\end{aligned} \tag{6.46}$$

Thus, wavevector representation $\phi(k)$ also has the Gaussian shape.

On comparing $\psi(x)$ and $\phi(k)$ with (6.42), we find the variance in the coordinate representation to be

$$s_x^2 = \frac{1}{2\alpha^2}, \tag{6.47}$$

while the variance in the wavevector representation is

$$s_k^2 = 2\alpha^2. \tag{6.48}$$

So the product of the standard deviations is

$$s_x s_k = \frac{1}{\sqrt{2}\,\alpha}\sqrt{2}\,\alpha = 1. \tag{6.49}$$

Using de Broglie's equation

$$p = \hbar k, \tag{6.50}$$

to replace k by p in ϕ yields a function in which the standard deviation is

$$s_p = \hbar s_k. \tag{6.51}$$

Consequently,

$$s_p s_x = \hbar s_k s_x = \hbar, \tag{6.52}$$

the standard deviation in the resulting momentum representation times the standard deviation in the coordinate representation equals Planck's constant divided by 2π. We have here a form of the momentum– coordinate *uncertainty principle*.

According to (1.46), the probability density of the particle in space is (6.44) times its complex conjugate:

$$\psi * \psi = N^*N\,\mathrm{e}^{-2\alpha^2 x^2}. \tag{6.53}$$

Similarly, the probability density over the wavevector is (6.46) times its complex conjugate:

$$\phi * \phi = \frac{N^*N}{2\alpha^2}\,\mathrm{e}^{-k^2/(2\alpha^2)}. \tag{6.54}$$

If we let Δx be the standard deviation in the distribution over x and Δk the standard deviation in the distribution over k, we now find that

$$(\Delta x)^2 (\Delta k)^2 = \frac{1}{4\alpha^2}\,\alpha^2 = \frac{1}{4}, \tag{6.55}$$

whence

$$\Delta x\,\Delta k = \frac{1}{2}. \tag{6.56}$$

Multiplying (6.56) by $\hbar$ and introducing the increment form of de Broglie's equation leads to

$$(\hbar\,\Delta k)\,\Delta x = \Delta p\,\Delta x = \frac{\hbar}{2}. \tag{6.57}$$

The uncertainty in momentum Δp times the uncertainty in position Δx equals $\hbar/2$, when the distributions are Gaussian. Abnormalities in the distributions would tend to make the product larger.

6.5. An Arbitrary Superposition of Simple Planar Waves

The exponential in the inverse Fourier transform integral can be the spatial factor in the function governing translation at a given velocity and momentum. The transform itself represents any linear combination of these exponentials. So introducing the appropriate temporal factor into the integral yields a general planar-wave function.

Free motion in a uniform beam perpendicular to the yz plane is described by formula (1.13), with wavevector $\mathbf{k}$ containing only an x component. At time $t = 0$, the function appears as in (6.43), while at other times it becomes

$$A\,\mathrm{e}^{i(kx - \omega t)}. \tag{6.58}$$

A general translatory state consists of a mixture of pure states. According to the superposition principle, the pure states combine additively. If the constituents with wavevectors between k and $k + \mathrm{d}k$ are present at the amplitude

$$A = \frac{\mathrm{d}k}{2\pi}\int_{-\infty}^{\infty} f(\xi)\,\mathrm{e}^{-ik\xi}\,\mathrm{d}\xi, \tag{6.59}$$

the resulting wave function is

$$\Psi(x, 0) = \frac{1}{(2\pi)^{\frac{1}{2}}}\int_{-\infty}^{\infty} \mathrm{e}^{ikx}\,\mathrm{d}k\left[\frac{1}{(2\pi)^{\frac{1}{2}}}\int_{-\infty}^{\infty} f(\xi)\,\mathrm{e}^{-ik\xi}\,\mathrm{d}\xi\right] \tag{6.60}$$

initially and

$$\Psi(x, t) = \frac{1}{(2\pi)^{\frac{1}{2}}}\int_{-\infty}^{\infty} \mathrm{e}^{i(kx - \omega t)}\,\mathrm{d}k\left[\frac{1}{(2\pi)^{\frac{1}{2}}}\int_{-\infty}^{\infty} f(\xi)\,\mathrm{e}^{-ik\xi}\,\mathrm{d}\xi\right] \tag{6.61}$$

when the time equals t. At a given instant of time, the real part of a pulse may appear as indicated in Figure 6.2.

On comparing (6.60) with Fourier integral (6.27), we see that

$$f(\xi) = \Psi(\xi, 0). \tag{6.62}$$

Equation (6.61) can therefore be rewritten in the form

$$\Psi(x, t) = \frac{1}{(2\pi)^{\frac{1}{2}}}\int_{-\infty}^{\infty} \mathrm{e}^{i(kx - \omega t)}\,\mathrm{d}k\left[\frac{1}{(2\pi)^{\frac{1}{2}}}\int_{-\infty}^{\infty} \Psi(\xi, 0)\,\mathrm{e}^{-ik\xi}\,\mathrm{d}\xi\right]. \tag{6.63}$$

This represents how an arbitrary planar wave $\Psi(x, 0)$ propagates in free space.

For general translatory motion in three dimensions, we replace the first exponential by expression (1.13) and each single integration by the appropriate triple integration. We then obtain

$$\Psi(\mathbf{r}, t) = \frac{1}{(2\pi)^{3/2}}\iiint \mathrm{e}^{i(\mathbf{k}\cdot\mathbf{r} - \omega t)}\,\mathrm{d}^3\mathbf{k}\left[\frac{1}{(2\pi)^{3/2}}\iiint \Psi(\xi, 0)\,\mathrm{e}^{-i\mathbf{k}\cdot\xi}\,\mathrm{d}^3\xi\right]. \tag{6.64}$$

In determining the dependence of ω on k, we need to consider a motion in which the constituent wavevectors are directed along one axis, about a certain magnitude with a peaked distribution law. The one-dimensional wave (6.63) is adequate.

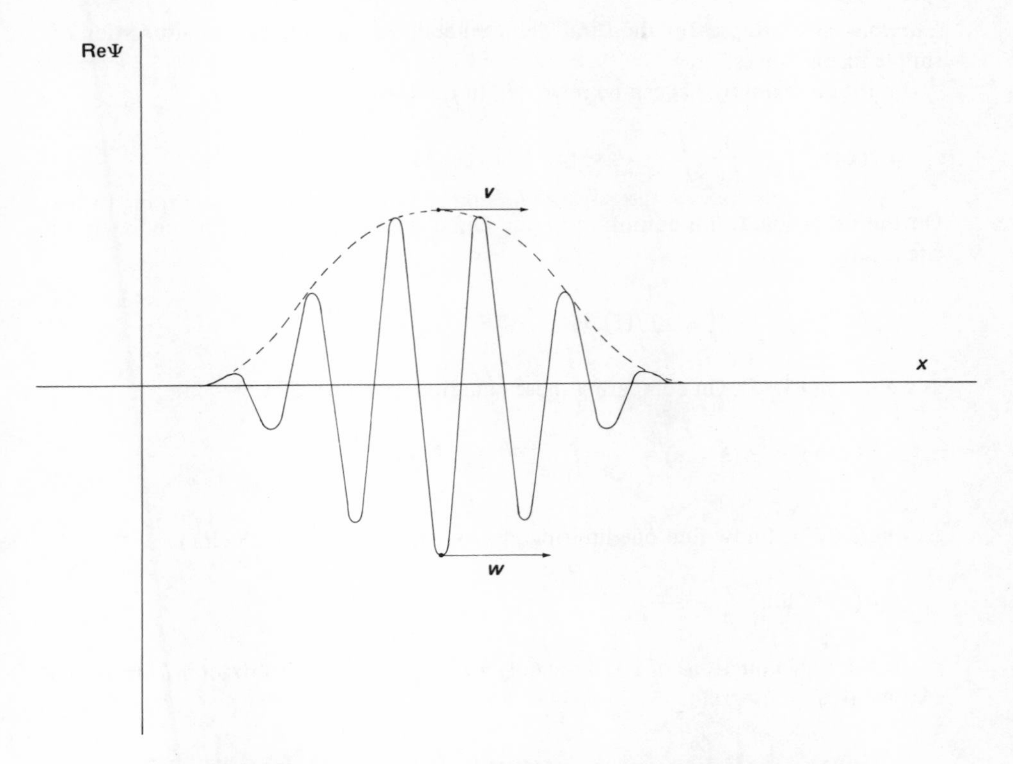

Fig. 6.2. The real part, at a given time, of a wave function formed by superposing pure, unidirectional, translational states.

Note particularly how the Fourier transform of the initial function

$$\phi(k) = \frac{1}{(2\pi)^{\frac{1}{2}}} \int_{-\infty}^{\infty} \Psi(\xi, 0)\, \mathrm{e}^{-ik\xi}\, \mathrm{d}\xi \tag{6.65}$$

determines the distribution over k of the simple planar wavelets. Indeed from formula (6.63), we obtain

$$\Psi(x, t) = \frac{1}{(2\pi)^{\frac{1}{2}}} \int_{-\infty}^{\infty} \phi(k)\, \mathrm{e}^{i(kx - \omega t)}\, \mathrm{d}k. \tag{6.66}$$

Letting u be the deviation of k from a properly weighted mean, designated $\bar{k}$,

$$u = k - \bar{k}, \tag{6.67}$$

converts (6.66) to

$$\Psi(x, t) = \frac{\mathrm{e}^{i\bar{k}x}}{(2\pi)^{\frac{1}{2}}} \int_{-\infty}^{\infty} \phi(u + \bar{k})\, \mathrm{e}^{i(ux - \omega t)}\, \mathrm{d}u. \tag{6.68}$$

Example 6.4. Represent the Dirac delta generalized function as a superposition of simple planar waves.

Fourier integral (6.27) can be rewritten in the form

$$f(x) = \int_{-\infty}^{\infty} \frac{1}{2\pi} \int_{-\infty}^{\infty} e^{ik(x - \xi)}\, dk\, f(\xi)\, d\xi.$$

On the other hand, the definition of the Dirac delta function leads immediately to the result

$$f(x) = \int_{-\infty}^{\infty} \delta(\xi - x) f(\xi)\, d\xi,$$

as we saw in (5.45). On comparing these equations, we find that

$$\delta(x - \xi) = \delta(\xi - x) = \frac{1}{2\pi} \int_{-\infty}^{\infty} e^{ik(x - \xi)}\, dk.$$

Example 6.5. Show that one limiting process yielding the Dirac delta is

$$\delta(x) = \lim_{\epsilon \to 0} \frac{1}{\pi} \frac{\epsilon}{x^2 + \epsilon^2}.$$

In the final equations of Example 6.4, set ξ equal to zero, subtract $\epsilon|k|$ from the exponent, and integrate:

$$\delta_\epsilon(x) = \frac{1}{2\pi} \left[\int_{-\infty}^{0} \exp(ikx + \epsilon k)\, dk + \int_{0}^{\infty} \exp(ikx - \epsilon k)\, dk \right]$$

$$= \frac{1}{2\pi} \left[\frac{\exp(ikx + \epsilon k)}{ix + \epsilon} \bigg|_{-\infty}^{0} + \frac{\exp(ikx - \epsilon k)}{ix - \epsilon} \bigg|_{0}^{\infty} \right]$$

$$= \frac{1}{2\pi} \left(\frac{1}{ix + \epsilon} - \frac{1}{ix - \epsilon} \right) = \frac{1}{2\pi} \frac{ix - \epsilon - ix - \epsilon}{-x^2 - \epsilon^2}$$

$$= \frac{1}{\pi} \frac{\epsilon}{x^2 + \epsilon^2}.$$

In the limit when ϵ goes to zero, the initial integrals reduce to the form for $\delta(x)$; consequently, the limit of the final expression is $\delta(x)$.

6.6. The De Broglie Angular Frequency

A wave function for a definite energy state does not reveal how its angular frequency relates to the particle's energy; but a mixture of planar waves possessing a peaked envelope, as Figure 6.2 shows, contains the information. Such an envelope moves approximately as a unit at a speed v equal to the derivative of ω with respect to k, around the dominant k. Integration of this relationship and introduction of the de Broglie formula produce a result that agrees with Einstein's equation for a photon.

No particle can be constrained to follow a classical trajectory exactly. Each given particle can be made to travel along a corridor of equivalent points only with a statistical distribution of positions and velocities or momenta, varying with time. The probability of finding the particle in an element of path dx at time t is given by $\Psi^* \Psi \, \mathrm{d}x$.

To make the uncertainty (or strictly, the variance) in location small, we must prepare the system with a strongly peaked $\Psi^* \Psi$. The particle is then generally found not far from the maximum and the position x_{max} of this point tends to move at the velocity of the particle. Classically, this velocity is p/μ, where p is the momentum, μ the mass. We thus have

$$x_{max} = vt + A \simeq \frac{p}{\mu} t + A. \tag{6.69}$$

Each pure motion contributing to a state possesses an independent angular frequency ω and wavevector k. In any small neighborhood in which the potential is practically constant, ω can only vary with k. Since there is no apparent cause for irregularities to exist in the variation, we assume the relationship to be analytical. Expanding ω about the dominant k, which is given by the weighted average of the k's, and is designated $\bar{k}$, yields

$$\omega = \omega(\bar{k}) + \left(\frac{\mathrm{d}\omega}{\mathrm{d}k}\right)_{\bar{k}} (k - \bar{k}) + \frac{1}{2}\left(\frac{\mathrm{d}^2\omega}{\mathrm{d}k^2}\right)_{\bar{k}} (k - \bar{k})^2 + \ldots$$

$$= \bar{\omega} + \bar{\omega}' u + \frac{1}{2} \bar{\omega}'' u^2 + \ldots \simeq \bar{\omega} + \bar{\omega}' u. \tag{6.70}$$

A prime on a symbol indicates that a differentiation with respect to k has been introduced; a bar over a symbol indicates evaluation at $k = \bar{k}$. Variable u is the difference $k - \bar{k}$ as in (6.67).

Over time intervals that are not too long, the linear approximation in (6.70) is valid and Equation (6.68) becomes

$$\Psi(x, t) \simeq \frac{e^{i(\bar{k}x - \bar{\omega}t)}}{(2\pi)^{\frac{1}{2}}} \int_{-\infty}^{\infty} \phi(u + \bar{k}) \, e^{iu(x - \bar{\omega}'t)} \, \mathrm{d}u$$

$$= \frac{e^{i(\bar{k}x - \bar{\omega}t)}}{(2\pi)^{\frac{1}{2}}} \, e^{-i\bar{k}(x - \bar{\omega}'t)} \int_{-\infty}^{\infty} \phi(u + \bar{k}) \, e^{i(u + \bar{k})(x - \bar{\omega}'t)} \, \mathrm{d}(u + \bar{k})$$

$$= e^{-i\bar{\omega}t + i\bar{k}\bar{\omega}'t} \, \Psi(x - \bar{\omega}'t, 0). \tag{6.71}$$

In the last step, (6.66) is employed to identify $1/(2\pi)^{\frac{1}{2}}$ times the integral.

The exponential in the resulting form merely acts to change the phase of the wave. Function $\Psi(x - \bar{\omega}'t, 0)$ is the original wave translated by distance $\bar{\omega}'t$. Now, a given point on this function obeys the equation

$$x - \bar{\omega}'t = B \tag{6.72}$$

with B constant. Differentiating (6.72) yields the velocity at which the envelope of the wave moves, the *group velocity*,

$$v = \frac{\mathrm{d}x}{\mathrm{d}t} = \bar{\omega}' = \left(\frac{\mathrm{d}\omega}{\mathrm{d}k}\right)_{\bar{k}}. \tag{6.73}$$

But since the particle tends to move with the envelope, we identify v with the velocity of the particle, as in (6.69), and write

$$v = \frac{p}{\mu}. \tag{6.74}$$

Combining (6.73), (6.74) with de Broglie's equation

$$p = \hbar k \tag{6.75}$$

leads to

$$\frac{\mathrm{d}\omega}{\mathrm{d}k} = \frac{\hbar k}{\mu}, \tag{6.76}$$

whence

$$\omega = \frac{\hbar k^2}{2\mu} + C. \tag{6.77}$$

Multiplying (6.77) by $\hbar$ and reintroducing (6.75) gives us

$$\hbar\omega = \frac{\hbar^2 k^2}{2\mu} + \hbar C = \frac{p^2}{2\mu} + \hbar C, \tag{6.78}$$

Term $p^2/2\mu$ is the kinetic energy while $\hbar C$ is interpreted as the potential energy (the particle is moving in a region of constant potential energy). The sum is thus the total energy E:

$$\hbar\omega = E. \tag{6.79}$$

There is no reason for a variation in potential to alter this form. Consequently, we assume that it is valid in general.

Note that differentiating Equation (6.76) yields

$$\frac{d^2\omega}{dk^2} = \frac{\hbar}{\mu} \tag{6.80}$$

and

$$\frac{d^n\omega}{dk^n} = 0 \qquad \text{when} \quad n > 2. \tag{6.81}$$

For a nonrelativistic particle, one obeying (6.74) with a constant mass, series (6.70) terminates at the quadratic term.

Example 6.6. Show that the group velocity v of a wave system at a certain angular frequency ω equals the corresponding phase velocity w only if w does not vary with ω.

From (1.16), the angular frequency ω is related to wavevector k by the formula

$$\omega = wk.$$

If w is constant, differentiating ω yields the result

$$\frac{d\omega}{dk} = w\frac{dk}{dk} = w,$$

which reduces (6.73) to

$$v = w.$$

Any variation in w would introduce another term into the derivative $d\omega/dk$ and would destroy the equality of v with w.

6.7. Variations of a Wave Function and Observable Properties with Time

The proportionality of ω to E enables us to construct a Schrödinger equation for mixed-energy states and then, a general formula for the temporal rate of change in expectation values.

Governing the behavior of a particle with a definite energy, near a point in a field of varying potential, is Equation (4.20). On comparing this with (4.21), we find that

$$\frac{\partial\Psi_\pm}{\partial t} = -i\omega\Psi_\pm. \tag{6.82}$$

Adding the two constituents of the state function,

$$\Psi = \Psi_+ + \Psi_-, \tag{6.83}$$

differentiating the sum with respect to t, and combining with (6.82) yields

$$\frac{\partial \Psi}{\partial t} = -i\omega\Psi. \tag{6.84}$$

Substituting (6.79) into (6.84) then leads to

$$\frac{\partial \Psi}{\partial t} = -i\frac{E}{\hbar}\Psi. \tag{6.85}$$

From Section 5.6, the Schrödinger equation can be written as

$$H\Psi = E\Psi \tag{6.86}$$

where H is the Hamiltonian operator. Eliminating energy E from (6.85) and (6.86) gives us

$$\frac{\partial \Psi}{\partial t} = -\frac{i}{\hbar}H\Psi, \tag{6.87}$$

a *time-dependent Schrödinger equation*. For a particle in a scalar potential V, the operator has the form

$$H = -\frac{\hbar^2}{2\mu}\nabla^2 + V. \tag{6.88}$$

Note that (6.87) is the same for each pure energy state. Consequently, it also applies, as it stands, to any arbitrary superposition of possible states. Indeed, it governs the temporal development of any system whose pure states are governed by (6.85) and (6.86).

From (5.39), the expectation value for a property corresponding to operator A is

$$\langle A \rangle = \int_R \Psi^* A \Psi \, \mathrm{d}^3\mathbf{r}. \tag{6.89}$$

Let us differentiate (6.89) with respect to time:

$$\frac{\mathrm{d}}{\mathrm{d}t}\langle A \rangle = \int_R \left(\frac{\partial \Psi^*}{\partial t} A\Psi + \Psi^* A \frac{\partial \Psi}{\partial t} + \Psi^* \frac{\partial A}{\partial t}\Psi \right) \mathrm{d}^3\mathbf{r}. \tag{6.90}$$

Then substitute (6.87) and its complex conjugate

$$\frac{\partial \Psi^*}{\partial t} = \frac{i}{\hbar}(H\Psi)^* \tag{6.91}$$

into the resulting integral:

$$\frac{\mathrm{d}}{\mathrm{d}t}\langle A\rangle = \frac{i}{\hbar}\int_R [(H\Psi)^*(A\Psi) - \Psi^* AH\Psi]\, \mathrm{d}^3\mathbf{r} + \left\langle\frac{\partial A}{\partial t}\right\rangle. \tag{6.92}$$

Under conditions that make H Hermitian, we have

$$\int_R (H\Psi)^*(A\Psi)\, \mathrm{d}^3\mathbf{r} = \int_R \Psi^* HA\Psi\, \mathrm{d}^3\mathbf{r}. \tag{6.93}$$

So (6.92) reduces to

$$\frac{\mathrm{d}}{\mathrm{d}t}\langle A\rangle = \frac{i}{\hbar}\int_R \Psi^*(HA - AH)\Psi\, \mathrm{d}^3\mathbf{r} + \left\langle\frac{\partial A}{\partial t}\right\rangle. \tag{6.94}$$

The difference between a combination of two operators in the given order and in the inverse order is called the *commutator* of the operators. Thus, composite operator

$$HA - AH \equiv [H, A] \tag{6.95}$$

is the commutator of H and A. With this notation, (6.94) becomes

$$\frac{\mathrm{d}}{\mathrm{d}t}\langle A\rangle = \frac{i}{\hbar}\langle [H, A]\rangle + \left\langle\frac{\partial A}{\partial t}\right\rangle. \tag{6.96}$$

Example 6.7. What is the commutator of p_x and x?

From Sections 5.4 and 5.5, the operators for momentum p_x and coordinate x, in the coordinate representation, are

$$\frac{\hbar}{i}\frac{\partial}{\partial x} \qquad \text{and} \qquad x,$$

respectively. Hence

$$[p_x, x]\Psi = (p_x x - x p_x)\Psi = \left(\frac{\hbar}{i}\frac{\partial}{\partial x}x - x\frac{\hbar}{i}\frac{\partial}{\partial x}\right)\Psi$$

$$= \frac{\hbar}{i}\Psi + \frac{\hbar}{i}x\frac{\partial\Psi}{\partial x} - x\frac{\hbar}{i}\frac{\partial\Psi}{\partial x} = \frac{\hbar}{i}\Psi$$

and

$$[p_x, x] = \frac{\hbar}{i}.$$

6.8. Generalizing Newton's Laws

Classical equations presume that at each instant of time a particle is at a definite position with a definite velocity, momentum, and rate of change of momentum, or force. But in quantum mechanics, probability reigns; only eigenvalues and expectation values are precise, or sharp. Nevertheless, generalizations of the Newtonian relationships can be constructed. The expectation value for a linear momentum equals the mass of the given particle multiplied by the rate of change in expectation value for its position. The expectation value for a force equals the temporal rate of change in expectation value for the corresponding momentum.

Let us consider a particle of mass μ moving in a scalar potential field V. Equations (6.86), (6.87), and (6.88) then govern its behavior. If the wave function for the particle is appreciable only in a small region, the operators that we are going to employ are all Hermitian and the time rates of change are given by (6.96).

The temporal derivative of expectation value for position along the x axis involves $[H, x]$. Substituting explicit forms for the operators into the commutator and reducing leads to

$$[H, x] = Hx - xH$$

$$= -\frac{\hbar^2}{2\mu}\left(\frac{\partial^2}{\partial x^2} + \frac{\partial^2}{\partial y^2} + \frac{\partial^2}{\partial z^2}\right)x + Vx + x\frac{\hbar^2}{2\mu}\left(\frac{\partial^2}{\partial x^2} + \frac{\partial^2}{\partial y^2} + \frac{\partial^2}{\partial z^2}\right) - xV$$

$$= -\frac{\hbar^2}{2\mu}\left(2\frac{\partial}{\partial x}\right) = \frac{\hbar}{i\mu}\frac{\hbar}{i}\frac{\partial}{\partial x} = \frac{\hbar}{i\mu}p_x. \tag{6.97}$$

In the last step, equality (5.58) has been employed. Since x is kept constant in the partial differentiation with respect to t, we also have

$$\frac{\partial x}{\partial t} = 0. \tag{6.98}$$

Substituting these results into (6.96), with A being x, and multiplying by μ gives us

$$\mu\frac{\mathrm{d}}{\mathrm{d}t}\langle x\rangle = \mu\left(\frac{i}{\hbar}\left\langle\frac{\hbar}{i\mu}p_x\right\rangle + \langle 0\rangle\right) = \langle p_x\rangle. \tag{6.99}$$

Thus, the expectation value for a Cartesian component of momentum equals the particle mass multiplied by the time rate of change of the expectation value for the corresponding coordinate in the isolated wave.

The rate of change of expectation value for the x component of momentum involves $[H, p_x]$. Substituting explicit forms for the operators into the commutator and reducing again yields a single term:

$$[H, p_x] = Hp_x - p_x H = -\frac{\hbar^2}{2\mu}\left(\frac{\partial^2}{\partial x^2} + \frac{\partial^2}{\partial y^2} + \frac{\partial^2}{\partial z^2}\right)\frac{\hbar}{i}\frac{\partial}{\partial x} +$$

$$+ V\frac{\hbar}{i}\frac{\partial}{\partial x} + \frac{\hbar}{i}\frac{\partial}{\partial x}\frac{\hbar^2}{2\mu}\left(\frac{\partial^2}{\partial x^2} + \frac{\partial^2}{\partial y^2} + \frac{\partial^2}{\partial z^2}\right) - \frac{\hbar}{i}\frac{\partial}{\partial x}V$$

$$= -\frac{\hbar}{i}\frac{\partial V}{\partial x}. \tag{6.100}$$

Furthermore, we have

$$\frac{\partial p_x}{\partial t} = 0. \tag{6.101}$$

So when A is p_x, formula (6.96) becomes

$$\frac{d}{dt}\langle p_x\rangle = \frac{i}{\hbar}\left\langle -\frac{\hbar}{i}\frac{\partial V}{\partial x}\right\rangle + \langle 0\rangle = -\left\langle\frac{\partial V}{\partial x}\right\rangle = \langle F_x\rangle. \tag{6.102}$$

Similar equations hold for the other components. Consequently,

$$\frac{d}{dt}\langle \mathbf{p}\rangle = -\langle \nabla V\rangle = \langle \mathbf{F}\rangle, \tag{6.103}$$

the rate of change in expectation value for the momentum of a wave packet equals the expectation value for the force acting on the particle.

Equation (6.103) does not reduce quantum theory to Newtonian mechanics, however, because (6.71) is approximate. In practice, a wave packet tends to spread with time and so the packet does not propagate as an unchanging entity. Nevertheless, this equation, when it was derived in 1927 by Paul Ehrenfest, contributed greatly to the general acceptance of quantum methods. We call (6.103) *Ehrenfest's theorem*.

6.9. Scalar and Vector Potentials

Since a field can carry both energy and momentum, it may contribute to the energy and to the momentum of a particle that interacts with it. The particle then appears to be dressed by the neighboring parts of the field. The energy effect is measured by an interaction coefficient and a scalar potential, the momentum effect by a coefficient and a vector potential.

Consider a particle of mass μ and velocity $\mathbf{v}$ moving in some field. If the potential energy resulting from interaction with the field is V, the total energy of the particle is

$$E = \frac{1}{2}\mu v^2 + V. \tag{6.104}$$

If the particle were free, its linear momentum would be $\mu\mathbf{v}$. But interaction with the field contributes to the inertia. This interaction alters the momentum of the particle by an amount depending on the field and an interaction coefficient q:

$$\mathbf{p} = \mu\mathbf{v} + q\mathbf{A}. \tag{6.105}$$

Consequently,

$$\mu^2 v^2 = \mu\mathbf{v}\cdot\mu\mathbf{v} = (\mathbf{p} - q\mathbf{A})\cdot(\mathbf{p} - q\mathbf{A}) = \Sigma\,(p_j - qA_j)\,(p_j - qA_j). \tag{6.106}$$

Substituting (6.106) into (6.104) leads to the Hamiltonian expression

$$E = \frac{1}{2\mu}\,\Sigma\,(p_j - qA_j)\,(p_j - qA_j) + V \equiv H, \tag{6.107}$$

which rearranges to

$$\Sigma\,(p_j - qA_j)\,(p_j - qA_j) = 2\mu(E - V). \tag{6.108}$$

In the coordinate representation, the operator for the jth component of momentum, from (5.58), (5.59), (5.60), is

$$\frac{\hbar}{i}\frac{\partial}{\partial x_j} = p_j. \tag{6.109}$$

But qA_j is generally independent of these components, being a function of position and possibly of time only. The corresponding operator then has the same form. Subtracting this from the left side of (6.109) and its equivalent from the right side yields

$$\frac{\hbar}{i}\frac{\partial}{\partial x_j} - qA_j = p_j - qA_j. \tag{6.110}$$

Let us combine each side of (6.110) with itself, sum over the repeated index, let the result act on Ψ, and reduce using (6.108). The result is

$$\Sigma\left(\frac{\hbar}{i}\frac{\partial}{\partial x_j} - qA_j\right)\left(\frac{\hbar}{i}\frac{\partial}{\partial x_j} - qA_j\right)\Psi$$

$$= \Sigma\,(p_j - qA_j)\,(p_j - qA_j)\,\Psi = 2\mu(E - V)\,\Psi, \tag{6.111}$$

a *Schrödinger equation* for motion in a field described by $\mathbf{A}$ and a part of V.

The potential energy may be expressed as an interaction coefficient times a scalar function ϕ. When the coefficient is q, we have

$$V = q\phi. \tag{6.112}$$

The Hamiltonian then becomes

$$\begin{aligned} H &= \frac{1}{2\mu}\,\Sigma\,(p_j - qA_j)\,(p_j - qA_j) + q\phi \\ &= \frac{1}{2\mu}\,(\mathbf{p} - q\mathbf{A})\cdot(\mathbf{p} - q\mathbf{A}) + q\phi \\ &= \frac{1}{2\mu}\,[\mathbf{p}\cdot\mathbf{p} - q\,(\mathbf{p}\cdot\mathbf{A} + \mathbf{A}\cdot\mathbf{p}) + q^2\,\mathbf{A}\cdot\mathbf{A}] + q\phi. \end{aligned} \tag{6.113}$$

6.10. The Lorentz Force Law

The statistical force acting on a charged particle equals the temporal rate of change of its statistical momentum. This rate can be obtained from (6.96) when the Hamiltonian is known.

Consider the particle of mass μ and interaction coefficient q moving as a compact wave packet through a field described by $\mathbf{A}$ and ϕ. Operator (6.113) then governs the motion. From (6.109),

$$p_j = \frac{\hbar}{i}\frac{\partial}{\partial x_j} = \frac{\hbar}{i}\,\nabla_j, \tag{6.114}$$

the momentum vector is represented by

$$\mathbf{p} = \frac{\hbar}{i}\,\nabla. \tag{6.115}$$

Since time t does not enter explicitly into $\mathbf{p}$, we have

$$\left\langle\frac{\partial\mathbf{p}}{\partial t}\right\rangle = 0. \tag{6.116}$$

Furthermore

$$\begin{aligned} \frac{i}{\hbar}\,[q\phi, \mathbf{p}]\,\Psi &= \frac{i}{\hbar}\,[q\phi, \frac{\hbar}{i}\,\nabla]\,\Psi = q\phi\,\nabla\Psi - \nabla(q\phi\Psi) \\ &= q\phi\,\nabla\Psi - q\phi\,\nabla\Psi - q\,\Psi\,\nabla\phi = -q(\nabla\phi)\Psi \end{aligned} \tag{6.117}$$

and

$$\frac{i}{\hbar}\left[\frac{1}{2\mu}(p_j - qA_j)(p_j - qA_j), \mathbf{p}\right]\Psi$$

$$= \frac{i}{\hbar}\left[\frac{1}{2\mu}\left(\frac{\hbar}{i}\nabla_j - qA_j\right)\left(\frac{\hbar}{i}\nabla_j - qA_j\right), \frac{\hbar}{i}\nabla\right]\Psi$$

$$= \frac{1}{2\mu}\left\{\left(\frac{\hbar}{i}\nabla_j - qA_j\right)\left(\frac{\hbar}{i}\nabla_j - qA_j\right)\nabla - \nabla\left[\left(\frac{\hbar}{i}\nabla_j - qA_j\right)\left(\frac{\hbar}{i}\nabla_j - qA_j\right)\right]\right\}\Psi$$

$$= \frac{1}{2\mu}\left[\left(\frac{\hbar}{i}\nabla_j - qA_j\right)\left(\frac{\hbar}{i}\nabla_j - qA_j\right)\nabla\Psi - \left(\frac{\hbar}{i}\nabla_j - qA_j\right)\left(\frac{\hbar}{i}\nabla_j - qA_j\right)\nabla\Psi - \right.$$

$$\left. - \left(\frac{\hbar}{i}\nabla_j - qA_j\right)(-q\,\nabla A_j)\Psi - (-q\,\nabla A_j)\left(\frac{\hbar}{i}\nabla_j - qA_j\right)\Psi\right]$$

$$= \frac{q}{2\mu}\left[(p_j - qA_j)\,\nabla A_j + \nabla A_j(p_j - qA_j)\right]\Psi. \tag{6.118}$$

To obtain the rate of change of expectation value for the momentum, we let the general operator in (6.96) be (6.115), the Hamiltonian be (6.113), and employ results (6.116), (6.117), and (6.118):

$$\frac{\mathrm{d}}{\mathrm{d}t}\langle\mathbf{p}\rangle = \frac{q}{2\mu}\langle\Sigma\,[(p_j - qA_j)(\nabla A_j) + (\nabla A_j)(p_j - qA_j)]\,\rangle - q\,\langle\nabla\phi\rangle$$

$$= \frac{1}{2}q\;\Sigma\,\langle(v_j\,\nabla A_j + \nabla A_j v_j)\rangle - q\,\langle\nabla\phi\rangle. \tag{6.119}$$

In the last step, a rearrangement of (6.105),

$$p_j - qA_j = \mu v_j, \tag{6.120}$$

has been introduced.

When the wave packet is small enough to make each ∇A_j and $\nabla\phi$ approximately constant within the integrals, Equation (6.119) reduces to

$$\frac{\mathrm{d}}{\mathrm{d}t}\langle\mathbf{p}\rangle = q\,\Sigma\,\langle v_j\rangle\,\nabla A_j - q\,\nabla\phi. \tag{6.121}$$

If we set

$$\langle\mathbf{v}\rangle = \mathbf{v}, \qquad \langle\mathbf{p}\rangle = \mathbf{p}, \tag{6.122}$$

we then obtain

$$\frac{\mathrm{d}}{\mathrm{d}t}\,\mathbf{p} = q\,\Sigma\,\nabla(v_jA_j - \phi) = q\,\nabla(\mathbf{v}\cdot\mathbf{A} - \phi). \tag{6.123}$$

The chain rule tells us that

$$\frac{\mathrm{d}}{\mathrm{d}t} = \frac{\partial}{\partial t} + \mathbf{v}\cdot\nabla, \tag{6.124}$$

while differentiating (6.105) yields

$$\frac{\mathrm{d}}{\mathrm{d}t}\,\mathbf{p} = \frac{\mathrm{d}}{\mathrm{d}t}\,(\mu\mathbf{v} + q\mathbf{A}) = \mu\,\frac{\mathrm{d}\mathbf{v}}{\mathrm{d}t} + q\left(\frac{\partial\mathbf{A}}{\partial t} + \mathbf{v}\cdot\nabla\mathbf{A}\right). \tag{6.125}$$

These formulas convert (6.123) to

$$\begin{aligned}\mu\,\frac{\mathrm{d}\mathbf{v}}{\mathrm{d}t} &= q\left(-\nabla\phi - \frac{\partial\mathbf{A}}{\partial t}\right) + q(\nabla\mathbf{v}\cdot\mathbf{A} - \mathbf{v}\cdot\nabla\mathbf{A})\\ &= q\left(-\nabla\phi - \frac{\partial\mathbf{A}}{\partial t}\right) + q\mathbf{v}\times(\nabla\times\mathbf{A}).\end{aligned} \tag{6.126}$$

In the last equality, formula

$$\mathbf{A}\times(\mathbf{B}\times\mathbf{C}) = \mathbf{B}(\mathbf{A}\cdot\mathbf{C}) - (\mathbf{A}\cdot\mathbf{B})\mathbf{C} \tag{6.127}$$

has been employed.

If we make the identifications

$$\mathbf{E} = -\nabla\phi - \frac{\partial\mathbf{A}}{\partial t}, \tag{6.128}$$

$$\mathbf{B} = \nabla\times\mathbf{A}, \tag{6.129}$$

Equation (6.126) reduces to

$$\mu\,\frac{\mathrm{d}\mathbf{v}}{\mathrm{d}t} = q(\mathbf{E} + \mathbf{v}\times\mathbf{B}). \tag{6.130}$$

Equation (6.130) is a form of the classical *Lorentz law* when **E** is the conventional electric intensity, **B** the conventional magnetic induction. Functions ϕ and **A** are the scalar and vector electromagnetic potentials when q is the electric charge on the given particle.

From the way it has been derived, we see that (6.130) has only statistical validity. Certain details of propagation through a field are obtained only from Schrödinger's equation (6.111) or generalizations therefrom.

6.11. Phase Changes along a Given Path through a Field

Some of these details become evident when we determine the effects of ϕ and **A** on Ψ along a particular path.

The pure energy states of a particle in an electromagnetic field are governed by (6.111), which can be rearranged to

$$\left[\frac{1}{2\mu}\Sigma\left(\frac{\hbar}{i}\frac{\partial}{\partial x_j} - qA_j\right)\left(\frac{\hbar}{i}\frac{\partial}{\partial x_j} - qA_j\right) + V\right]\Psi = E\Psi. \tag{6.131}$$

The assumed symmetry with respect to time and the dependence of coefficient ω on energy E gave us (6.85), whence

$$E\Psi = -\frac{\hbar}{i}\frac{\partial\Psi}{\partial t}. \tag{6.132}$$

Eliminating energy E from (6.131) and (6.132) gives

$$\left[\frac{1}{2\mu}\Sigma\left(\frac{\hbar}{i}\frac{\partial}{\partial x_j} - qA_j\right)\left(\frac{\hbar}{i}\frac{\partial}{\partial x_j} - qA_j\right) + V\right]\Psi = -\frac{\hbar}{i}\frac{\partial\Psi}{\partial t}. \tag{6.133}$$

Since (6.133) is the same for each pure energy state, it applies to any arbitrary superposition of possible states. Thus, it governs the development in time of any system whose pure states are governed by (6.131). We call (6.133) the *time-dependent Schrödinger equation* for movement of a charged particle in a field described by potentials **A** and a factor in V. The equation may be abbreviated as

$$H\Psi = -\frac{\hbar}{i}\frac{\partial\Psi}{\partial t}. \tag{6.134}$$

Let us introduce the transformation

$$\Psi = \Psi_1 \exp\left(-\frac{i}{\hbar}\int_0^t V\,\mathrm{d}t\right) \tag{6.135}$$

in which the integration proceeds along a *definite path* between a given initial and a variable final point. Since the exponential is an explicit function of t alone, we have

$$H\Psi = H\Psi_1 \exp\left(-\frac{i}{\hbar}\int_0^t V\,\mathrm{d}t\right) \tag{6.136}$$

and

$$-\frac{\hbar}{i}\frac{\partial\Psi}{\partial t} = -\frac{\hbar}{i}\Psi_1\left[\exp\left(-\frac{i}{\hbar}\int_0^t V\,\mathrm{d}t\right)\right]\left(-\frac{i}{\hbar}V\right) - \frac{\hbar}{i}\frac{\partial\Psi_1}{\partial t}\exp\left(-\frac{i}{\hbar}\int_0^t V\,\mathrm{d}t\right)$$

$$= \left(V\Psi_1 - \frac{\hbar}{i}\frac{\partial\Psi_1}{\partial t}\right)\exp\left(-\frac{i}{\hbar}\int_0^t V\,\mathrm{d}t\right). \tag{6.137}$$

Substituting (6.136) and (6.137) into (6.134), canceling the exponential factors, and rearranging leads to

$$(H - V)\Psi_1 = -\frac{\hbar}{i}\frac{\partial\Psi_1}{\partial t} \tag{6.138}$$

or

$$\frac{1}{2\mu}\Sigma\left(\frac{\hbar}{i}\frac{\partial}{\partial x_j} - qA_j\right)\left(\frac{\hbar}{i}\frac{\partial}{\partial x_j} - qA_j\right)\Psi_1 = -\frac{\hbar}{i}\frac{\partial\Psi_1}{\partial t}. \tag{6.139}$$

Note how the transformation removes V from the equation.

Next let us introduce the transformation

$$\Psi_1 = \Psi_0 \exp\left(\frac{iq}{\hbar}\int_0^l \mathbf{A}\cdot\mathrm{d}\mathbf{r}\right) \tag{6.140}$$

in which the integration also proceeds along the definite path from a given initial to a variable final point. Now, the exponential is an explicit function of position alone and

$$-\frac{\hbar}{i}\frac{\partial\Psi_1}{\partial t} = -\frac{\hbar}{i}\frac{\partial\Psi_0}{\partial t}\exp\left(\frac{iq}{\hbar}\int_0^l \mathbf{A}\cdot\mathrm{d}\mathbf{r}\right), \tag{6.141}$$

$$\left(\frac{\hbar}{i}\frac{\partial}{\partial x_j} - qA_j\right)\Psi_1 = \left(\frac{\hbar}{i}\frac{\partial\Psi_0}{\partial x_j} - qA_j\Psi_0\right)\exp\left(\frac{iq}{\hbar}\int_0^l \mathbf{A}\cdot\mathrm{d}\mathbf{r}\right) +$$

$$+\frac{\hbar}{i}\Psi_0\left[\exp\left(\frac{iq}{\hbar}\int_0^l \mathbf{A}\cdot\mathrm{d}\mathbf{r}\right)\right]\left(\frac{iq}{\hbar}A_j\right)$$

$$= \frac{\hbar}{i}\frac{\partial\Psi_0}{\partial x_j}\exp\left(\frac{iq}{\hbar}\int_0^l \mathbf{A}\cdot\mathrm{d}\mathbf{r}\right), \tag{6.142}$$

and

$$\left(\frac{h}{i}\frac{\partial}{\partial x_j} - qA_j\right)\left[\frac{\hbar}{i}\frac{\partial \Psi_0}{\partial x_j}\exp\left(\frac{iq}{\hbar}\int_0^l \mathbf{A}\cdot\mathbf{dr}\right)\right]$$

$$= -\hbar^2 \frac{\partial}{\partial x_j}\frac{\partial \Psi_0}{\partial x_j}\exp\left(\frac{iq}{\hbar}\int_0^l \mathbf{A}\cdot\mathbf{dr}\right). \tag{6.143}$$

Substituting these results into (6.139) and canceling the exponential factor leaves

$$-\frac{\hbar^2}{2\mu}\nabla^2\Psi_0 = -\frac{\hbar}{i}\frac{\partial \Psi_0}{\partial t}, \tag{6.144}$$

the Schrödinger equation for a free particle. The wave function Ψ for movement along a given path through a field is related to the solution Ψ_0 for the trajectory in absence of the field by the transformation

$$\Psi = \Psi_0 \exp\left[-\frac{i}{\hbar}\left(\int_0^t V\,\mathrm{d}t - q\int_0^l \mathbf{A}\cdot\mathbf{dr}\right)\right]. \tag{6.145}$$

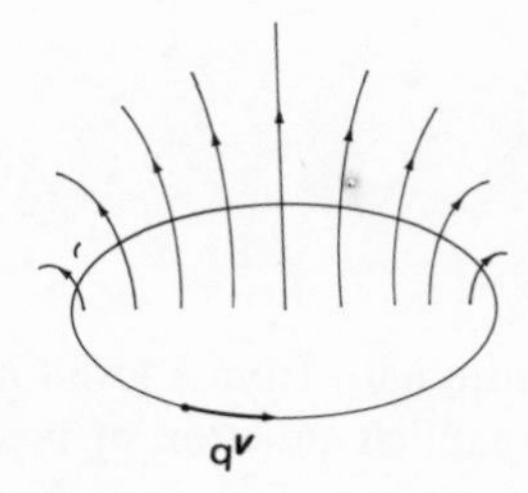

Fig. 6.3. Charged particle moving along a loop enclosing magnetic flux.

6.12. Quantization of Magnetic Flux

Equation (6.145) and the condition that a wave function describing motion through space be single valued imply that flux is quantized.

Consider a particle moving along a closed path that encloses magnetic flux, as Figure 6.3 shows. Suppose that the wave function for a steady state when there is no charge on the particle is Ψ_0. Function Ψ in (6.145) then describes the corresponding steady state when the particle carries charge q.

Because Ψ_0 is a possible eigenfunction, it is single valued. So for Ψ to return to its initial value when its coordinates complete the circuit, the exponent in

(6.145) must be $2\pi i$ multiplied by an integer n. Since the first integral in the exponent is zero, we have

$$-\frac{2\pi i}{h}\left(-q \oint \mathbf{A} \cdot d\mathbf{r}\right) = 2\pi i n, \tag{6.146}$$

whence

$$\oint \mathbf{A} \cdot d\mathbf{r} = n\frac{h}{q}. \tag{6.147}$$

But from Stokes' law and Equation (6.129), we obtain

$$\oint \mathbf{A} \cdot d\mathbf{r} = \int_S \nabla \times \mathbf{A} \cdot d\mathbf{S} = \int_S \mathbf{B} \cdot d\mathbf{S} = \text{flux enclosed}. \tag{6.148}$$

Equation (6.147) therefore states that when a steady current encloses a flux, the flux must be an integral multiple of h/q.

Now, a current can be set up in a ring of appropriate material and the ring cooled down until it becomes a superconductor. Experimenters find that the flux enclosed is then quantized in units of

$$\frac{h}{2e}, \tag{6.149}$$

where e is the magnitude of charge on an electron. This result indicates that superconducting electrons move in pairs.

6.13. Effect of a Field on Two-Slit Diffraction

Ehrenfest's theorem tells us that the field intensities determine how given particles move on the average through a particular field. But the potentials determine how and where interferences occur. Because **A** and/or ϕ may differ from zero in regions where **B** and **E** are zero, a field can exert an infleunce in a region where its intensities are zero.

Consider a diffraction system in which a homogeneous beam strikes a perpendicular mask with two parallel slits, which allow parts to pass through, as Figure 1.1 shows. The wave function for movement along the first path from source to screen in absence of a field contains the spatial factor

$$\psi_{10} = A_1\, e^{iks_1}, \tag{6.150}$$

while that for the second path is

$$\psi_{20} = A_2\, e^{iks_2}. \tag{6.151}$$

Here s_1 is the distance traveled, measured along the first path, while s_2 is the distance traveled, measured along the second path. The temporal factor for each of the functions is

$$e^{-i\omega_0 t}. \tag{6.152}$$

According to (6.145), a field transforms (6.150) and (6.151) to

$$\psi_1 = A_1 \exp i\left(ks_1 - \frac{S_1}{\hbar}\right) \tag{6.153}$$

and

$$\psi_2 = A_2 \exp i\left(ks_2 - \frac{S_2}{\hbar}\right) \tag{6.154}$$

if

$$S = \int V\,dt - q\int \mathbf{A}\cdot d\mathbf{r}. \tag{6.155}$$

When the screen is far from the mask, the paths leaving the slits and going to a given point on the screen are approximately parallel. The difference in distance is then the projection of the distance between the slits d on the longer path, as Figure 1.2 illustrates:

$$s_2 - s_1 = d \sin\theta. \tag{6.156}$$

On the screen, the difference in phase varies with the path length difference $s_2 - s_1$ and with the integral difference $S_2 - S_1$. If s_2 and s_1 are labeled s_{20} and s_{10} when there is no field, then a given difference in the exponents in (6.153) and (6.154) appears where

$$ks_2 - \frac{S_2}{\hbar} - \left(ks_1 - \frac{S_1}{\hbar}\right) = ks_{20} - ks_{10} \tag{6.157}$$

or

$$k(s_2 - s_1) - \frac{1}{\hbar}(S_2 - S_1) = k(s_{20} - s_{10}). \tag{6.158}$$

Introducing (6.156), letting θ_0 be the deflection angle in absence of field,

$$kd \sin\theta - \frac{1}{\hbar}(S_2 - S_1) = kd \sin\theta_0, \tag{6.159}$$

and rearranging leads to

$$\sin\theta - \sin\theta_0 = \frac{1}{\hbar kd}(S_2 - S_1) \tag{6.160}$$

or

$$\Delta\sin\theta = \frac{1}{\hbar kd}\Delta S. \tag{6.161}$$

When the deflection angle is small,

$$\sin\theta \simeq \theta, \tag{6.162}$$

Equation (6.161) reduces to

$$\Delta\theta = \frac{1}{\hbar kd}\Delta S. \tag{6.163}$$

Introducing de Broglie's formula for momentum converts (6.163) to

$$\Delta\theta = \frac{1}{pd}\Delta S. \tag{6.164}$$

Here θ is the angle of diffraction, p the momentum of a particle in the beam, d the distance between the slit centers, and S the expression in (6.155).

Since each fringe arises from a definite difference in phase, (6.164) describes the angular displacement of a fringe caused by the field. Note that this displacement does not involve Planck's constant. Merely altering the magnitude of $\hbar$ has no effect on $\Delta\theta$.

The influences of potentials V and $\mathbf{A}$ on details of movement were first emphasized by Yakir Aharonov and David Bohm in 1959. As a consequence, they maintained that potentials play a fundamental role in quantum mechanics which they do not play in classical mechanics.

6.14. Magnetic and Electric Aharonov–Bohm Effects

Wherever diffraction of charged particles occurs, and separated rays travel along different paths from a source to a particular spot on a screen, the difference in phasing at the final point depends on the flux enclosed and on the potential-multiplied-by-time difference, between the paths. The amount scattered through a certain angle is modulated by the resulting superposition, reaching a maximum when the wavelets arrive in phase and dropping to zero when they arrive π radians out of phase.

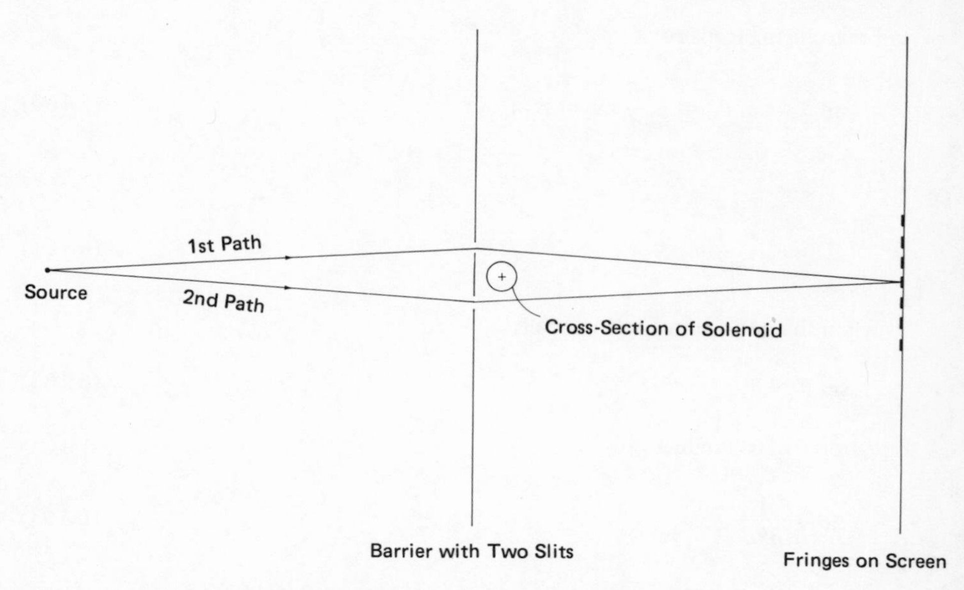

Fig. 6.4. Setup for observing the magnetic Aharonov–Bohm effect.

Consider a homogeneous beam of similar charged particles striking a perpendicular barrier in which two parallel identical slits are cut. Equivalent parts of the beam then pass through each slit to a distant screen.

In the experiment sketched in Figure 6.4, a long shielded solenoid is placed behind the barrier between the beamlets, parallel to the slits, and a current is passed through the solenoid. Equation (6.155) then yields

$$\Delta S = S_2 - S_1 = -q \oint \mathbf{A}\cdot d\mathbf{r} = -q \int_S \nabla \times \mathbf{A}\cdot d\mathbf{S}$$

$$= -q \int_S \mathbf{B}\cdot d\mathbf{S} = -q \text{ (flux enclosed)}. \tag{6.165}$$

Substituting this result into (6.164) gives us

$$\Delta\theta = -\frac{q}{pd}\int_S \mathbf{B}\cdot d\mathbf{S}. \tag{6.166}$$

Now, the magnetic field is concentrated within the cyclindrical windings. At each end, it spreads out. In the region through which the charged particles pass, it is very weak, particularly when the solenoid is relatively long. In these circumstances, almost no force acts on the particles, and there is little deflection in the average particle motion. Nevertheless, the fringes appear displaced by angle (6.166).

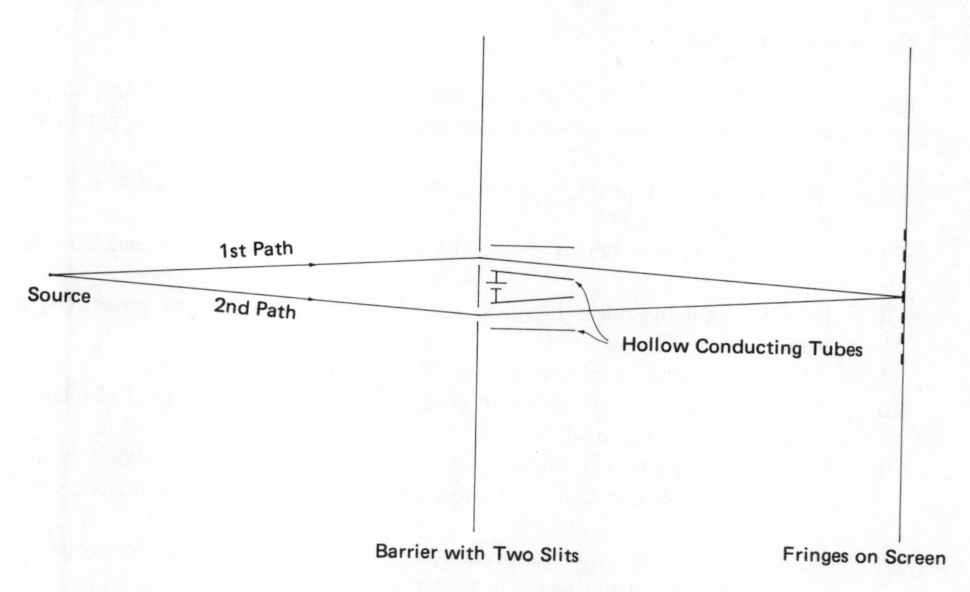

Fig. 6.5. System that exhibits the electric Aharonov–Bohm effect.

In the setup described by Figure 6.5, a metallic tube surrounds each beamlet as it emerges from its slit. A potential difference is imposed between the tubes, as indicated. The field intensity is then small all along each path. But Equation (6.155) produces

$$\Delta S = S_2 - S_1 = \int q\phi_2 \, \mathrm{d}t - \int q\phi_1 \, \mathrm{d}t$$

$$= q \int (\phi_2 - \phi_1) \, \mathrm{d}t, \tag{6.167}$$

which converts (6.164) to

$$\Delta\theta = \frac{q}{pd} \int \Delta\phi \, \mathrm{d}t. \tag{6.168}$$

Since the electric field intensity is almost zero all along both paths, no appreciable electric force acts on the charged particles. As a consequence, the imposed potential difference between the two tubes causes little deflection in the average particle motion, but it does displace the fringes by angle (6.168).

Experimental results do support formulas (6.166) and (6.168). Now, an observable influence of V or $\mathbf{A}$ that is not attributable to a field intensity acting in the given system is called an *Aharonov–Bohm effect.*

Discussion Questions

6.1. Why is the most general allowed wave function for a given system a linear superposition of the possible eigenfunctions with respect to a pertinent operator?
6.2. When is the superposition in Question 6.1 (a) a sum, (b) an integral, (c) a sum plus an integral?
6.3. What can a series (a) of sines, (b) of cosines, (c) of sines and cosines represent?
6.4. How is a single-valued function expressed by (a) a Fourier series, (b) a Fourier integral?
6.5. Explain what transform inverts the effect of a Fourier transform.
6.6. How does instability broaden an energy level? How is the mean life related to the width of such a level?
6.7. What is the Fourier transform of a Gaussian wave packet? What is the product of the standard deviations in position and momentum for such a mixed state?
6.8. How does Fourier integral theorem (6.27) determine how an arbitrary superposition of planar waves propagates?
6.9. Why can the Dirac delta be represented as a superposition of planar waves?
6.10. How can subtracting $\epsilon|k|$ from the exponent in the representation of the Dirac delta enable one to carry out the integration?
6.11. Why may the de Broglie angular frequency depend on the wavevector? Why is the dependence analytical? When is the dependence linear?
6.12. When does propagation in free space merely translate the wave and shift the phase?
6.13. What determines the group velocity of a wave system? When does the group velocity differ from the phase velocity?
6.14. Why is the energy of a de Broglie wave proportional to its frequency? In what context was this relationship first discovered?
6.15. Construct and establish a Schrödinger equation that describes mixed-energy states.
6.16. How does a person formulate a Schrödinger equation for the development of a system in time?
6.17. Why does the rate of change in an expectation value involve the Hamiltonian operator?
6.18. How does the expression for this rate simplify when H is Hermitian with respect to Ψ, $A\Psi$, and the given boundary conditions?
6.19. When is operator H not Hermitian with respect to Ψ and $x\Psi$?
6.20. Why are classical properties represented by expectation values?
6.21. Explain how Newtonian laws apply to a wave packet.
6.22. How does a field contribute (a) to the energy of a particle, (b) to the momentum, and inertia, of a particle?

6.23. Why are the operators for the components of momentum unchanged by the presence of the field term $q\mathbf{A}$?

6.24. Derive the Schrödinger equation for motion of a particle in an interacting field.

6.25. How do the scalar and vector potentials appear in the Hamiltonian operator?

6.26. How does a person obtain $[H, \mathbf{p}]$?

6.27. What does $\langle \partial \mathbf{p}/\partial t \rangle$ equal?

6.28. In the expression for $d\mathbf{p}/dt$, when can $\nabla \mathbf{A}$ and $\nabla \phi$ be factored out of the expectation value integrals?

6.29. With what are the classical velocity and classical momentum of a particle identified?

6.30. Establish the rule

$$\frac{\mathrm{d}}{\mathrm{d}t} = \frac{\partial}{\partial t} + \mathbf{v} \cdot \nabla .$$

6.31. Establish the formula

$$\mathbf{A} \times (\mathbf{B} \times \mathbf{C}) = \mathbf{B}(\mathbf{A} \cdot \mathbf{C}) - (\mathbf{A} \cdot \mathbf{B})\mathbf{C}.$$

6.32. How are the electric and magnetic field intensities defined? How are they related to the scalar and vector potentials?

6.33. How does the Lorentz force law arise in quantum mechanics?

6.34. Explain how the scalar and vector potentials affect the phase of the wave function.

6.35. Under what limitations can a person transform the wave function to remove (a) the scalar potential, (b) the vector potential, from Schrödinger's time-dependent equation?

6.36. Explain how magnetic flux is quantized.

6.37. What evidence is there for concluding that superconducting electrons travel in pairs?

6.38. Describe the paths taken by particles traveling from a single source through two parallel slits to a screen. What determine the change in phase along each path and the resulting difference at a given point on the screen?

6.39. What kind of influences can a field exert in a region where its intensities vanish?

6.40. Why is the Aharonov-Bohm fringe deflection angle independent of the magnitude of Planck's constant?

Problems

6.1. A particle confined between $x = 0$ and $x = a$ at $V = 0$ is governed by the state function $\psi = Ax(a - x)$. Express ψ as a Fourier series in which each term is an eigenfunction.

6.2. For the particle in Problem 6.1, obtain the probability that the system is in each eigenstate. Then determine the average energy.

6.3. If the mean life of a charged pion is 2.60×10^{-8} s, what is the width of the state?

6.4. What modulated wave is obtained when two similarly directed planar de Broglie waves equal in amplitude but differing in wavevector and angular frequency by Δk and $\Delta\omega$ are superposed? At what velocity does the modulation factor travel?

6.5. The phase velocity of surface waves on a nonviscous fluid is given by the formula $w = (g/k)^{\frac{1}{2}}$, in which g is the acceleration due to gravity and k is the wavevector. Find an expression for the group velocity of such waves. Calculate the group velocity of water waves for which the mean wavelength is 15.0 cm.

6.6. Use the representation of the Dirac delta as a superposition of planar waves to establish

$$\delta(x) = \lim_{n\to\infty} \frac{\sin nx}{\pi x}.$$

6.7. Determine wave function $\Psi(x, t)$ when the wavevector representation for the state of a system of free particles is

$$\phi = (2\pi)^{\frac{1}{2}}\, \delta(k - \bar{k}).$$

6.8. Verify the commutator formulas

$$[A, B + C] = [A, B] + [A, C],$$

$$[A, BC] = [A, B]\, C + B\, [A, C].$$

6.9. Show that if A does not explicitly contain t and if the commutator of H with A is zero, the derivative $(\mathrm{d}/\mathrm{d}t)\,\langle A^2\rangle$ is zero.

6.10. Let S be some function of the independent variables and introduce the transformation

$$\Psi_1 = \Psi \exp\left(\frac{i}{h} S\right)$$

into the general time-dependent Schrödinger equation.

6.11. What Fourier series represent

$$\text{(a)}\quad f(x) = x \qquad \text{when} \quad -\pi < x < \pi,$$

$$\text{(b)}\quad F(x) = x^2 \qquad \text{when} \quad -\pi < x < \pi?$$

6.12. In the range $0 < \varphi < \pi$, a one-dimensional rotator is governed by the wave function

$$\Phi = A\varphi(\pi - \varphi).$$

Express Φ as a sum of eigenfunctions assuming (a) Φ is even about the point $\varphi = 0$, (b) Φ is odd about $\varphi = 0$.

6.13. What is the mean life of the particle corresponding to a resonance 120 MeV in width?

6.14. How does the phase velocity of a de Broglie wave vary with the potential of the particle?

6.15. A given phase of a sinusoidal capillary wave moves on the surface of a liquid at the speed $w = (\gamma k/\rho)^{\frac{1}{2}}$ if γ is the surface tension and ρ the density of the liquid. How fast does a group of such waves move? Calculate the group velocity on the surface of water at 20°C when the mean wavelength λ is 0.100 cm. The surface tension of water is 72.75 dyne cm^{-1} at 20°C.

6.16. Employ the representation of the Dirac delta as a superposition of planar waves in establishing

$$\delta(x) = \lim_{\epsilon \to 0} \frac{1}{(\pi\epsilon)^{\frac{1}{2}}} e^{-x^2/\epsilon}$$

6.17. If a particle were located with certainty at point $\mathbf{r} = \bar{\mathbf{r}}$ at time $t = 0$, what would be its wavevector representation at that time? How would the probabilities of the different linear momenta compare?

6.18. Verify the commutator formulas

$$[AB, C] = A[B, C] + [A, C]B,$$

$$[A, [B, C]] + [B, [C, A]] + [C, [A, B]] = 0.$$

6.19. If J_z is the operator for angular momentum around the z axis, when is $\mathrm{d}\langle J_z\rangle/\mathrm{d}t$ zero?

6.20. Determine how a gauge transformation

$$A' = A + \nabla f, \qquad \phi' = \phi - \frac{\partial f}{\partial t}$$

alters the phase of the wave function.

References

Books

Arfken, G.: 1970, *Mathematical Methods for Physicists*, 2nd edn, Academic Press, New York, pp. 643–688.

Beard, D. B., and Beard, G. B.: 1970, *Quantum Mechanics with Applications*, Allyn and Bacon, Boston, pp. 12–65.

Gottfried, K.: 1966, *Quantum Mechanics, Volume I: Fundamentals*, Benjamin, New York, pp. 16–76.

Merzbacher, E.: 1970, *Quantum Mechanics*, 2nd edn, Wiley, New York, pp. 13–52.

Park, D.: 1974, *Introduction to the Quantum Theory*, McGraw-Hill, New York, pp. 60–88.

White, R. L.: 1966, *Basic Quantum Mechanics*, McGraw-Hill, New York, pp. 124–146.

Articles

Abbott, L. F., and Wise, M. B.: 1981, 'Dimension of a Quantum-Mechanical Path', *Am. J. Phys.* **49**, 37–39.

Aharonov, Y., and Carmi, G.: 1974, 'Quantum-Related Reference Frames and the Local Physical Significance of Potentials', *Found. Phys.* **4**, 75–81.

Amand, J., St, and Uritam, R. A.: 1973, 'A Simple Illustration of the Uncertainty Relation in Hadronic Collisions', *Am. J. Phys.* **41**, 650–653.

Berry, M. V.: 1980, 'Exact Aharonov-Bohm Wavefunction Obtained by Applying Dirac's Magnetic Phase Factor', *Eur. J. Phys.* **1**, 240–244.

Boyer, T. H.: 1972, 'Misinterpretation of the Aharonov–Bohm Effect', *Am. J. Phys.* **40**, 56–59.

Brown, L. S.: 1973, 'Classical Limit of the Hydrogen Atom', *Am. J. Phys.* **41**, 525–530.

Buch, L. H., and Denman, H. H.: 1974, 'Solution of the Schrödinger Equation for Some Electric Field Problems', *Am. J. Phys.* **42**, 304–309.

Cohen, B. L.: 1965, 'A Simple Treatment of Potential Barrier Penetration', *Am. J. Phys.* **33**, 97–98.

Cohn, J.: 1972, 'Quantum Theory in the Classical Limit', *Am. J. Phys.* **40**, 463–467.
Cook, L. F.: 1980, 'Einstein and $\Delta t\, \Delta E$', *Am. J. Phys.* **48**, 142–145.
Crawford, F. S.: 1982, 'Can a Wave Packet be Wider than it is Long?', *Am. J. Phys.* **50**, 199.
Crawford, F. S.: 1982, 'Elementary Derivation of the Magnetic Flux Quantum', *Am. J. Phys.* **50**, 514–516.
Danos, M.: 1982, 'Bohm–Aharonov Effect: The Quantum Mechanics of the Electrical Transformer', *Am. J. Phys.* **50**, 64–66.
Diu, B.: 1980, 'Plane Waves and Wave Packets in Elementary Quantum Mechanics Problems', *Eur. J. Phys.* **1**, 231–240.
Eck, J. S., and Thompson, W. J.: 1977, 'Dissipative Forces and Quantum Mechanics', *Am. J. Phys.* **45**, 161–163.
Erlichson, H.: 1970, 'Aharonov–Bohm Effect – Quantum Effects on Charged Particles in Field-Free Regions', *Am. J. Phys.* **38**, 162–173.
Espinosa, J. M.: 1982, 'Physical Properties of de Broglie's Phase Waves', *Am. J. Phys.* **50**, 357–362.
Garwin, L. J., and Garwin, R. L.: 1977, 'On the Strength of the Dirac Monopole', *Am. J. Phys.* **45**, 164–165.
Greenberger, D. M.: 1980, 'Comment on "Nonspreading Wave Packets"', *Am. J. Phys.* **48**, 256.
Harris, R. A., and Strauss, H. L.: 1978, 'Paradoxes from the Uncertainty Principle', *J. Chem. Educ.* **55**, 374–375.
Hill, R. N.: 1973, 'A Paradox Involving the Quantum Mechanical Ehrenfest Theorem', *Am. J. Phys.* **41**, 736–738.
Holze, D. H., and Scott, W. T.: 1964, 'Relationships of Wave Packet Variances', *Am. J. Phys.* **32**, 853–856.
Hrasko, P.: 1977, 'Quasiclassical Quantization of the Magnetic Charge', *Am. J. Phys.* **45**, 838–840.
Janossy, L.: 1974, 'The Physical Interpretation of Wave Mechanics. II', *Found. Phys.* **4**, 445–452.
Klein, J. R.: 1980, 'Do Free Quantum-Mechanical Wave Packets Always Spread?, *Am. J. Phys.* **48**, 1035–1037.
Konopinski, E. J.: 1978, 'What the Electromagnetic Vector Potential Describes', *Am. J. Phys.* **46**, 499–502.
Krivchenkov, V. D., and Kukin, V. D.: 1972, 'Equidistant Spacing of the Energy Levels of the One-Dimensional Schrödinger Equation', *Soviet Phys. Doklady* **17**, 444–446.
Lai, C. S.: 1979, 'Spreading of Wave Packets in the Coordinate Representation', *Am. J. Phys.* **47**, 766–768.
Lieber, M.: 1975, 'Quantum Mechanics in Momentum Space: An Illustration', *Am. J. Phys.* **43**, 486–491.
Lubkin, E.: 1971, 'A Simple Picture for Dirac's Charge-Pole Quantization Law', *Am. J. Phys.* **39**, 94–96.
Marshall, A. G., and Comisarow, M. B.: 1975, 'Fourier Transform Methods in Spectroscopy', *J. Chem. Educ.* **52**, 638–641.
Paul, D.: 1980, 'Dispersion Relation for de Broglie Waves', *Am. J. Phys.* **48**, 283–284.
Pena, de la, L.: 1980, 'Conceptually Interesting Generalized Heisenberg Inequality', *Am. J. Phys.* **48**, 775–776.
Peres, A.: 1980, 'Measurement of Time by Quantum Clocks', *Am. J. Phys.* **48**, 552–557.
Peslak, J., Jr: 1979, 'Comparison of Classical and Quantum Mechanical Uncertainties', *Am. J. Phys.* **47**, 39–45.
Roy, C. L., and Sannigrahi, A. B.: 1979, 'Uncertainty Relation between Angular Momentum and Angle Variable', *Am. J. Phys.* **47**, 965–967.
Roychoudhuri, C.: 1978, 'Heisenberg's Microscope – A Misleading Illustration', *Found. Phys.* **8**, 845–849.

Santana, P. H. A., and Rosato, A.: 1973, 'Use of the Laplace Transform Method to Solve the One-Dimensional Periodic-Potential Problem', *Am. J. Phys.* **41**, 1138–1144.

Semon, M. D.: 1982, 'Experimental Verification of an Aharonov–Bohm Effect in Rotating Reference Frames', *Found. Phys.* **12**, 49–57.

Snygg, J.: 1977, 'Wave Functions Rotated in Phase Space', *Am. J. Phys.* **45**, 58–60.

Sorkin, R.: 1979, 'On the Failure of the Time-Energy Uncertainty Principle', *Found. Phys.* **9**, 123–128.

Stevens, K. W. H.: 1980, 'A Note on Quantum Mechanical Tunnelling', *Eur. J. Phys.* **1**, 98–101.

Taylor, P. L.: 1969, 'Wave Mechanics and the Concept of Force', *Am. J. Phys.* **37**, 29–33.

Weichel, H.: 1976, 'The Uncertainty Principle and the Spectral Width of a Laser Beam', *Am. J. Phys.* **44**, 839–840.

Williams, D. N.: 1979, 'New Mathematical Proof of the Uncertainty Relation', *Am. J. Phys.* **47**, 606–607.

Chapter 7

Angular Motion in a Spherically Symmetric Field

7.1. Conditions on the Angular Eigenfunctions

We have seen how rotation of a linear system of particles is modeled by a single particle of reduced mass μ traveling around the center of mass of the system. The probability per unit volume that the model particle is distance r from the center, with colatitude θ from the axis of rotation and azimuthal angle φ about this axis, is given by the product $[\Psi(r, \theta, \varphi)]^* \Psi(r, \theta, \varphi)$.

When the rotation is free, all points at a given r and θ are equivalent. When the motion at the given r and θ exhibits a single definite angular momentum, the corresponding wave function satisfies both (2.15) and (2.16). Integrating these equations leads to formula (2.17).

But pure rotation about an axis is accompanied by back-and-forth motion through the middle plane of rotation. We have taken this plane to be the xy plane, so that the pure rotation is described by the $\Phi(\varphi)$ factor, the back-and-forth movement by the $\Theta(\theta)$ factor, in Ψ. The product of Θ and Φ is labeled Y.

The simplest back-and-forth movement involves the superposition of two equivalent oppositely traveling waves in each small region. While wavevector $\mathbf{k}$ is then not definite, the scalar $\mathbf{k} \cdot \mathbf{k}$ is. As a consequence, such movement is governed by the second-order Schrödinger equation. Since this equation can also govern any unidirectional rotation that might be present, it is a fundamental condition that an angular eigenfunction $Y(\theta, \varphi)$ in Ψ must satisfy.

Definiteness in the probability density does not imply that Ψ has only one value at each point. But in diffraction experiments, the relative phases of different wavelets reaching a given point on a screen are significant. The components do add to yield a definite Ψ, insofar as a person can tell. Consequently, we assume that a wave function describing motion *through space* is single valued.

7.2. The Homogeneous Polynomial Factor

Free angular motion about a point, with a definite l or J, is described, when the point is made the origin, by a ψ that is the product of a homogeneous polynomial in x, y, z, and a function of distance r from the origin. The polynomial must satisfy Laplace's equation.

Consider a model particle of reduced mass

$$\mu = \frac{m_1 m_2}{m_1 + m_2}, \tag{7.1}$$

with the coordinates of Figures 2.2 and 4.3. Let it move in a central field, for which the potential energy is

$$V = V(r). \tag{7.2}$$

Any state for which the square of wavevector **k** is a definite function is governed by the Schrödinger equation

$$\nabla^2 \psi + k^2 \psi = 0. \tag{7.3}$$

Condition (7.2) implies that k^2 is independent of θ and φ. The argument in Section 4.5 tells us that the radial variable r then separates from the angular variables:

$$\psi = R(r)\, Y(\theta, \varphi). \tag{7.4}$$

Indeed from (4.55), we obtain

$$-\frac{1}{Y} r^2 \nabla^2 Y = B, \tag{7.5}$$

with B a constant.

Equation (7.5) does not involve wavevector k in any way. Hence the angular functions for

$$k = 0 \tag{7.6}$$

are also those for any well-behaved k; *Laplace's equation*

$$\nabla^2 \psi = 0, \tag{7.7}$$

obtained from (7.3) by setting k equal to zero, yields the various Y's for the central field motion.

A general analytic form for the solution of (7.7) is a power series in the rectangular coordinates. Terms in this series can be grouped according to degree and each set

$$\psi = \sum_{j,k} A_{jk} x^{l-j-k} y^j z^k \tag{7.8}$$

considered separately.

Indeed, the operator

$$\nabla^2 = \frac{\partial^2}{\partial x^2} + \frac{\partial^2}{\partial y^2} + \frac{\partial^2}{\partial z^2} \tag{7.9}$$

reduces the degree of each term by 2. So, Equation (7.7) only imposes conditions on sets (7.8). It cannot introduce conditions linking different sets.

In the double summation in (7.8), we let

$$j = 0, 1, \ldots, l, \tag{7.10}$$

$$k = 0, 1, \ldots, l, \tag{7.11}$$

with

$$j + k \leqslant l. \tag{7.12}$$

When the exponent on x is $l, l - 1, \ldots, 2, 1, 0$, the number of terms are $1, 2, \ldots, l - 1, l, l + 1$, respectively. So the total number of terms equals

$$1 + 2 + \ldots + (l - 1) + l + (l + 1). \tag{7.13}$$

Action of ∇^2 on (7.8) leads to a polynomial of degree $l - 2$ with

$$1 + 2 + \ldots + (l - 1) \tag{7.14}$$

terms. Equation (7.7) tells us that this polynomial must be identically zero. Consequently, the A_{jk}'s have to satisfy sum (7.14) conditions. The number of independent polynomials of degree l is quantity (7.13) minus quantity (7.14); that is,

$$1 + 2 + \ldots + (l - 1) + l + (l + 1) - [1 + 2 + \ldots + (l - 1)]$$

$$= l + (l + 1) = 2l + 1. \tag{7.15}$$

Dividing any homogeneous polynomial solution by r^l yields a function Y of θ and φ. This function is a suitable angular factor for an eigenfunction of Schrödinger's equation for field (7.2), since the nature of k does not affect Equation (7.5). An acceptable solution of Laplace's equation is called a *harmonic*. Because each homogeneous polynomial solution exhibits a symmetry appropriate to a sphere centered on the origin, it is called a *spherical harmonic*. Function Y is the angular part of such a harmonic.

Standard forms for Y were found in Section 2.13. When expressed in rectangular coordinates, each of these involves a homogeneous polynomial in x, y, z, multiplied by a function of r. The degree of the polynomial equals quantum number l or J.

In general, dividing (7.8) by r^l and introducing (2.88), (2.89), (2.90) yields

$$Y = A_{00} \sin^l \theta \cos^l \varphi + \ldots . \tag{7.16}$$

Expanding each cos φ and sin φ in Y by the identity in Example 1.1 leads to terms in which the φ factors are

$$e^{il\varphi}, e^{i(l-1)\varphi}, \ldots . e^{-il\varphi}. \tag{7.17}$$

These correspond to quantum numbers

$$m = l, l-1, \ldots, -l \tag{7.18}$$

and to parameter l being quantum number l (or quantum number J for a rotator).

Example 7.1. Show that Laplace's equation does not restrict the linear terms in the power series solution.

When l is 1, polynomial (7.8) is

$$\psi = A_{00}x + A_{10}y + A_{01}z.$$

But since

$$\frac{\partial \psi}{\partial x} = A_{00}, \qquad \frac{\partial \psi}{\partial y} = A_{10}, \qquad \frac{\partial \psi}{\partial z} = A_{01},$$

and

$$\frac{\partial^2 \psi}{\partial x^2} = 0, \qquad \frac{\partial^2 \psi}{\partial y^2} = 0, \qquad \frac{\partial^2 \psi}{\partial z^2} = 0,$$

the Laplacian

$$\frac{\partial^2 \psi}{\partial x^2} + \frac{\partial^2 \psi}{\partial y^2} + \frac{\partial^2 \psi}{\partial z^2}$$

is zero regardless of what the constants equal.

Thus (7.7) does not restrict A_{00}, A_{10}, and A_{01}. Any one may be set equal to A and the others set equal to zero. Dividing the result by r then yields

$$Y_{p_x} = A\frac{x}{r}, \qquad Y_{p_y} = A\frac{y}{r}, \qquad Y_{p_z} = A\frac{z}{r}.$$

Since Equation (7.5) does not contain $k(r)$, these are independent angular functions for (7.3), the Schrödinger equation, when $V = V(r)$.

7.3. Solutions Derived from the Reciprocal of the Radius Vector

Eigenfunctions for free angular motion about the origin can also be expressed as r^{l+1} times a multiple spatial differentiating operator acting on $1/r$.

The wavevector magnitude for a particle depends on the nature of the potential. When this potential varies only with distance of the particle from a source point, as in (7.2), the magnitude k varies only with this distance. When the source point is made the origin of a spherical coordinate system, the angular coordinates can be separated from the radial coordinate in the Schrödinger equation. Since the angular equation (7.5) that results does not contain k, all angular eigenfunctions also appear as factors in the solutions of Laplace's equation, for which k is zero.

A particularly simple solution of Laplace's equation (7.7) is a function that varies inversely with the distance from a given point. When the point is the origin, we have

$$\psi = \frac{1}{r}. \tag{7.19}$$

Indeed, take the formula for r^2,

$$r^2 = x^2 + y^2 + z^2, \tag{7.20}$$

differentiate,

$$2r \frac{\partial r}{\partial x} = 2x, \tag{7.21}$$

and rearrange,

$$\frac{\partial r}{\partial x} = \frac{x}{r}. \tag{7.22}$$

Differentiate (7.19) twice, obtaining

$$\frac{\partial}{\partial x} \frac{1}{r} = -\frac{1}{r^2} \frac{\partial r}{\partial x} = -\frac{x}{r^3} \tag{7.23}$$

and

$$\frac{\partial^2}{\partial x^2} \frac{1}{r} = -\frac{r^3 - x[3r^2(x/r)]}{r^6} = \frac{3x^2 - r^2}{r^5}. \tag{7.24}$$

Similarly, we find that

$$\frac{\partial^2}{\partial y^2} \frac{1}{r} = \frac{3y^2 - r^2}{r^5}, \tag{7.25}$$

$$\frac{\partial^2}{\partial z^2} \frac{1}{r} = \frac{3z^2 - r^2}{r^5}, \tag{7.26}$$

whence

$$\nabla^2\left(\frac{1}{r}\right) = \left(\frac{\partial^2}{\partial x^2} + \frac{\partial^2}{\partial y^2} + \frac{\partial^2}{\partial z^2}\right)\frac{1}{r}$$

$$= \frac{3x^2 - r^2 + 3y^2 - r^2 + 3z^2 - r^2}{r^5}$$

$$= 0. \tag{7.27}$$

Thus, function (7.19) is a solution of

$$\nabla^2 \psi = \frac{\partial^2 \psi}{\partial x^2} + \frac{\partial^2 \psi}{\partial y^2} + \frac{\partial^2 \psi}{\partial z^2} = 0. \tag{7.28}$$

Any spatial derivative of a solution of (7.28) is also a solution. Let us choose an arbitrary axis along which $x^{(j)}$ is measured and make the operator

$$\frac{\partial}{\partial x^{(j)}} = \frac{\partial x}{\partial x^{(j)}}\frac{\partial}{\partial x} + \frac{\partial y}{\partial x^{(j)}}\frac{\partial}{\partial y} + \frac{\partial z}{\partial x^{(j)}}\frac{\partial}{\partial z}$$

$$= a\frac{\partial}{\partial x} + b\frac{\partial}{\partial y} + c\frac{\partial}{\partial z} \tag{7.29}$$

act on (7.28):

$$\left(a\frac{\partial}{\partial x} + b\frac{\partial}{\partial y} + c\frac{\partial}{\partial z}\right)\left(\frac{\partial^2 \psi}{\partial x^2} + \frac{\partial^2 \psi}{\partial y^2} + \frac{\partial^2 \psi}{\partial z^2}\right)$$

$$= \frac{\partial^2}{\partial x^2}\left(a\frac{\partial \psi}{\partial x} + b\frac{\partial \psi}{\partial y} + c\frac{\partial \psi}{\partial z}\right) +$$

$$+ \frac{\partial^2}{\partial y^2}\left(a\frac{\partial \psi}{\partial x} + b\frac{\partial \psi}{\partial y} + c\frac{\partial \psi}{\partial z}\right) +$$

$$+ \frac{\partial^2}{\partial z^2}\left(a\frac{\partial \psi}{\partial x} + b\frac{\partial \psi}{\partial y} + c\frac{\partial \psi}{\partial z}\right)$$

$$= \nabla^2 \frac{\partial \psi}{\partial x^{(j)}} = 0. \tag{7.30}$$

We see that expression

$$\frac{\partial \psi}{\partial x^{(j)}} \tag{7.31}$$

does satisfy Laplace's equation when ψ does. As a result, the derivative

$$\frac{\partial^l}{\partial x^{(1)}\,\partial x^{(2)}\ldots\partial x^{(l)}}\,\frac{1}{r} \tag{7.32}$$

is a solution.

Each term in the expansion of (7.32) is of degree $-(l+1)$ in the rectangular coordinates. Consequently, multiplying the sum by r^{l+1} and introducing substitutions (2.88), (2.89), (2.90) leads to a function of θ and φ alone. Since (7.5) does not contain k, this function is proportional to the angular factor. We write

$$Y(\theta,\varphi) = N(-1)^l\,r^{l+1}\,\frac{\partial^l}{\partial x^{(1)}\,\partial x^{(2)}\ldots\partial x^{(l)}}\,\frac{1}{r}. \tag{7.33}$$

The axes of $x^{(1)}, x^{(2)}, \ldots, x^{(l)}$ may point in any direction from the origin. These axes intersect a unit sphere centered on the origin at points called *Maxwell poles.*

Example 7.2. What operations on $1/r$ yield the p functions of Example 7.1?

For a p orbital, quantum number l is 1 and only a single axis exists. When this axis is the x axis, $x^{(1)}$ is x and formula (7.33) leads to

$$N(-1)\,r^2\,\frac{\partial}{\partial x}\,\frac{1}{r} = Nr^2\,\frac{x}{r^3} = N\frac{x}{r},$$

which is Y_{p_x}. When the only axis is the y axis, $x^{(1)}$ is y and formula (7.33) yields

$$N\frac{y}{r},$$

which is Y_{p_y}. When the only axis is the z axis, $x^{(1)}$ is z and the formula gives us

$$N\frac{z}{r},$$

which is Y_{p_z}. The axes for Y_{p_x}, Y_{p_y}, Y_{p_z} strike the unit sphere centered on the origin at $(1, 0, 0)$, $(0, 1, 0)$, and $(0, 0, 1)$, respectively.

7.4. Dependence on the Azimuthal Angle

Because unidirectional rotation about an axis is independent of the orthogonal back-and-forth motion, the rotation is described by a definite factor in the angular eigenfunction. Making the axis the z axis then allows variable φ to be separated from variable θ in the Schrödinger equation for Y. The resulting differential equation for φ is easily integrated.

As in Sections 4.5 and 7.2, we consider a model particle of mass μ in a potential that depends only on spherical coordinate r. The Schrödinger equation then appears as in (4.47). When the particle is in a definite energy state, its radial motion is independent of its angular motion and (4.47) reduces to (4.55), which yields (7.5) and (4.61). Employing expansion (4.54) in (4.61) leads to

$$\frac{1}{\sin\theta}\frac{\partial}{\partial\theta}\left(\sin\theta\frac{\partial Y}{\partial\theta}\right)+\frac{1}{\sin^2\theta}\frac{\partial^2 Y}{\partial\varphi^2}+l(l+1)Y=0. \tag{7.34}$$

We also suppose that pure circulation exists around one axis and label this axis the z axis. Presuming that this circulation is independent of the accompanying back-and-forth motion, as we have indicated, we set

$$Y=\Theta(\theta)\Phi(\varphi) \tag{7.35}$$

in (7.34) and obtain

$$\frac{\Phi}{\sin\theta}\frac{\mathrm{d}}{\mathrm{d}\theta}\left(\sin\theta\frac{\mathrm{d}\Theta}{\mathrm{d}\theta}\right)+\frac{\Theta}{\sin^2\theta}\frac{\mathrm{d}^2\Phi}{\mathrm{d}\varphi^2}+l(l+1)\Theta\Phi=0, \tag{7.36}$$

whence

$$\left[\frac{1}{\Theta\sin\theta}\frac{\mathrm{d}}{\mathrm{d}\theta}\left(\sin\theta\frac{\mathrm{d}\Theta}{\mathrm{d}\theta}\right)+l(l+1)\right]\sin^2\theta=-\frac{1}{\Phi}\frac{\mathrm{d}^2\Phi}{\mathrm{d}\varphi^2}=m^2, \tag{7.37}$$

where m^2 has been introduced by definition.

Since the left side of (7.37) is a function of θ alone, variation of φ cannot cause it to vary. Since the middle expression is a function of φ alone, variation of θ cannot cause it to vary. Since m^2 equals both of these expressions, it cannot vary when either φ or θ vary. Therefore, m^2 must be constant.

Rearranging the second equality in (7.37) yields the equation

$$\frac{\mathrm{d}^2\Phi}{\mathrm{d}\varphi^2}+m^2\Phi=0 \tag{7.38}$$

whose solution is

$$\Phi=e^{im\varphi} \tag{7.39}$$

if we allow m to be both positive and negative. This form agrees with (2.28). When m^2 is zero, an independent solution

$$\Phi=C\varphi \tag{7.40}$$

arises. But since (7.40) yields a multiple valued probability density, it is not acceptable.

For (7.39) to yield a definite probability density, m must be either integral or half integral. For Φ itself to be definite, as Section 7.1 indicates, this parameter must be integral. Later we will find that the limits are $-l$ and l, so we write

$$m = -l, -l + 1, \ldots, 0, \ldots, l. \tag{7.41}$$

We call m the magnetic quantum number.

Function (7.39) represents a wave traveling in the positive direction of φ when m is positive and in the negative direction when m is negative. Superposing oppositely traveling waves with

$$m = \pm |m| \tag{7.42}$$

yields the independent functions

$$\Phi = \sqrt{2} \cos |m|\varphi \tag{7.43}$$

and

$$\Phi = \sqrt{2} \sin |m|\varphi \tag{7.44}$$

as in Section 2.6.

7.5. Dependence on the Colatitude Angle

By multiply differentiating an equation obtained on differentiating a simple polynomial, we can construct a form of the remaining differential equation in θ. An explicit expression for each suitable solution can then be deduced.

The equation that $\Theta(\theta)$ has to satisfy is the overall equation in (7.37). Rearranging this,

$$\frac{1}{\sin\theta}\frac{\mathrm{d}}{\mathrm{d}\theta}\left(\sin\theta\,\frac{\mathrm{d}\Theta}{\mathrm{d}\theta}\right) + \left[l(l+1) - \frac{m^2}{\sin^2\theta}\right]\Theta = 0, \tag{7.45}$$

letting

$$w = \cos\theta \tag{7.46}$$

so

$$1 - w^2 = \sin^2\theta, \tag{7.47}$$

$$\mathrm{d}w = -\sin\theta\,\mathrm{d}\theta, \tag{7.48}$$

and setting

$$\Theta(\theta) = P(w) \tag{7.49}$$

leads to the differential equation

$$\frac{\mathrm{d}}{\mathrm{d}w}\left[(1 - w^2)\frac{\mathrm{d}P}{\mathrm{d}w}\right] + \left[l(l+1) - \frac{m^2}{1 - w^2}\right]P = 0. \tag{7.50}$$

Since the field in which the particle moves is central about the origin, there is nothing to distinguish the hemisphere above $\theta = \pi/2$ from the one below $\theta = \pi/2$. One might suppose that solutions could be constructed from simple expressions possessing such symmetry. A candidate is a power of (7.47), a derivative of such a power, or a power of (7.47) times such a derivative; therefore, we start with the polynomial

$$y = c(1 - w^2)^l, \tag{7.51}$$

in which l is an integer, differentiate,

$$\frac{\mathrm{d}y}{\mathrm{d}w} = cl(1 - w^2)^{l-1}(-2w) = -\frac{2lwy}{1 - w^2}, \tag{7.52}$$

and rearrange:

$$(1 - w^2)\frac{\mathrm{d}y}{\mathrm{d}w} + 2lwy = 0. \tag{7.53}$$

Then differentiate (7.53) $l + 1$ times to get

$$(1 - w^2)\frac{\mathrm{d}^{l+2}y}{\mathrm{d}w^{l+2}} + (l+1)(-2w)\frac{\mathrm{d}^{l+1}y}{\mathrm{d}w^{l+1}} + \frac{1}{2}(l+1)l(-2)\frac{\mathrm{d}^l y}{\mathrm{d}w^l} +$$

$$+ 2lw\frac{\mathrm{d}^{l+1}y}{\mathrm{d}w^{l+1}} + (l+1)2l\frac{\mathrm{d}^l y}{\mathrm{d}w^l} = 0 \tag{7.54}$$

or

$$(1 - w^2)\frac{\mathrm{d}^{l+2}y}{\mathrm{d}w^{l+2}} - 2w\frac{\mathrm{d}^{l+1}y}{\mathrm{d}w^{l+1}} + l(l+1)\frac{\mathrm{d}^l y}{\mathrm{d}w^l} = 0. \tag{7.55}$$

Letting

$$\frac{\mathrm{d}^l y}{\mathrm{d}w^l} = P \tag{7.56}$$

reduces (7.55) to

$$(1-w^2)\frac{d^2P}{dw^2}-2w\frac{dP}{dw}+l(l+1)P=0, \tag{7.57}$$

a result that agrees with (7.50) when $m = 0$.

Finally, differentiate (7.53) $l + m + 1$ times to obtain

$$(1-w^2)\frac{d^{l+m+2}y}{dw^{l+m+2}}+(l+m+1)(-2w)\frac{d^{l+m+1}y}{dw^{l+m+1}}+$$
$$+\frac{1}{2}(l+m+1)(l+m)(-2)\frac{d^{l+m}y}{dw^{l+m}}+2lw\frac{d^{l+m+1}y}{dw^{l+m+1}}+$$
$$+(l+m+1)2l\frac{d^{l+m}y}{dw^{l+m}}=0 \tag{7.58}$$

or

$$(1-w^2)\frac{d^{l+m+2}y}{dw^{l+m+2}}-2(m+1)w\frac{d^{l+m+1}y}{dw^{l+m+1}}+$$
$$+(l+m+1)(l-m)\frac{d^{l+m}y}{dw^{l+m}}=0. \tag{7.59}$$

Letting

$$(1-w^2)^{m/2}\frac{d^{l+m}y}{dw^{l+m}}=P \tag{7.60}$$

or

$$\frac{d^{l+m}y}{dw^{l+m}}=(1-w^2)^{-m/2}P \tag{7.61}$$

and differentiating,

$$\frac{d^{l+m+1}y}{dw^{l+m+1}}=(1-w^2)^{-m/2}\left(\frac{dP}{dw}+\frac{mw}{1-w^2}P\right), \tag{7.62}$$

$$\frac{d^{l+m+2}y}{dw^{l+m+2}}=(1-w^2)^{-m/2}\left[\frac{d^2P}{dw^2}+\frac{2mw}{1-w^2}\frac{dP}{dw}+\right.$$
$$\left.+\left(\frac{m}{1-w^2}+\frac{m(m+2)w^2}{(1-w^2)^2}\right)P\right], \tag{7.63}$$

yields expressions that reduce (7.59) to the form

$$(1 - w^2)\frac{d^2P}{dw^2} - 2w\frac{dP}{dw} + \left[l(l+1) - \frac{m^2}{1-w^2}\right]P = 0, \tag{7.64}$$

which is differential equation (7.50).

Solutions of the symmetry that we seek come from combining (7.60) and (7.51) in the expression

$$P_l^m(w) = (1 - w^2)^{m/2}\frac{d^m}{dw^m}P_l(w) \tag{7.65}$$

where

$$P_l(w) = c\frac{d^l}{dw^l}(1 - w^2)^l. \tag{7.66}$$

These solutions are all well behaved.

In them, parameter m is the *absolute value* of the magnetic quantum number. It is true that in Section 7.4 this constant was given both signs to allow (7.39) to represent both solutions of the φ differential equation. But such a trick is not needed for the θ equation. Parameter l is the integer

$$l = 0, 1, 2, \ldots \tag{7.67}$$

called the azimuthal quantum number.

Since $(1 - w^2)^l$ is a polynomial of degree $2l$, differentiating it l times, as in (7.66), yields a polynomial of degree l. Differentiating it m more times, as in (7.65), reduces the degree to $l - m$. The solution ceases to exist when integer m exceeds integer l; the upper limit on m is l. So in formula (7.41), where both signs are allowed, m runs from $-l$ to l.

With each m^2, increasing l from m to ∞ by integral steps causes the degree of

$$(P_l^m)(1 - w^2)^{-m/2} \tag{7.68}$$

to increase by integers from 0 to ∞. A sum of such polynomials is as flexible as a power series. The superposition can therefore represent any well-behaved function in the pertinent interval

$$-1 < w < 1. \tag{7.69}$$

Multiplying the sum by $(1 - w^2)^{m/2}$ does not destroy this flexibility. So we know a general superposition of the P_l^m yields each possible Θ for the given m^2.

When normalization constant c is chosen to be

$$c = \frac{(-1)^l}{2^l l!}, \tag{7.70}$$

expression (7.66) is called the *Legendre polynomial* of degree l. The corresponding $P_l^m(w)$ is called the *associated Legendre function*. Combining (7.65), (7.66), and (7.70) yields *Rodrigues' formula*

$$P_l^m(w) = \frac{(-1)^l}{2^l l!} (1 - w^2)^{m/2} \frac{d^{l+m}}{dw^{l+m}} (1 - w^2)^l. \tag{7.71}$$

The Legendre polynomials through the tenth degree are listed in Table 7.1.

TABLE 7.1.
Legendre polynomials

Symbol	Formula
$P_0(w)$	1
$P_1(w)$	w
$P_2(w)$	$\frac{1}{2}(3w^2 - 1)$
$P_3(w)$	$\frac{1}{2}(5w^3 - 3w)$
$P_4(w)$	$\frac{1}{8}(35w^4 - 30w^2 + 3)$
$P_5(w)$	$\frac{1}{8}(63w^5 - 70w^3 + 15w)$
$P_6(w)$	$\frac{1}{16}(231w^6 - 315w^4 + 105w^2 - 5)$
$P_7(w)$	$\frac{1}{16}(429w^7 - 693w^5 + 315w^3 - 35w)$
$P_8(w)$	$\frac{1}{128}(6435w^8 - 12\,012w^6 + 6930w^4 - 1260w^2 + 35)$
$P_9(w)$	$\frac{1}{128}(12\,155w^9 - 25\,740w^7 + 18\,018w^5 - 4620w^3 + 315w)$
$P_{10}(w)$	$\frac{1}{256}(46\,189w^{10} - 109\,395w^8 + 90\,090w^6 - 30\,030w^4 + 3465w^2 - 63)$

Example 7.3. Show that choice (7.70) makes

$$P_l(1) = 1 \qquad \text{and} \qquad P_l(-1) = (-1)^l.$$

On carrying out the indicated differentiations, we find that

$$\frac{d}{dw}(1 - w^2) = -2w$$

and

$$\frac{d^2}{dw^2}(1 - w^2)^2 = -4 + 12w^2.$$

Evaluating these expressions at $w = 1$ yields

$$\frac{d}{dw}(1 - w^2)\bigg|_{w=1} = -2 = (-1)(2)(1)$$

and

$$\frac{d^2}{dw^2}(1 - w^2)^2\bigg|_{w=1} = -4 + 12 = 8 = (-1)^2(2)^2(2).$$

A possible generalization of these results is

$$\frac{d^l}{dw^l}(1 - w^2)^l\bigg|_{w=1} = (-1)^l 2^l l!$$

with l any integer. By differentiating, we know that

$$\frac{d^{l+1}}{dw^{l+1}}(1 - w^2)^{l+1} = \frac{d^l}{dw^l}(l + 1)(1 - w^2)^l(-2w)$$

$$= (-2w)(l + 1)\frac{d^l}{dw^l}(1 - w^2)^l +$$

$$+ l(-2)(l + 1)\frac{d^{l-1}}{dw^{l-1}}(1 - w^2)^l.$$

The last term yields a sum out of which $1 - w^2$ factors. Consequently, this term does not contribute at $w = 1$ and

$$\frac{d^{l+1}}{dw^{l+1}}(1 - w^2)^{l+1}\bigg|_{w=1} = (-2)(l + 1)\frac{d^l}{dw^l}(1 - w^2)^l\bigg|_{w=1}.$$

The assumed formula reduces this to

$$(-2)(l+1)(-1)^l 2^l l! = (-1)^{l+1} 2^{l+1} (l+1)!$$

Thus, if the formula is true for integer l, it is true for integer $l + 1$. But the formula is known to be correct for $l = 1$ and $l = 2$; hence it has to be correct for $l = 3$, and so on. By induction, the generalization is valid. Similarly,

$$\frac{d^l}{dw^l}(1 - w^2)^l \bigg|_{w=-1} = 2^l l!$$

With choice (7.70) for coefficient c, formula (7.66) now yields

$$P_l(1) = \frac{(-1)^l}{2^l l!}(-1)^l 2^l l! = 1$$

and

$$P_l(-1) = \frac{(-1)^l}{2^l l!} 2^l l! = (-1)^l.$$

7.6. Normalizing the Angular Eigenfunctions

The function Ψ describing the state of a particle is conventionally defined so that $\Psi^*\Psi$ equals its probability density ρ. Since the integral of this density over all available space is 1, the corresponding integrals of $\Psi^*\Psi$ and $\psi^*\psi$ are set equal to 1.

In central field calculations, when the radial factor is normalized as (4.113) indicates, the angular factor is also normalized to 1 and

$$\int_0^{2\pi}\int_0^{\pi} Y^*Y \sin\theta \, d\theta \, d\varphi = 1 \tag{7.72}$$

or

$$\int_0^{2\pi} \Phi^*\Phi \, d\varphi \int_0^{\pi} \Theta^*\Theta \sin\theta \, d\theta = 1. \tag{7.73}$$

Since the Φ in (7.39) yields

$$\int_0^{2\pi} e^{-im\varphi} e^{im\varphi} \, d\varphi = \int_0^{2\pi} d\varphi = 2\pi, \tag{7.74}$$

Equation (7.49) is replaced by

$$\Theta(\theta) = NP(w) \tag{7.75}$$

and normalization factor N is chosen to make the second integral in (7.73) equal $1/(2\pi)$. With w defined by (7.46), the θ integral becomes

$$\int_0^{\pi} \Theta^* \Theta \sin\theta \, d\theta = N^2 \int_{-1}^{1} [P_l^m(w)]^2 \, dw = \frac{1}{2\pi}. \tag{7.76}$$

When m is 0, P_l^m is given by (7.66) and (7.70). The θ integral can then be evaluated in the following manner:

$$\begin{aligned}
\int_{-1}^{1} [P_l(w)]^2 \, dw &= c^2 \int_{-1}^{1} \left[\frac{d^l}{dw^l}(1-w^2)^l\right] \frac{d^l}{dw^l}(1-w^2)^l \, dw \\
&= (-1)^l c^2 \int_{-1}^{1} \left[\frac{d^{2l}}{dw^{2l}}(1-w^2)^l\right] (1-w^2)^l \, dw \\
&= c^2 (2l)! \int_{-1}^{1} (1-w^2)^l \, dw \\
&= c^2 (2l)! \int_{-1}^{1} (1-w)^l (1+w)^l \, dw \\
&= c^2 (2l)! \frac{l(l-1)\ldots 1}{(l+1)(l+2)\ldots 2l} \int_{-1}^{1} (1+w)^{2l} \, dw \\
&= c^2 (l!)^2 \left.\frac{(1+w)^{2l+1}}{2l+1}\right|_{-1}^{1} = \frac{(l!)^2}{2^{2l}(l!)^2} \frac{2^{2l+1}}{2l+1} \\
&= \frac{2}{2l+1}.
\end{aligned} \tag{7.77}$$

The second step involves integration by parts l times, the third step $2l$ differentiations in the first factor of the integrand, the fourth step a factoring, the fifth step integration by parts l times, and the sixth step integration of variable $(1 + w)$ to the $2l$th power.

When m is not zero, introduction of formula (7.65), followed by integration by parts, leads to

$$\begin{aligned}
\int_{-1}^{1} [P_l^m(w)]^2 \, dw &= \int_{-1}^{1} (1-w^2)^m \frac{d^m P_l}{dw^m} \frac{d^m P_l}{dw^m} \, dw \\
&= -\int_{-1}^{1} \frac{d}{dw}\left[(1-w^2)^m \frac{d^m P_l}{dw^m}\right] \frac{d^{m-1} P_l}{dw^{m-1}} \, dw.
\end{aligned} \tag{7.78}$$

Differentiating (7.53) $l+m$ times and multiplying by $(1-w^2)^{m-1}$ produces

$$(1-w^2)^m \frac{\mathrm{d}^{m+1}P_l}{\mathrm{d}w^{m+1}} - 2mw(1-w^2)^{m-1}\frac{\mathrm{d}^m P_l}{\mathrm{d}w^m} +$$

$$+(l+m)(l-m+1)(1-w^2)^{m-1}\frac{\mathrm{d}^{m-1}P_l}{\mathrm{d}w^{m-1}} = 0, \tag{7.79}$$

whence

$$\frac{\mathrm{d}}{\mathrm{d}w}\left[(1-w^2)^m \frac{\mathrm{d}^m P_l}{\mathrm{d}w^m}\right] = -(l+m)(l-m+1)(1-w^2)^{m-1}\frac{\mathrm{d}^{m-1}P_l}{\mathrm{d}w^{m-1}}. \tag{7.80}$$

Formula (7.80) reduces (7.78) to

$$\int_{-1}^{1}[P_l^m(w)]^2\,\mathrm{d}w$$

$$= (l+m)(l-m+1)\int_{-1}^{1}(1-w^2)^{m-1}\left(\frac{\mathrm{d}^{m-1}P_l}{\mathrm{d}w^{m-1}}\right)^2\mathrm{d}w$$

$$= (l+m)(l-m+1)\int_{-1}^{1}[P_l^{m-1}(w)]^2\,\mathrm{d}w. \tag{7.81}$$

Iterating with this formula, introducing (7.77),

$$\int_{-1}^{1}[P_l^m(w)]^2\,\mathrm{d}w = \frac{(l+m)!}{(l-m)!}\int_{-1}^{1}[P_l(w)]^2\,\mathrm{d}w$$

$$= \frac{(l+m)!}{(l-m)!}\,\frac{2}{2l+1}, \tag{7.82}$$

and substituting the result into (7.76) yields

$$N^2 = \frac{2l+1}{4\pi}\,\frac{(l-m)!}{(l+m)!}. \tag{7.83}$$

In setting up (7.76), we chose N to be real. Let us further identity N by making it negative when the magnetic quantum number m is odd and positive, and positive otherwise. Formula (7.35) then yields

$$Y_l^m(\theta,\varphi) = \sqrt{\frac{2l+1}{4\pi}\,\frac{(l-m)!}{(l+m)!}}\,(-1)^m\,\mathrm{e}^{im\varphi}\,P_l^m(\cos\theta) \tag{7.84}$$

for m, in (7.39), greater or equal to 0 and

$$Y_l^m(\theta, \varphi) = \sqrt{\frac{2l+1}{4\pi} \frac{(l-|m|)!}{(l+|m|)!}}\, e^{im\varphi} P_l^{|m|}(\cos\theta) \tag{7.85}$$

for the magnetic quantum number $m < 0$. Explicit expressions for small quantum numbers appear in Table 7.2.

TABLE 7.2
Normalized spherical harmonics

Symbol	Formula
Y_0^0	$\sqrt{\frac{1}{4\pi}}$
Y_1^0	$\sqrt{\frac{3}{4\pi}} \cos\theta = \sqrt{\frac{3}{4\pi}} \frac{z}{r}$
$Y_1^{\pm 1}$	$\mp \sqrt{\frac{3}{8\pi}} e^{\pm i\varphi} \sin\theta = \mp \sqrt{\frac{3}{8\pi}} \frac{x \pm iy}{r}$
Y_2^0	$\sqrt{\frac{5}{16\pi}} (3\cos^2\theta - 1) = \sqrt{\frac{5}{16\pi}} \frac{3z^2 - r^2}{r^2}$
$Y_2^{\pm 1}$	$\mp \sqrt{\frac{15}{8\pi}} e^{\pm i\varphi} \cos\theta \sin\theta = \mp \sqrt{\frac{15}{8\pi}} \frac{(x \pm iy)z}{r^2}$
$Y_2^{\pm 2}$	$\sqrt{\frac{15}{32\pi}} e^{\pm 2i\varphi} \sin^2\theta = \sqrt{\frac{15}{32\pi}} \frac{(x \pm iy)^2}{r^2}$
Y_3^0	$\sqrt{\frac{7}{16\pi}} (5\cos^3\theta - 3\cos\theta) = \sqrt{\frac{7}{16\pi}} \frac{z(5z^2 - 3r^2)}{r^3}$
$Y_3^{\pm 1}$	$\mp \sqrt{\frac{21}{64\pi}} e^{\pm i\varphi} \sin\theta\, (5\cos^2\theta - 1) = \mp \sqrt{\frac{21}{64\pi}} \frac{(x \pm iy)(5z^2 - r^2)}{r^3}$
$Y_3^{\pm 2}$	$\sqrt{\frac{105}{32\pi}} e^{\pm 2i\varphi} \sin^2\theta \cos\theta = \sqrt{\frac{105}{32\pi}} \frac{(x \pm iy)^2 z}{r^3}$
$Y_3^{\pm 3}$	$\mp \sqrt{\frac{35}{64\pi}} e^{\pm 3i\varphi} \sin^3\theta = \mp \sqrt{\frac{35}{64\pi}} \frac{(x \pm iy)^3}{r^3}$

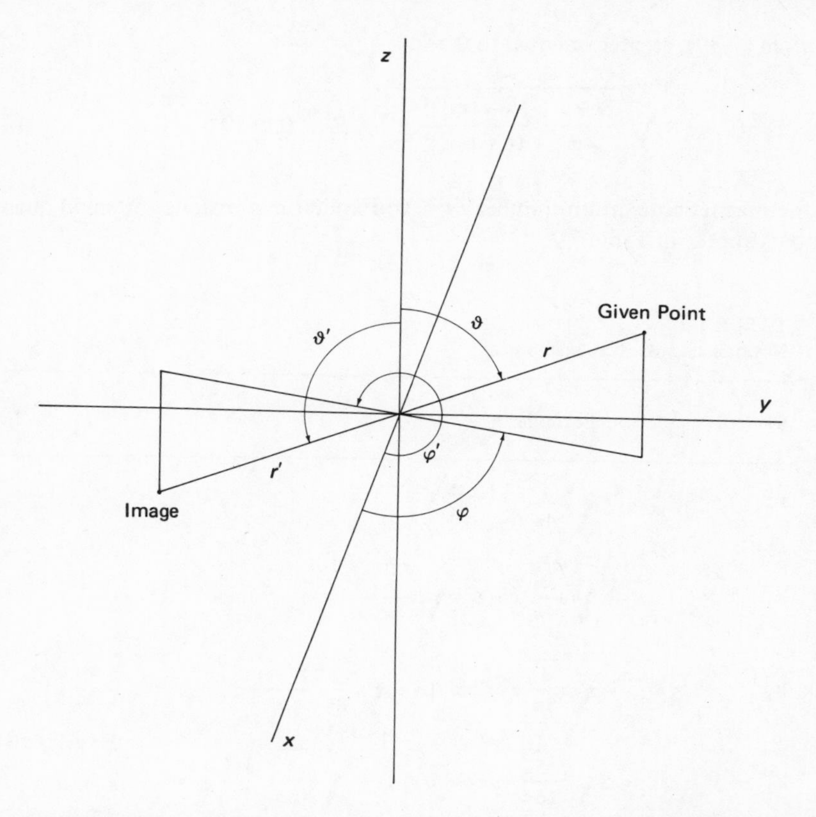

Fig. 7.1. Spherical coordinates of a point and the reflection of this point through the origin.

Example 7.4. How does inversion through the origin affect a spherical harmonic?

Reflection of a given point through an origin does not alter the distance of the point from the origin. But the reflection changes the angle that the radius vector makes with the z axis into its supplement and adds π to the angle between the projection of the radius vector on the xy plane and the x axis. From Figure 7.1, the initial coordinates are related to the final coordinates by the equations

$$r = r', \qquad \theta = \pi - \theta', \qquad \varphi = \varphi' - \pi.$$

A reorientation of a function produces the same expression as the inverse operation acting on the coordinates. But the inverse of a reflection is the same reflection. Therefore, reflecting (7.39) produces the expression

$$\Phi(\varphi') = e^{im\varphi'} = e^{im(\pi + \varphi)} = e^{im\pi}\, e^{im\varphi} = (-1)^m\, e^{im\varphi}$$

$$= (-1)^m\, \Phi(\varphi).$$

Furthermore, since

$$w = \cos\theta,$$

the transformed w equals negative w,

$$w' = \cos\theta' = \cos(\pi - \theta) = -\cos\theta = -w,$$

and the transformed associated Legendre function of (7.65) has the form

$$P_l^m(w') = c(1 - w'^2)^{m/2} \frac{\mathrm{d}^{l+m}}{\mathrm{d}w'^{l+m}} (1 - w'^2)^l$$

$$= c(1 - w^2)^{m/2} \frac{\mathrm{d}^{l+m}}{(-1)^{l+m}\,\mathrm{d}w^{l+m}} (1 - w^2)^l = (-1)^{l+m} P_l^m(w).$$

From (7.35) and (7.75) the transformed spherical harmonic is

$$(Y_l^m)' = NP_l^m(w')\,\Phi(\varphi'),$$

while the original spherical harmonic is

$$Y_l^m = NP_l^m(w)\,\Phi(\varphi).$$

The relationships we have obtained reduce the transformed expression as follows:

$$(Y_l^m)' = N(-1)^{l+m} P_l^m(w)(-1)^m\, \mathrm{e}^{im\varphi} = (-1)^l\, Y_l^m.$$

We see that reflecting the spherical harmonic through the origin has the same effect as multiplying the harmonic by $(-1)^l$. So when l is even, the eigenfunction has even parity; when l is odd, the eigenfunction has odd parity.

Example 7.5. Explain why reorientation of a function produces the same effect as the corresponding inverse reorientation of the coordinates on which the function depends.

Consider a typical reorientation R that changes radius vector $\mathbf{r}$ to radius vector $\mathbf{r}'$ and function $\psi(\mathbf{r})$ to function $\psi'(\mathbf{r})$:

$$R\mathbf{r} = \mathbf{r}',$$

$$R\psi(\mathbf{r}) = \psi'(\mathbf{r}).$$

When both the coordinates and the function are subjected to this reorientation, the mathematical expression is not altered:

$$R\psi(R\mathbf{r}) = \psi(\mathbf{r}).$$

Applying the inverse operation to the radius vector in parentheses on each side yields

$$R\psi(R^{-1}R\mathbf{r}) = \psi(R^{-1}\mathbf{r}),$$

whence

$$R\psi(\mathbf{r}) = \psi(R^{-1}\mathbf{r}).$$

7.7. Mutual Orthogonality of the Angular Eigenfunctions

The spherical harmonics we have obtained describe completely independent physical states. As a consequence, we expect them to be orthogonal to each other. If the functions are also normalized, we would have

$$\int_0^{2\pi}\int_0^{\pi} Y_{l_1}^{m_1\,*}\, Y_{l_2}^{m_2} \sin\theta \, \mathrm{d}\theta \, \mathrm{d}\varphi = \delta_{m_1 m_2}\, \delta_{l_1 l_2}. \tag{7.86}$$

From (7.35), (7.39), and (7.75), an angular eigenfunction for definite l and m is

$$Y_l^m = \mathrm{e}^{im\varphi} N P_l^m. \tag{7.87}$$

When N is chosen to satisfy (7.83), the Y_l^m are normalized to 1 and formula (7.86) holds for $m_1 = m_2$, $l_1 = l_2$.

When $m_1 \neq m_2$, the φ factor yields

$$\int_0^{2\pi} \Phi_1^* \Phi_2 \, \mathrm{d}\varphi = \int \mathrm{e}^{-im_1\varphi}\, \mathrm{e}^{im_2\varphi} \, \mathrm{d}\varphi = \int \mathrm{e}^{i(m_2 - m_1)\varphi} \, \mathrm{d}\varphi$$
$$= \left. \frac{\mathrm{e}^{i(m_2 - m_1)\varphi}}{i(m_2 - m_1)} \right|_0^{2\pi} = 0. \tag{7.88}$$

The double integral in (7.86) is then zero, as the formula states.

The Legendre function in the θ factor satisfies (7.64). When l is l_1 and when l is l_2, we have

$$\frac{\mathrm{d}}{\mathrm{d}w}\left[(1 - w^2)\frac{\mathrm{d}P_{l_1}^m}{\mathrm{d}w}\right] + \left[l_1(l_1 + 1) - \frac{m^2}{1 - w^2}\right] P_{l_1}^m = 0 \tag{7.89}$$

and

$$\frac{\mathrm{d}}{\mathrm{d}w}\left[(1 - w^2)\frac{\mathrm{d}P_{l_2}^m}{\mathrm{d}w}\right] + \left[l_2(l_2 + 1) - \frac{m^2}{1 - w^2}\right] P_{l_2}^m = 0. \tag{7.90}$$

Multiplying (7.89) by $P_{l_2}^m$, (7.90) by $P_{l_1}^m$, and subtracting yields

$$\frac{d}{dw}\left[(1-w^2)\left(P_{l_2}^m \frac{dP_{l_1}^m}{dw} - P_{l_1}^m \frac{dP_{l_2}^m}{dw}\right)\right] +$$

$$+ [l_1(l_1+1) - l_2(l_2+1)]\, P_{l_1}^m P_{l_2}^m = 0. \tag{7.91}$$

Let us integrate the differential form of Equation (7.91) from $w = -1$ to $w = 1$. The first term in the resulting expression vanishes because of factor $(1 - w^2)$. The quantity in brackets in the second term can then be canceled, leaving

$$\int_{-1}^{1} P_{l_1}^m(w)\, P_{l_2}^m(w)\, dw = 0. \tag{7.92}$$

Therefore,

$$\int_0^{\pi} \Theta_{l_1}^m(\theta)\, \Theta_{l_2}^m(\theta) \sin\theta\, d\theta = 0. \tag{7.93}$$

This relationship makes integral (7.86) zero whenever $m_1 = m_2$ and $l_1 \neq l_2$. Formula (7.86) is thus valid in general.

Note that Legendre functions with different m's are usually not orthogonal. However, the spherical harmonics with different m's are orthogonal because the φ factors are orthogonal.

7.8. Spherical-Harmonic Analysis of a Pure Planar Wave

A particle that moves freely along a straight line a given distance from a given point possesses a certain angular momentum with respect to the point. A beam of particles moving freely along parallel straight lines contains a distribution of angular momenta. Such a beam can be described as a superposition of partial waves having various angular momentum quantum numbers with respect to the given point.

Let us place the origin of a Cartesian system at the point and orient the axes so the z axis is parallel to the direction of motion as Figure 7.2 shows. Let us assume that the beam is homogeneous and uniform over the significant cross section. The argument leading to (1.13) then applies with wavevector $\mathbf{k}$ possessing only a z component. When the amplitude is taken to be 1, we have

$$\psi = e^{ikz} = e^{ikr\cos\theta} = e^{ikrw}, \tag{7.94}$$

where

$$w = \cos\theta \tag{7.95}$$

as before.

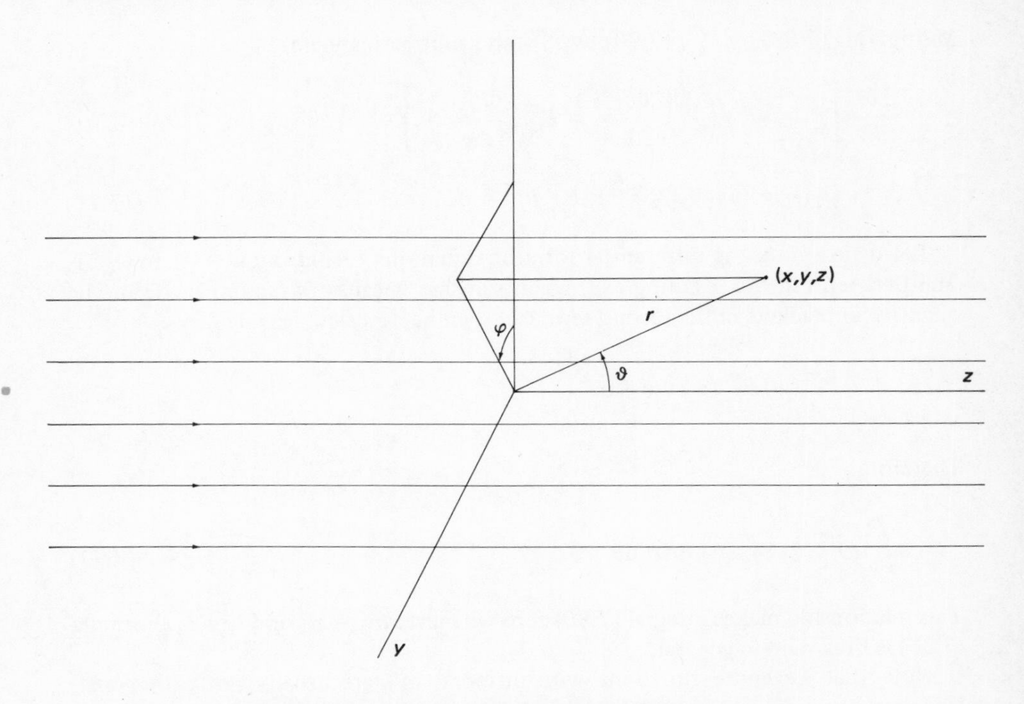

Fig. 7.2. Coordinates for a point in the homogeneous beam.

Now, the degree of Legendre polynomial

$$P_l(\cos\theta) = P_l(w) \tag{7.96}$$

varies by integers from 0 to ∞ as l increases by unit steps from 0. As a consequence, a sum of such polynomials is as flexible as a power series. Exponential (7.94) can therefore be represented as a series of such polynomials. We write

$$e^{ikz} = e^{ikr\cos\theta} = \Sigma\, a_m j_m(kr) P_m(\cos\theta) \tag{7.97}$$

or

$$\Sigma\, a_m j_m(kr) P_m(w) = e^{ikrw}. \tag{7.98}$$

Let us multiply both sides of (7.98) by $P_l(w)$ and integrate over the range of w:

$$\Sigma\, a_m j_m(kr) \int_{-1}^{1} P_l(w) P_m(w)\, dw = \int_{-1}^{1} e^{ikrw} P_l(w)\, dw. \tag{7.99}$$

According to (7.92), the integral on the left is zero except when lower index m

is l. Then (7.77) tells us that the integral equals $2/(2l+1)$. Multiplying through by $(2l+1)/2$ therefore yields

$$a_l j_l(kr) = \frac{2l+1}{2} \int_{-1}^{1} e^{ikrw} P_l(w)\, dw. \tag{7.100}$$

We next define

$$a_l = (2l+1)\, i^l \tag{7.101}$$

and set

$$kr = z, \tag{7.102}$$

for simplicity. Note that this z is not a Cartesian coordinate. Then

$$j_l(z) = \frac{1}{2i^l} \int_{-1}^{1} e^{izw} P_l(w)\, dw. \tag{7.103}$$

Expression $j_l(z)$ is called the *spherical Bessel function* of order l.

A descending power series for this function is obtained by successive partial integrations. Implementing the first term and neglecting the higher terms coming from the remaining integral leads to the asymptotic representation

$$\begin{aligned} j_l(z) &= \frac{1}{2i^l}\left[\frac{e^{izw}}{iz} P_l(w) \Big|_{-1}^{1} - \int_{-1}^{1} \frac{e^{izw}}{iz} \frac{d}{dw} P_l(w)\, dw \right] \\ &\simeq \frac{1}{2i^l} \frac{1}{iz} [e^{iz} - (-1)^l e^{-iz}] = \frac{e^{-il\pi/2} (e^{iz} - e^{il\pi} e^{-iz})}{2iz} \\ &= \frac{\sin\left(z - \frac{1}{2} l\pi\right)}{z}, \end{aligned} \tag{7.104}$$

valid at large z.

7.9. Schrödinger Equation for Radial Constituents of Free Motion

Each term of sum (7.98) for a homogeneous beam separates into a radial factor and an angular factor. The Schrödinger equation governing the beam can be similarly separated and a differential equation for the spherical Bessel function constructed.

For particles in free space, wavevector k in Equation (4.47) is constant. The argument in Section 4.5 is then applicable and equations (4.63), (4.53) yield

$$-\frac{1}{r^2}\frac{d}{dr}\left(r^2\frac{dR}{dr}\right)+\frac{l(l+1)}{r^2}R=k^2R \tag{7.105}$$

or

$$-\frac{1}{(kr)^2}\frac{d}{d(kr)}\left[(kr)^2\frac{dR}{d(kr)}\right]+\frac{l(l+1)}{(kr)^2}R=R. \tag{7.106}$$

Setting

$$kr=z \tag{7.107}$$

as in (7.102), we have

$$-\frac{1}{z^2}\frac{d}{dz}\left(z^2\frac{dR}{dz}\right)+\frac{l(l+1)}{z^2}R=R. \tag{7.108}$$

Let us represent the solution of (7.108) as an integral transform of a function f to be determined:

$$R=z^\lambda\int e^{izw}f(w)\,dw. \tag{7.109}$$

Substituting (7.109) into (7.108) and carrying out the indicated operations leads to the relationship

$$z^{\lambda-2}[-\lambda(\lambda+1)+l(l+1)]\int e^{izw}f(w)\,dw -$$

$$-2(\lambda+1)z^{\lambda-1}\int iw\,e^{izw}f(w)\,dw+z^\lambda\int w^2\,e^{izw}f(w)\,dw$$

$$=z^\lambda\int e^{izw}f(w)\,dw. \tag{7.110}$$

To make the coefficient of $z^{\lambda-2}$ vanish, we set

$$\lambda=l. \tag{7.111}$$

Multiplying through by $-i$, writing $iz\,e^{izw}$ as $(d/dw)\,e^{izw}$, and canceling $z^{\lambda-1}$ then yields

$$\int f(w)\left[-2(l+1)w+(1-w^2)\frac{d}{dw}\right]e^{izw}\,dw=0. \tag{7.112}$$

Let us integrate the second term by parts and impose limits that make the integrated part vanish:

$$\int_{-1}^{1} \left\{ f(w)\,[-2(l+1)w] - \frac{\mathrm{d}}{\mathrm{d}w}\,[f(w)\,(1-w^2)] \right\} \mathrm{e}^{izw}\,\mathrm{d}w + + f(w)\,(1-w^2)\,\mathrm{e}^{izw} \Big|_{-1}^{1} = 0. \tag{7.113}$$

We are left with the equation

$$-\frac{\mathrm{d}}{\mathrm{d}w}\,[(1-w^2)\,f(w)] = 2(l+1)w\,f(w) \tag{7.114}$$

that is satisfied when

$$f(w) = c(1-w^2)^l. \tag{7.115}$$

Solution (7.109) then becomes

$$R = cz^l \int_{-1}^{1} \mathrm{e}^{izw}\,(1-w^2)^l\,\mathrm{d}w. \tag{7.116}$$

Now, Rodrigues' formula for Legendre's polynomial converts (7.103) to

$$j_l(z) = \frac{(-1)^l}{2^l\,l!}\,\frac{1}{2i^l} \int_{-1}^{1} \mathrm{e}^{izw}\,\frac{\mathrm{d}^l}{\mathrm{d}w^l}\,(1-w^2)^l\,\mathrm{d}w. \tag{7.117}$$

If we integrate this expression by parts l times,

$$j_l(z) = \frac{(-1)^l}{2^l\,l!}\,\frac{1}{2i^l}\,(-1)^l\,(iz)^l \int_{-1}^{1} \mathrm{e}^{izw}\,(1-w^2)^l\,\mathrm{d}w = \frac{z^l}{2^{l+1}\,l!} \int_{-1}^{1} \mathrm{e}^{izw}\,(1-w^2)^l\,\mathrm{d}w, \tag{7.118}$$

we see that the solution we have constructed of (7.109) is a constant times the spherical Bessel function. Consequently, differential equation (7.108) describes how the spherical Bessel function changes throughout its range.

Discussion Questions

7.1. How is it that different linear molecules and hydrogen-like atoms exhibit the same angular variations in their wave functions?

7.2. What conditions govern factor $Y(\theta, \varphi)$ in the Ψ for these two-dimensional free rotators? Why is Y independent of the formula for wavevector k?

7.3. How can Y appear as a homogeneous polynomial in x, y, z divided by r to a power equal to the degree of the polynomial? What conditions must the coefficients in the polynomial satisfy?

7.4. To what Schrödinger equation are

$$N \frac{1}{r}$$

and

$$N \frac{\partial^l}{\partial x^{(1)}\, \partial x^{(2)} \ldots \partial x^{(l)}} \frac{1}{r}$$

solutions?

7.5. Explain why

$$N r^{l+1} \frac{\partial^l}{\partial x^{(1)}\, \partial x^{(2)} \ldots \partial x^{(l)}} \frac{1}{r}$$

is the angular factor in each definite-l solution to the Schrödinger equation for a central field.

7.6. What operations on $1/r$ yield the standard real d angular factors?

7.7. Explain how θ and φ variables are separated in the Schrödinger equation for Y.

7.8. Explain which solutions of the differential equation for Φ are acceptable.

7.9. From the appropriate exponential function derive the differential equation for $\Phi(\varphi)$.

7.10. How do we handle the transcendental functions in the differential equation for Θ?

7.11. With what symmetry must the physically significant solutions of the differential equation for Θ be consistent? How do we seek such solutions?

7.12. Why are the limits on the magnetic quantum number $-l$ and l?

7.13. Explain how the integral of the square of $P_l^m(w)$ is obtained. How are spherical harmonics normalized?

7.14. Explain why reorientation of a function is equivalent to the inverse operation acting on the coordinates.

7.15. To what do the spherical coordinates of a point transform when the point is reflected through the origin?

7.16. How does inversion affect a spherical harmonic? What parity does a spherical harmonic possess?

7.17. Why are functions for completely independent states orthogonal?

7.18. Explain what (a) Φ factors, (b) Θ factors are orthogonal to each other.

7.19. Explain how a homogeneous beam is made up of particles with various angular momenta with respect to a given point in its path.

7.20. How can a planar wave be broken down into angular and radial factors with respect to an arbitrary point?

7.21. Define the spherical Bessel function. Determine the asymptotic behavior of this function.

7.22. How can the spherical Bessel function be involved in describing a pure planar wave?

7.23. How is the Schrödinger equation for translatory motion separated into a radial equation and an angular equation with respect to a given point?

7.24. How do we construct an integral transform suitable for solving a linear ordinary differential equation?

7.25. Develop a suitable integral transform and use it to solve the radial equation for translation.

7.26. What differential equation defines a spherical Bessel function? Explain.

Problems

7.1. What relationships must exist among the coefficients for

$$ax^2 + by^2 + cz^2 + dxy + eyz + fzx$$

to be a harmonic function? Show that the homogeneous polynomial factor in each standard rotational eigenfunction for $l = 2$ satisfies these conditions.

7.2. Determine the harmonic represented by

$$Nr^3 \frac{\partial^2}{\partial x\, \partial y} \frac{1}{r}.$$

7.3. Express $Y_{d_{3z^2-r^2}}$ as a constant times a power of r times a derivative of $1/r$.

7.4. Employ properties of $(1 - w^2)^l$ in showing that $P_l(w)$ has l distinct real roots.

7.5. Find the multiplier that rotates Y_l^m by angle α around the z axis.

7.6. Determine how expression (7.33) behaves on inversion through the origin.

7.7. Show that the square root of two times (a) the real part, (b) the imaginary part, of Y_l^m is an eigenfunction of the Schrödinger equation for the particle in a central field.

7.8. Show how the conventional quantum states in Table 7.2 superpose to form the states described by

$$\sqrt{\frac{21}{32\pi}}\,\frac{x(5z^2 - r^2)}{r^3},\quad \sqrt{\frac{21}{32\pi}}\,\frac{y(5z^2 - r^2)}{r^3},\quad \sqrt{\frac{105}{16\pi}}\,\frac{z(x^2 - y^2)}{r^3},$$

$$\sqrt{\frac{105}{4\pi}}\,\frac{xyz}{r^3},\quad \sqrt{\frac{35}{32\pi}}\,\frac{x(x^2 - 3y^2)}{r^3},\quad \sqrt{\frac{35}{32\pi}}\,\frac{y(y^2 - 3x^2)}{r^3}.$$

7.9. What superposition of Y_1^0 and Y_1^1 yields the same probability distribution over θ as Y_0^0?

7.10. By substitution, show that

$$A\,\frac{\sin\left(z - \frac{1}{2}l\pi\right)}{z} \quad \text{and} \quad B\,\frac{\cos\left(z - \frac{1}{2}l\pi\right)}{z}$$

are asymptotic solutions of (7.108), valid at large z.

7.11. What relationships must exist among the coefficient for

$$ax^3 + by^3 + cz^3 + dx^2y + ex^2z + fy^2z + gy^2x + hz^2x + jz^2y + kxyz$$

to be a harmonic function? Show that the homogeneous polynomial factor in each Y_3^m meets these conditions.

7.12. If primed Cartesian axes are obtained from the unprimed Cartesian axes by a 45° clockwise rotation about the z axis, what spherical harmonic is

$$Nr^3 \frac{\partial^2}{\partial x' \, \partial y'} \frac{1}{r}?$$

7.13. Determine $Y_{f_{xyz}}$ as a constant times a power of r times a derivative of $1/r$.

7.14. Use Rodrigues. formula to calculate the colatitude factor for the state existing when $l = 2$ and $m = 1$. Then form the angular wave function Y_2^1 for the state.

7.15. Rotate the spherical harmonic $N(yz/r^2)$ 45° counterclockwise around the x axis.

7.16. Determine how homogeneous polynomial (7.8) and the corresponding Y behave on inversion through the origin.

7.17. Show how the real f orbitals, Y_3^0 and the forms listed in Problem 7.8, superpose to form

$$\sqrt{\frac{7}{16\pi}} \frac{x(5x^2 - 3r^2)}{r^3}.$$

7.18. Superpose the real f orbitals to form

$$\sqrt{\frac{105}{16\pi}} \frac{x(y^2 - z^2)}{r^3}.$$

7.19. What probability density distribution over angles does (a) a p_z orbital, (b) a p_x orbital, (c) a p_y orbital yield? What angular distribution arises from placing one electron in a p_x orbital, one electron in a p_y orbital, and one electron in a p_z orbital?

7.20. Show that

$$\frac{(-1)^l(-1)^m}{2^l l!} \frac{(l+m)!}{(l-m)!} (1 - w^2)^{-m/2} \frac{\mathrm{d}^{l-m}}{\mathrm{d}w^{l-m}} (1 - w^2)^l$$

behaves as a $P_l^m(w)$ in (7.82) and in (7.92). Therefore, this expression also represents $P_l^m(w)$.

References

Books

Arfken, G.: 1970, *Mathematical Methods for Physicists*, 2nd edn, Academic Press, New York, pp. 534–608.

Eyring, H., Walter, J., and Kimball, G. E.: 1944, *Quantum Chemistry*, Wiley, New York, pp. 48–91.

Schiff, L. I.: 1968, *Quantum Mechanics*, 3rd edn, McGraw-Hill, New York, pp. 66–69.

Ziock, K.: 1969, *Basic Quantum Mechanics*, Wiley, New York, pp. 73–103.

Articles

Bordass, W. T., and Linnett, J. W.: 1970, 'A New Way of Presenting Atomic Orbitals', *J. Chem. Educ.* **47**, 672–675.
Bragg, L. E.: 1970, 'Legendre's Equation for Undergraduates', *Am. J. Phys.* **38**, 641–643.
Essen, H.: 1978, 'Quantization and Independent Coordinates', *Am. J. Phys.* **46**, 983–988.
Ley-Koo, E.: 1972, 'On the Expansion of a Plane Wave in Spherical Waves', *Am. J. Phys.* **40**, 1538–1539.
Miyakawa, K.: 1969, 'Legendre's Polynomials in Undergraduate Courses', *Am. J. Phys.* **37**, 924–925.
Ramamurti, G., Ranganathan, K., and Ganesan, L. R.: 1972, 'Solutions of Legendre's Equation – Simple Proof', *Am. J. Phys.* **40**, 913.
Whippman, M. L.: 1966, 'Orbital Angular Momentum in Quantum Mechanics', *Am. J. Phys.* **34**, 656–659.

Chapter 8

Operators for Angular Momentum and Spin

8.1. Rotational States, Spin States, and Suitable Composites

When the eigenstates for a mode, or modes, of motion in a given system form a simple sequence or ladder, these states may be interconverted by applying an operator that moves the system up or down the ladder. In Section 5.7, such operators were constructed for the states of a harmonic oscillator; in Section 5.11, step-up and step-down operators were formulated for the radial motion of a particle in a Coulombic field.

The rotational states of a molecule with given energy, the orbital states of an electron with given principal and azimuthal quantum numbers, and the analogous nuclear states also form simple sequences. So we are led to develop operators that shift such systems up or down the pertinent ladders. Properties of the basic rotational operators will be obtained from the effects of small changes in angle on possible wave functions. Combinations of the basic operators that shift the magnetic quantum number will then be deduced.

An elementary particle exhibits a mass, a charge, and a magnetic moment. Orientation of the moment behaves as orientation of the orbital angular momentum of the model particle of mass μ. So we will assume that analogous operators govern the moment's orientation and that a *spin* similar to a molecule's rotation is present. But since this spin does not involve movement through space, the restriction to integral units of angular momentum does not apply; half integral units are also allowed. However, other fractions do not appear, since they would violate the overall up-down symmetry which the particle can exhibit.

Common are composite systems of particles presenting a net spin and angular momentum. The shift operators can be applied to the resultant states. Furthermore, each such state may be obtained by assigning a spin and an angular momentum state to each constituent particle in a number of ways. The shift operators can be applied to any one of these.

Following the law for combining independent motions, the functions for each assignment are combined multiplicatively. Each product is multiplied by a coefficient and added to the others to yield the function for the composite system. Applying the shift operators to such a sum leads to conditions on the coefficients.

8.2. The Azimuthal-Angle Operator Governing Angular Momentum

The basic operator for angular momentum assumes its simplest, most explicable form when it is expressed in terms of the angle of rotation about the pertinent axis. So let us review the construction of this form first.

A model for rotatory behavior is the particle of mass μ traveling with a given energy in a given circulatory direction within a binding field. The axis for such motion generally passes through the center of the field. Let us call this axis the z axis and label the distance from the center to the particle r, the colatitude angle for the position of the particle θ, and the azimuthal angle φ, as Figure 2.2 shows.

For free rotation about the center, the potential is a function of r alone. In a given energy state, with a definite angular momentum, all points at a fixed r and θ are then equivalent. By symmetry, multiplying $\mathrm{d}\varphi$ by a factor multiplies the effect on the differential wave function $\mathrm{d}\Psi$ by the factor. Also in the pure state, each equivalent part of Ψ itself produces the same effect on $\mathrm{d}\Psi$. Letting the angular constituent of Ψ be Y, as before, we have

$$\mathrm{d}Y = \frac{\partial Y}{\partial \theta}\,\mathrm{d}\theta + iMY\,\mathrm{d}\varphi. \tag{8.1}$$

Constant M is considered to be real so that the rotation in one direction about the axis proceeds with a fixed probability density.

We assume that Y is a smoothly varying function, satisfying Equation (2.12) in the form

$$\mathrm{d}Y = \frac{\partial Y}{\partial \theta}\,\mathrm{d}\theta + \frac{\partial Y}{\partial \varphi}\,\mathrm{d}\varphi. \tag{8.2}$$

On comparing (8.1) and (8.2), we see that

$$\frac{\partial}{\partial \varphi} Y = iMY, \tag{8.3}$$

whence

$$\frac{\hbar}{i}\frac{\partial}{\partial \varphi} Y = M\hbar Y. \tag{8.4}$$

From (2.26), $M\hbar$ equals the angular momentum around the axis of rotation. As a consequence, the differentiating operator on the left of (8.4) is called the *angular momentum operator*

$$J_z = \frac{\hbar}{i}\frac{\partial}{\partial \varphi} \tag{8.5}$$

and Equation (8.4) is rewritten in the form

$$J_z Y = M\hbar Y. \tag{8.6}$$

Introducing a variation of V with θ does not alter the symmetry with respect to φ and so does not alter the argument. When V varies with φ because of the presence of other particles in the central field, we may suppose that the effect is to mix the rotational configuration with other rotational configurations consistent with the states in which the other particles move through. Equation (8.5) is presumed to apply to each of the constituent rotational states.

Thus, we consider (8.5) to be general. When the axis for the rotation is not the z axis, we may label it z' and write

$$J_{z'} = \frac{\hbar}{i} \frac{\partial}{\partial \varphi'}, \tag{8.7}$$

where φ' is the azimuthal angle around the new axis. In a composite system, the axis for each particle's angular momentum varies with time; at any given instant, we must employ the pertinent axis as the z' axis in (8.7).

The axis for a particle's spin is still different. If we put two primes on the corresponding coordinates, the basic operator is

$$J_{z''} \tag{8.8}$$

with algebraic properties similar to those of J_z.

There is, however, an additional freedom. Because the orientation motion of a particle does not involve displacement of parts in space, the argument for integral units of $\hbar$ does not apply. Half integral units in $\hbar$ of spin are allowed. Since, by symmetry, the maximum negative angular momentum must have the same absolute value as the maximum positive angular momentum, other fractional units do not appear. (Recall the argument in Section 2.5.)

The angular momenta associated with the orbital motions in general combine with the angular momenta associated with the spins to yield a total angular momentum about a z axis fixed in an inertial frame. The basic operator for this total presumably also has the properties to be derived.

8.3. Cartesian Angular-Momentum Operators

A change in variables transforms the azimuthal-angle operator for angular momentum around the z axis to rectangular coordinates. Rotating these coordinates then yields operators for angular momentum around the x axis and the y axis.

From Figure 2.2 and the definitions of the sine and cosine, we obtain the transformation equations

$$x = r \sin \theta \cos \varphi, \tag{8.9}$$

$$y = r \sin \theta \sin \varphi, \tag{8.10}$$

$$z = r \cos \theta. \tag{8.11}$$

With the chain rule

$$\frac{\partial}{\partial\varphi} = \frac{\partial x}{\partial\varphi}\frac{\partial}{\partial x} + \frac{\partial y}{\partial\varphi}\frac{\partial}{\partial y} + \frac{\partial z}{\partial\varphi}\frac{\partial}{\partial z}, \tag{8.12}$$

we then find that

$$\begin{aligned}\frac{\partial}{\partial\varphi} &= -r\sin\theta\sin\varphi\frac{\partial}{\partial x} + r\sin\theta\cos\varphi\frac{\partial}{\partial y}\\ &= -y\frac{\partial}{\partial x} + x\frac{\partial}{\partial y}.\end{aligned} \tag{8.13}$$

Using (8.13) to eliminate $\partial/\partial\varphi$ from (8.5) leads to

$$J_z = \frac{\hbar}{i}\left(x\frac{\partial}{\partial y} - y\frac{\partial}{\partial x}\right). \tag{8.14}$$

Rotating the coordinate system by replacing x with y, y with z, z with x yields

$$J_x = \frac{\hbar}{i}\left(y\frac{\partial}{\partial z} - z\frac{\partial}{\partial y}\right), \tag{8.15}$$

while replacing x with z, y with x, z with y gives us

$$J_y = \frac{\hbar}{i}\left(z\frac{\partial}{\partial x} - x\frac{\partial}{\partial z}\right). \tag{8.16}$$

We thus obtain expressions for J_x, J_y, and J_z which can be applied simultaneously, with a given orientation of the axes.

8.4. Expressions Governing Rotational Standing Waves

Because the wave function for rotation is required to be smoothly varying in θ as well as with φ, pure unidirectional circulation around a rotational axis must be accompanied by some wobbling motion through the equatorial plane. This additional motion is represented by a standing wave.

Since such a wave consists of two oppositely traveling waves, it is not the eigenfunction of an angular momentum operator. Indeed when Y_+ is the eigenfunction of J_z with the eigenvalue $|M|\hbar$ and Y_- is the oppositely traveling wave, we have

$$J_z Y_+ = |M|\hbar\, Y_+, \tag{8.17}$$

$$J_z Y_- = -|M|\hbar\, Y_-, \tag{8.18}$$

and

$$J_z(Y_+ + Y_-) = |M|\hbar\, Y_+ - |M|\hbar\, Y_-. \tag{8.19}$$

But applying the operator J_z again yields

$$J_z^2(Y_+ + Y_-) = |M|\hbar J_z Y_+ - |M|\hbar J_z Y_- = (M\hbar)^2 Y_+ + (M\hbar)^2 Y_-$$

$$= M^2\hbar^2(Y_+ + Y_-). \tag{8.20}$$

The standing wave $Y_+ + Y_-$ is an eigenfunction of J_z^2 with the square of the angular momentum around the z axis as the eigenvalue.

Consequently, we consider the standing wave that accompanies a traveling wave around the z axis to be the eigenfunction of the corresponding contribution to the square of angular momentum; that is, of

$$J_x^2 + J_y^2. \tag{8.21}$$

The complete eigenfunction for rotation is the product of the independent constituent functions and is here represented by Y.

Since operator

$$J_z = \frac{\hbar}{i}\frac{\partial}{\partial\varphi} \tag{8.22}$$

cannot change the dependence of Y on θ, $J_z Y$ is the same eigenfunction of $J_x^2 + J_y^2$ as Y is. Operation by J_z yields a function with the same quantum numbers M and J, if J is the additional number needed to determine the overall state. Since they affect the overall state in a similar way, then neither J_x nor J_y can change the quantum number J. This number will be defined further later.

Example 8.1. By operator methods, obtain the eigenfunctions of J_z that contribute to $xR(r)$.

Let operator (8.5) act on the given function once,

$$J_z xR(r) = \frac{\hbar}{i}\frac{\partial}{\partial\varphi}(r\sin\theta\cos\varphi)R(r) = \frac{\hbar}{i}(-r\sin\theta\sin\varphi)R$$

$$= \frac{\hbar}{i}(-y)R = \hbar i y R,$$

and again,

$$J_z^2 xR(r) = \frac{\hbar}{i}\frac{\partial}{\partial\varphi}(\hbar i r\sin\theta\sin\varphi R) = \hbar^2 r\sin\theta\cos\varphi R = \hbar^2 xR(r).$$

The standing wave is an eigenfunction of J_z^2 with the eigenvalue $\hbar^2$.

On comparing the result with (8.20), we see that here $|M| = 1$ and that $Y_+ + Y_-$ is proportional to $xR(r)$. To preserve normalization, we let the proportionality constant be $\sqrt{2}$:

$$Y_+ + Y_- = \sqrt{2}\, xR(r).$$

On comparing the first equation in this example with (8.19), we also see that

$$Y_+ - Y_- = \sqrt{2}\, iyR(r).$$

Consequently,

$$Y_+ = \frac{\sqrt{2}}{2}(x + iy)R(r) = \frac{1}{\sqrt{2}} \sin\theta\, e^{i\varphi}\, rR(r)$$

and

$$Y_- = \frac{\sqrt{2}}{2}(x - iy)R(r) = \frac{1}{\sqrt{2}} \sin\theta\, e^{-i\varphi}\, rR(r).$$

8.5. Angular Momentum Shift Operators

How the states for a given J are related can be obtained from properties of operators that shift the M of a wave traveling about the z axis by ± 1. We will first construct useful commutators, then derive the pertinent expressions and properties.

From (8.15) and (8.16), differentiating operator J_y followed by differentiating operator J_x has the effect of

$$J_x J_y = \frac{\hbar}{i}\left(y\frac{\partial}{\partial z} - z\frac{\partial}{\partial y}\right)\frac{\hbar}{i}\left(z\frac{\partial}{\partial x} - x\frac{\partial}{\partial z}\right)$$

$$= -\hbar^2\left(yz\frac{\partial^2}{\partial z\,\partial x} + y\frac{\partial}{\partial x} - xy\frac{\partial^2}{\partial z^2} - z^2\frac{\partial^2}{\partial x\,\partial y} + zx\frac{\partial^2}{\partial y\,\partial z}\right), \tag{8.23}$$

while J_x followed by J_y acts as

$$J_y J_x = \frac{\hbar}{i}\left(z\frac{\partial}{\partial x} - x\frac{\partial}{\partial z}\right)\frac{\hbar}{i}\left(y\frac{\partial}{\partial z} - z\frac{\partial}{\partial y}\right)$$

$$= -\hbar^2\left(yz\frac{\partial^2}{\partial z\,\partial x} - z^2\frac{\partial^2}{\partial x\,\partial y} - xy\frac{\partial^2}{\partial z^2} + zx\frac{\partial^2}{\partial y\,\partial z} + x\frac{\partial}{\partial y}\right). \tag{8.24}$$

Subtracting (8.24) from (8.23) yields

$$J_x J_y - J_y J_x = \hbar^2\left(x\frac{\partial}{\partial y} - y\frac{\partial}{\partial x}\right) = i\hbar J_z, \tag{8.25}$$

Similarly,

$$J_yJ_z - J_zJ_y = i\hbar J_x, \tag{8.26}$$

$$J_zJ_x - J_xJ_z = i\hbar J_y. \tag{8.27}$$

Relationships (8.25), (8.26), and (8.27) form the *commutation rules* which J_x, J_y, and J_z obey.

Of particular interest are the complex operators

$$J_+ = J_x + iJ_y, \tag{8.28}$$

$$J_- = J_x - iJ_y. \tag{8.29}$$

Constructing the commutator of J_z with each and using the commutation rules to reduce yields

$$\begin{aligned} J_zJ_+ - J_+J_z &= J_zJ_x + iJ_zJ_y - J_xJ_z - iJ_yJ_z = i\hbar J_y + i(-i\hbar J_x) \\ &= \hbar(J_x + iJ_y) = \hbar J_+ \end{aligned} \tag{8.30}$$

and

$$\begin{aligned} J_zJ_- - J_-J_z &= J_zJ_x - iJ_zJ_y - J_xJ_z + iJ_yJ_z = i\hbar J_y + i(i\hbar J_x) \\ &= -\hbar(J_x - iJ_y) = -\hbar J_-. \end{aligned} \tag{8.31}$$

Next, let the product $J_zJ_\pm$ act on a Y describing rotation in one direction around the z axis, or its analog for the intrinsic state or the composite state. Then employ (8.30), (8.31), and (8.6) to obtain an eigenvalue equation

$$\begin{aligned} J_z(J_\pm Y) &= (J_\pm J_z \pm \hbar J_\pm)Y = J_\pm M\hbar Y \pm \hbar J_\pm Y \\ &= (M \pm 1)\hbar(J_\pm Y). \end{aligned} \tag{8.32}$$

Since operators J_x and J_y do not alter the quantum number J, according to the argument in Section 8.4, then neither can operators J_+ and J_-. So eigenfunction $J_\pm Y$ in (8.32) must be some coefficient times a normalized Y for quantum numbers J, $M \pm 1$. Thus, J_+ and J_- are *step-up* and *step-down operators* shifting quantum number M by $+1$ and -1, respectively, while leaving quantum number J unchanged.

The coefficient will be determined later. For its calculation, we will need some additional relationships. Now, operator (8.21) can be expressed as the sum of the product of two shift operators and the term $\hbar J_z$:

$$\begin{aligned} J_x^2 + J_y^2 &= J_x^2 + J_y^2 + i(J_xJ_y - J_yJ_x) - i(J_xJ_y - J_yJ_x) \\ &= (J_x - iJ_y)(J_x + iJ_y) + \hbar J_z = J_-J_+ + \hbar J_z. \end{aligned} \tag{8.33}$$

The product J_-J_+ acting on Y yields an eigenfunction of J_z,

$$J_z(J_-J_+Y) = (J_-J_z - \hbar J_-)J_+Y = J_-(J_+J_z + \hbar J_+)Y - \hbar J_-J_+Y$$
$$= J_-J_+M\hbar Y + \hbar J_-J_+Y - \hbar J_-J_+Y = M\hbar(J_-J_+Y), \quad (8.34)$$

with the same eigenvalue $M\hbar$ that Y has.

Since operators J_x and J_y do not alter the total quantum number, J_+ and J_- do not, either. Therefore, the eigenfunction J_-J_+Y also has the same label J as Y itself. Furthermore, term $\hbar J_z$ acting on Y yields a function with the same M and J. So $J_x^2 + J_y^2$ acting on Y yield a function that can at most differ from Y in its magnitude and phase; it yields a number times Y.

Example 8.2. Explain the operator equation

$$\mathbf{J} \times \mathbf{J} = i\hbar\mathbf{J}.$$

On expanding the left side,

$$\mathbf{J} \times \mathbf{J} = \begin{vmatrix} \hat{\mathbf{x}} & \hat{\mathbf{y}} & \hat{\mathbf{z}} \\ J_x & J_y & J_z \\ J_x & J_y & J_z \end{vmatrix}$$
$$= (J_yJ_z - J_zJ_y)\hat{\mathbf{x}} + (J_zJ_x - J_xJ_z)\hat{\mathbf{y}} + (J_xJ_y - J_yJ_x)\hat{\mathbf{z}},$$

and the last factor on the right side,

$$\mathbf{J} = J_x\hat{\mathbf{x}} + J_y\hat{\mathbf{y}} + J_z\hat{\mathbf{z}},$$

we see that the formula merely implies that Equations (8.25), (8.26), and (8.27) are valid.

Example 8.3. Apply J_+ to the eigenfunction $zR(r)$.

Let operators (8.15) and (8.16) act on the given function:

$$J_xzR(r) = \frac{\hbar}{i}\left(y\frac{\partial}{\partial z} - z\frac{\partial}{\partial y}\right)zR = \frac{\hbar}{i}\left(yR + yzR'\frac{z}{r} - z^2R'\frac{y}{r}\right)$$
$$= -i\hbar yR(r),$$

$$J_yzR(r) = \frac{\hbar}{i}\left(z\frac{\partial}{\partial x} - x\frac{\partial}{\partial z}\right)zR = \frac{\hbar}{i}\left(z^2R'\frac{x}{r} - xR - xzR'\frac{z}{r}\right)$$
$$= i\hbar xR(r).$$

Next, multiply the second result by i and add to the first result:

$$J_+zR(r) = (J_x + iJ_y)zR(r) = -\hbar(x + iy)R(r)$$

$$= \sqrt{2}\,\hbar\left[-\frac{1}{\sqrt{2}}(x + iy)R(r)\right].$$

We thus obtain $\sqrt{2}\,\hbar$ times the negative of the Y_+ found in Example 8.1. The given function has $M = 0$; the derived function has $M = 1$.

8.6. The Integral or Half-Integral Nature of Quantum Number M

From the commutation rules, we have deduced that complex operator $J_\pm$ shifts the M of an angular eigenfunction by 1 unit. Let us now establish that these M's are arranged symmetrically about zero and that they are integral or half-integral.

With a free rotator, there is nothing to distinguish a positive azimuthal angle change $d\varphi$ from a negative one of the same magnitude. So if differential changes in θ and φ from a given point produce a change in Y_1,

$$dY_1 = \frac{\partial Y_1}{\partial \theta}\,d\theta + iM_1 Y_1\,d\varphi, \tag{8.35}$$

variations $d\theta$ and $-d\varphi$ from an equivalent point produce the same change in a possible state function Y_2 of the same value at the point:

$$dY_1 = dY_2 = \frac{\partial Y_2}{\partial \theta}\,d\theta + iM_2 Y_2\,(-d\varphi)$$

$$= \frac{\partial Y_2}{\partial \theta}\,d\theta + i(-M_2)Y_2\,d\varphi. \tag{8.36}$$

But for (8.35) and (8.36) to be both true, we must have

$$M_1 = -M_2. \tag{8.37}$$

For each allowable positive quantum number

$$M_1 = |M|, \tag{8.38}$$

there is a negative quantum number

$$M_2 = -|M|. \tag{8.39}$$

According to Equation (8.32), successive M's for a given J differ by 1. Two of these M's related as in (8.37) must satisfy the equation

$$M_1 = M_2 + n = -M_1 + n, \tag{8.40}$$

where n is an integer. Solving for M_1, we find that

$$M_1 = \frac{n}{2}. \tag{8.41}$$

When n is even for one pair of M's in a set, it is even for all of the M's and the M's are integral; when n is odd, the M's are half-integral.

The integral M's yield single-valued Y's, as we have already seen in Chapter 3. These are generally acceptable. Each half-integral M, on the other hand, would yield a double-valued Φ. So such an M may occur only for motions involving one or more spins.

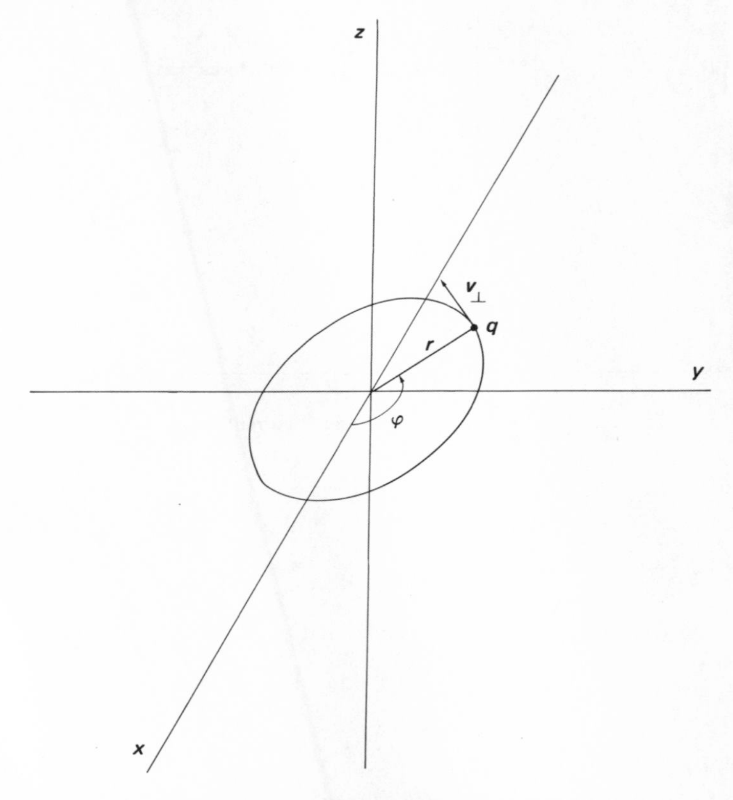

Fig. 8.1. Model particle carrying charge q around a small circle.

8.7. Magnetic Energy Associated with an Angular Momentum

A particle with spin, a particle with charge and orbital angular momentum, and an array of spinning and/or charged particles with a total angular momentum possess magnetic moments. Each such moment tends to interact with a magnetic field in a quantized manner.

Let us consider the given system in a magnetic field. A model for a pertinent constituent of motion (or behavior) is then a particle carrying charge q at the appropriate speed $v_{\perp}$ around a small circle enclosing some of the magnetic field lines. The plane of the circle is set perpendicular to the average course of these lines where the plane meets the lines. At a given instant, we also place the origin of a coordinate system at the center of the circle and draw the x and y axes in its plane, as Figure 8.1 shows.

Let us also suppose that the magnetic lines converge in parallel planes. The coordinate system is turned so that the y axis lies in one of these planes, as Figure 8.2

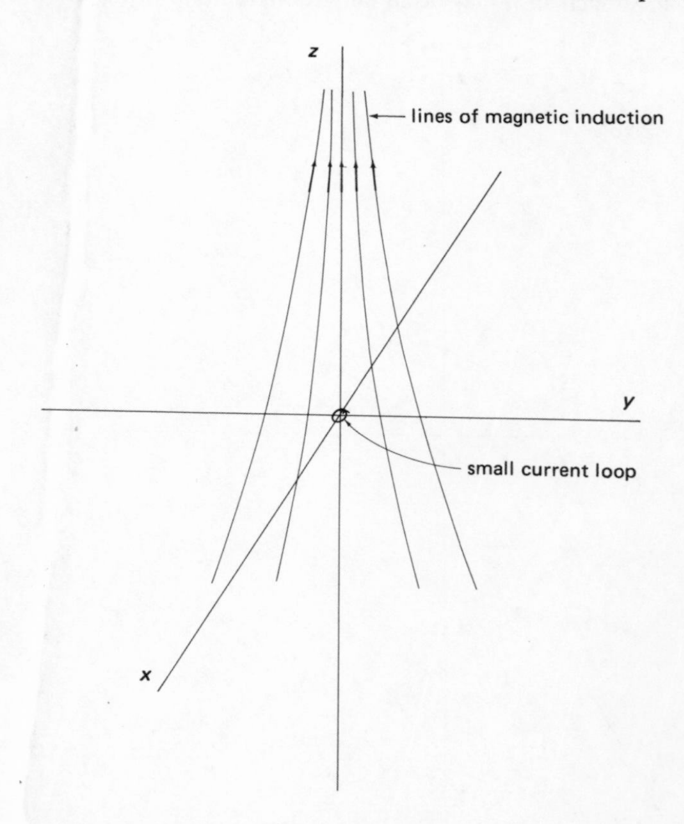

Fig. 8.2. Axes and momentary position of the circular orbit in a two-dimensional converging magnetic field.

illustrates. Since magnetic lines have no source or sink, the divergence of **B** is zero,

$$\nabla \cdot \mathbf{B} = 0. \tag{8.42}$$

In the approximation that B_x is zero, we have

$$\frac{\partial B_y}{\partial y} + \frac{\partial B_z}{\partial z} = 0 \tag{8.43}$$

or

$$\frac{\partial B_y}{\partial y} = -\frac{\partial B_z}{\partial z}. \tag{8.44}$$

Near the origin, we may consider that

$$\frac{\partial B_z}{\partial z} \simeq f(z). \tag{8.45}$$

Equation (8.44) then integrates to

$$B_y = -\frac{\partial B_z}{\partial z} y. \tag{8.46}$$

Furthermore, the radial component of B acting on the charge q is

$$B_r = B_y \sin \varphi = -\frac{\partial B_z}{\partial z} y \sin \varphi. \tag{8.47}$$

Since coordinate y is the projection of r on the y axis,

$$y = r \sin \varphi, \tag{8.48}$$

Equation (8.47) can be rewritten in the form

$$B_r = -\frac{\partial B_z}{\partial z} r \sin^2 \varphi. \tag{8.49}$$

The Lorentz force acting on the charge is

$$\mathbf{F} = q\mathbf{v} \times \mathbf{B}. \tag{8.50}$$

Employing only the component of velocity

$$v_\perp \hat{\varphi} \tag{8.51}$$

that involves the pertinent pure quantum-mechanical motion and resolving the magnetic field in the cylindrical coordinate system,

$$\mathbf{B} = B_z\hat{\mathbf{z}} + B_r\hat{\mathbf{r}} + B_\varphi\hat{\varphi}, \tag{8.52}$$

converts (8.50) to

$$\mathbf{F} = qv_\perp B_r(-\hat{\mathbf{z}}) = qv_\perp r \sin^2 \varphi \frac{\partial B_z}{\partial z} \hat{\mathbf{z}}. \tag{8.53}$$

Since the angular momentum $\mu v_\perp r$ of the rotating system is constant and the average of $\sin^2 \varphi$ is $\frac{1}{2}$, the mean $\mathbf{F}$, over an integral number of revolutions, is

$$\bar{\mathbf{F}} = qv_\perp r \frac{1}{2} \frac{\partial B_z}{\partial z} \hat{z} = \mathsf{M}_z \frac{\partial B_z}{\partial z} \hat{\mathbf{z}}. \tag{8.54}$$

Here *magnetic moment* M_z has been introduced as the current times the area in the loop:

$$\mathsf{M}_z = \frac{qv_\perp}{2\pi r} \pi r^2 = \frac{1}{2} qv_\perp r. \tag{8.55}$$

Because the angular momentum is quantized in units of $\hbar$,

$$\mu v_\perp r = M\hbar, \tag{8.56}$$

the magnetic moment M_z of the model is quantized as follows:

$$\mathsf{M}_z = \frac{q}{2\mu} \mu v_\perp r = M \frac{q\hbar}{2\mu}. \tag{8.57}$$

For an electron moving in an atom, M is replaced by m, q by $-e$, and μ by the rest mass of the electron m_e:

$$\mathsf{M}_z = -m \frac{e\hbar}{2m_e}. \tag{8.58}$$

The unit of magnetic moment is the *Bohr magneton*

$$\frac{e\hbar}{2m_e} = 9.2741 \times 10^{-24} \text{ J tesla}^{-1}. \tag{8.59}$$

The *magnetic energy* of a current loop or magnetic dipole in field $\mathbf{B}$ equals the force opposing the effect of the field times the displacement, integrated over a path into the field from a point where the field vanishes. Using (8.54), we write

$$\Delta E = -\int \bar{\mathbf{F}} \cdot \mathrm{ds} = -\int \mathsf{M}_z \frac{\partial B_z}{\partial z}\, \mathrm{d}z = -\mathsf{M}_z B_z. \tag{8.60}$$

In general, we would have

$$\Delta E = -\mathbf{M} \cdot \mathbf{B}, \tag{8.61}$$

where ΔE is the energy required to move the system with magnetic moment $\mathbf{M}$ to the point where the field is $\mathbf{B}$ from a reference point where no field exists.

If the orbital energy of an electron with quantum numbers n, l, m is E_0 when the field vanishes, a magnetic field B changes this energy to

$$E = E_0 + m \frac{e\hbar}{2m_e} B \tag{8.62}$$

according to (8.58) and (8.60). The effect on the spin energy is added to this.

We see that the separation between levels with different magnetic quantum numbers is altered by the field. From (4.115), interaction with the electromagnetic field leads to transitions for which

$$\Delta m = 0 \qquad \text{and} \qquad \Delta m = \pm 1. \tag{8.63}$$

The first condition leads to no change in a photon energy; the second condition leads to a change calculable from (8.62). The resulting perturbations in the spectrum were first observed by Pieter Zeeman and are called *Zeeman effects.*

8.8. Observing Spatial Quantization in a Beam

Interaction of a magnetic moment with an induction field $\mathbf{B}$ is quantized as we have seen. If in addition, the field is nonhomogeneous, with an appreciable $\partial B/\partial z$ around the given molecule, atom, or neutralized particle, a force proportional to the projected moment appears. If this force acts for a fixed length of time, the particle will be deflected an amount that is proportional to the force and to the moment. Such deflections are studied in the experiment first carried out by Otto Stern and Walther Gerlach.

Essential features of a setup appear in Figure 8.3. The material to be studied is vaporized, collimated by slits, and then passed through a velocity filter. The filter may consist of successive rotating disks containing holes that are in line only for particles moving at the chosen velocity. The beam then enters the magnetic field. Nonhomogeneity is achieved by employing shaped pole pieces.

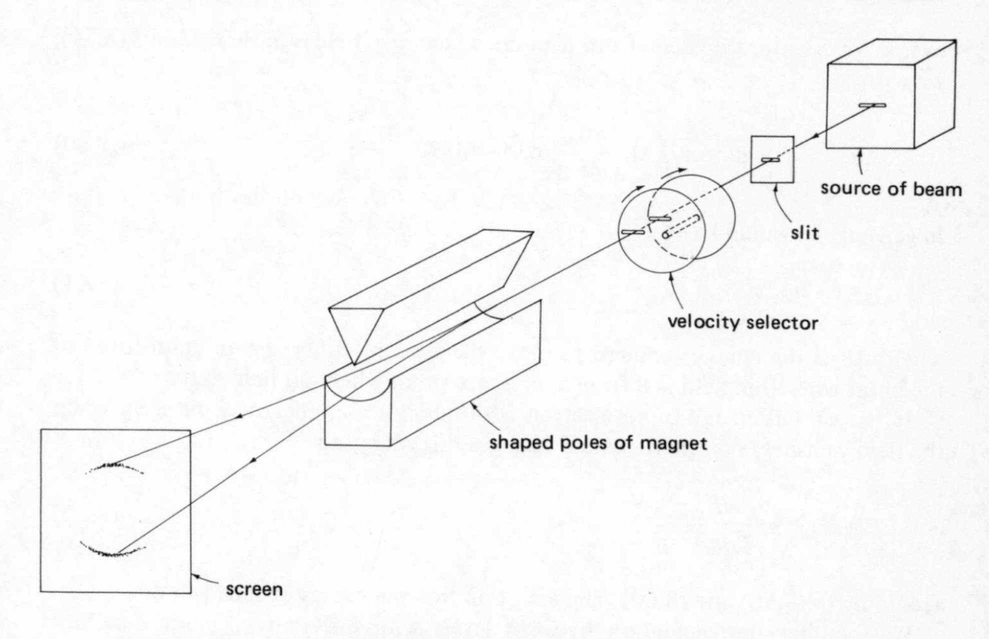

Fig. 8.3. Arrangement of parts in a Stern–Gerlach apparatus.

As each particle moves through the magnetic field, it is subjected to the force given by formula (8.54). The magnetic moment is proportional to quantum number M, following equation (8.57). As a consequence, Newton's second law applies in the form

$$\frac{d^2 z}{dt^2} = kM, \tag{8.64}$$

in which k is a constant. This equation integrates to yield

$$z = \frac{1}{2} kMt^2, \tag{8.65}$$

where t is the time spent by the particle in the field and z is the vertical displacement resulting. This displacement is projected onto the screen by further travel of the particle.

Since M is either integral or half-integral, particles from the homogeneous beam should appear in narrow bands on the screen. Such patterns are indeed observed. Thus, direct evidence is available that the spatial quantization we have postulated does occur.

Since J is the maximum M and since the M values are one integer apart and symmetric about zero, there are $2J + 1$ different M's and $2J + 1$ Stern-Gerlach bands. When J is an integer, the number of bands is odd; when J is half an odd integer, the number of bands is even.

A beam of silver or hydrogen atoms yields two bands. In these atoms there is no net orbital angular momentum – just the spin angular momentum of an odd electron. We presume that this spin is governed by a quantum number s analogous to l and J. The projection of the spin in the direction of the imposed field is governed by an m_s analogous to m and M. From Section 8.6 we know that a negative m_s exists for each positive m_s and that neighboring m_s's differ by 1.

The observed projections of the angular momentum of the spinning electron are therefore

$$\frac{1}{2}\hbar \quad \text{and} \quad -\frac{1}{2}\hbar. \tag{8.66}$$

Zeeman studies show that the first state possesses the magnetic moment

$$-\frac{e\hbar}{2m_e}, \tag{8.67}$$

the second state the magnetic moment

$$\frac{e\hbar}{2m_e}, \tag{8.68}$$

twice what one would expect for a charge $-e$ with angular momenta given by (8.66).

This fact suggests that interpreting particle spin as rotation of the particle like a rigid body is wrong. We have to consider the intrinsic angular momentum of the particle and its magnetic moment as properties.

8.9. Eigenoperators and Eigenfunctions for Spin

Over a continuous range in a coordinate of a particle, an infinite number of independent sinusoidal waves can be constructed. When the potential of the particle is constant within the given interval, an infinite number of these sinusoidal waves meet the boundary conditions and are state functions. Introducing changes in the potential alters the eigenfunctions but does not reduce the total number to a finite value.

But in a spin system, only a finite number of states exist. The proton, the neutron, and the electron each exhibit just two states in a given nonhomogeneous field. As a consequence, the corresponding eigenfunctions cannot be regarded as completely continuous functions of one or more coordinates, as rotational eigenfunctions can. The pertinent operators do, however, obey the algebra of rotational operators.

Let us consider a system exhibiting a magnetic moment and spin. We assume that the spin is governed by an operator **S** whose components satisfy commutation relations like (8.25), (8.26), (8.27). These are summarized in the statements

$$\mathbf{S} = S_x\hat{\mathbf{x}} + S_y\hat{\mathbf{y}} + S_z\hat{\mathbf{z}}, \tag{8.69}$$

$$\mathbf{S} \times \mathbf{S} = i\hbar\mathbf{S}. \tag{8.70}$$

Furthermore, let us formally construct eigenfunction χ_k and the corresponding eigenvalue equation

$$S_z\chi_k(\tau) = m_s\hbar\,\chi_k(\tau) \tag{8.71}$$

in which m_s is a quantum number one unit from its nearest neighbors, with extreme values $\pm s$, while τ is a hypothesized spin coordinate. We also assume that the eigenfunctions form an orthogonal, normalized set

$$\int \chi_j^*\chi_k \,\mathrm{d}\tau = \delta_{jk}, \tag{8.72}$$

and that each operator is Hermitian,

$$\int \chi_j^* S_i \chi_k \,\mathrm{d}\tau = \left[\int \chi_k^* S_i \chi_j \,\mathrm{d}\tau\right]^* = \int (S_i\chi_j)^*\chi_k \,\mathrm{d}\tau. \tag{8.73}$$

Since these relationships are similar to those for the rotational angular-momentum operators, they lead to the same algebra. And since the eigenfunctions of S_z describe the possible orientations with respect to the z axis, the general spin function is a superposition of these eigenfunctions:

$$\chi = \Sigma\, a_j\chi_j. \tag{8.74}$$

The total probability that the system would exhibit any of the allowed eigenvalues is unity; as a consequence, the superposition is normalized to 1. Using orthonormality condition (8.72) to reduce the integral, we obtain

$$\begin{aligned} 1 &= \int \chi^*\chi \,\mathrm{d}\tau = \int (\Sigma\, a_j\chi_j)^*(\Sigma\, a_k\chi_k)\,\mathrm{d}\tau \\ &= \Sigma\, a_j^* a_j. \end{aligned} \tag{8.75}$$

Because the only term involving the ith state is $a_i^* a_i$, this term is interpreted as the probability that the system is in the ith state.

Letting S_z act on both sides of (8.74) yields the sum

$$S_z\chi = S_z\;\Sigma\, a_k\chi_k = \Sigma\, a_k S_z\chi_k = \Sigma\, a_k m_s(k)\hbar\,\chi_k. \tag{8.76}$$

In the last equality, Equation (8.71) has been employed. Multiplying the result by the complex conjugate of (8.74) and integrating gives us

$$\int \chi^* S_z \chi \, \mathrm{d}\tau = \sum_j \sum_k a_j^* a_k m_s(k)\hbar \int \chi_j^* \chi_k \, \mathrm{d}\tau = \sum_j a_j^* a_j m_s(j)\hbar, \tag{8.77}$$

if we make use of the orthonormality condition on the eigenfunctions.

Since coefficient $a_i^* a_i$ in the ith term is the weight of eigenvalue $m_s(i)\hbar$, the integral

$$\int \chi^* S_z \chi \, \mathrm{d}\tau \tag{8.78}$$

is the average z-component of spin for the state described by χ. We call (8.78) the *expectation value* for S_z. Since there is no fundamental difference between the z, x, and y axes, we interpret

$$\int \chi^* S_x \chi \, \mathrm{d}\tau \tag{8.79}$$

as the expectation value for S_x and

$$\int \chi^* S_y \chi \, \mathrm{d}\tau \tag{8.80}$$

as the expectation value for S_y.

When a system exists in a single-spin state with respect to the z axis, the weight of one eigenvalue m_s is 1 and the weight of each other eigenvalue is 0. Now, the coordinate axes may be rotated so the axis for this spin state points in an arbitrary direction in space. We thus obtain

$$\int \chi^* S' \chi \, \mathrm{d}\tau = m_s' \hbar \tag{8.81}$$

where S' is the spin operator with respect to the axis for the unique spin. (See Figure 8.4.)

An electron exists in two independent spin states corresponding to

$$m_s = \pm \frac{1}{2}. \tag{8.82}$$

Eigenvalue equation (8.71) then takes on the forms

$$S_z \chi_+ = +\frac{1}{2} \hbar \chi_+, \tag{8.83}$$

$$S_z \chi_- = -\frac{1}{2} \hbar \chi_-, \tag{8.84}$$

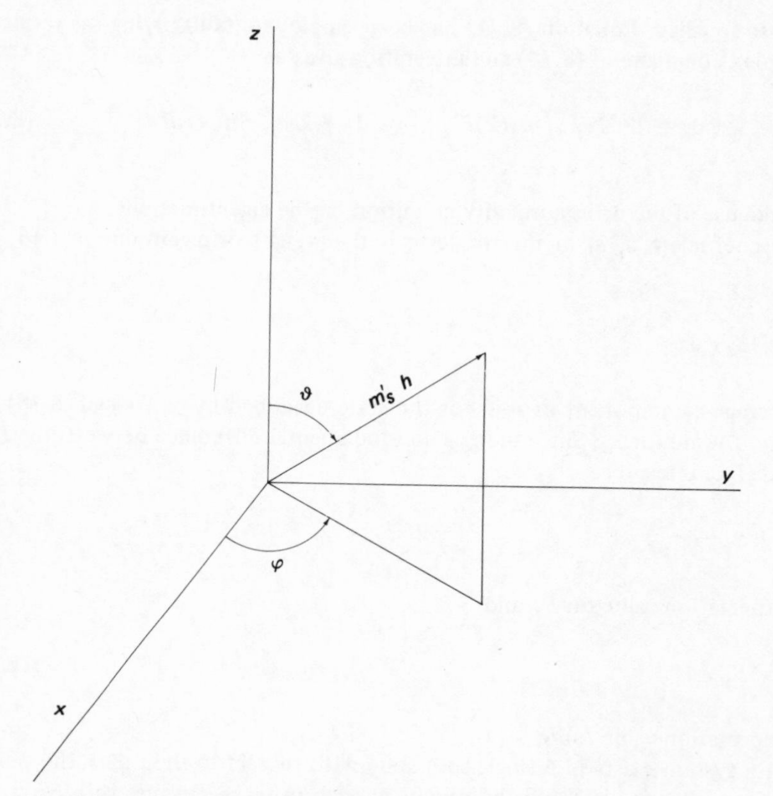

Fig. 8.4. General orientation of spin $m_s'\hbar$ for a system.

while (8.74) becomes

$$\chi = a_+\chi_+ + a_-\chi_-. \tag{8.85}$$

Sum (8.75) is now

$$a_+^*a_+ + a_-^*a_- = 1. \tag{8.86}$$

If a_+ and a_- were real, formula (8.86) would correspond to the condition on the cosine and sine of an arbitrary angle. Making the coefficients complex corresponds to multiplying each by the exponential of i times another arbitrary angle. So we set

$$a_+ = e^{i\delta} \cos \frac{1}{2}\theta, \tag{8.87}$$

$$a_- = e^{i(\delta + \varphi)} \sin \frac{1}{2}\theta, \tag{8.88}$$

without losing generality. In Example 8.5 we will find that angles θ and φ correspond to those in Figure 8.4.

Example 8.4. Determine the effects of S_x and S_y on χ_+ and χ_-.

Equations (8.69) and (8.70) imply that S_x, S_y, and S_z obey the commutation rules. Consequently, formula (8.32) applies. But this formula is satisfied and the necessary normalization conditions are met if

$$(S_x + iS_y)\chi_- = \hbar\chi_+,$$

$$(S_x + iS_y)\chi_+ = 0,$$

$$(S_x - iS_y)\chi_+ = \hbar\chi_-,$$

$$(S_x - iS_y)\chi_- = 0.$$

These equations combine to yield

$$S_x\chi_+ = \frac{1}{2}\hbar\chi_-,$$

$$S_x\chi_- = \frac{1}{2}\hbar\chi_+,$$

$$S_y\chi_+ = -\frac{1}{2i}\hbar\chi_- = \frac{i}{2}\hbar\chi_-,$$

$$S_y\chi_- = \frac{1}{2i}\hbar\chi_+ = -\frac{i}{2}\hbar\chi_+.$$

Example 8.5. Ascertain the significance of angles θ and φ in (8.87) and (8.88).

Substitute (8.85) into (8.79), (8.80), (8.78) and reduce using the formulas for the action of the components of **S** on χ_+, χ_- and formulas (8.87), (8.88):

$$\int\chi^* S_x\chi\,\mathrm{d}\tau = \int(a_+\chi_+ + a_-\chi_-)^* S_x(a_+\chi_+ + a_-\chi_-)\,\mathrm{d}\tau$$

$$= \int(a_+\chi_+ + a_-\chi_-)^*\left(\frac{1}{2}\hbar a_+\chi_- + \frac{1}{2}\hbar a_-\chi_+\right)\mathrm{d}\tau$$

$$= \frac{1}{2}\hbar a_+^* a_- + \frac{1}{2}\hbar a_-^* a_+$$

$$= \frac{1}{2}\hbar\,\mathrm{e}^{i\varphi}\cos\frac{1}{2}\theta\sin\frac{1}{2}\theta + \frac{1}{2}\hbar\,\mathrm{e}^{-i\varphi}\cos\frac{1}{2}\theta\sin\frac{1}{2}\theta$$

$$= \frac{1}{2}\hbar\sin\theta\cos\varphi,$$

$$\int \chi^* S_y \chi \, d\tau = \int (a_+\chi_+ + a_-\chi_-)^* \left(-\frac{1}{2i}\hbar a_+\chi_- + \frac{1}{2i}\hbar a_-\chi_+\right) d\tau$$

$$= \frac{1}{2i}\hbar\, e^{i\varphi} \cos\frac{1}{2}\theta \sin\frac{1}{2}\theta - \frac{1}{2i}\hbar\, e^{-i\varphi} \cos\frac{1}{2}\theta \sin\frac{1}{2}\theta$$

$$= \frac{1}{2}\hbar \sin\theta \sin\varphi,$$

$$\int \chi^* S_z \chi \, d\tau = \int (a_+\chi_+ + a_-\chi_-)^* \left(\frac{1}{2}\hbar a_+\chi_+ - \frac{1}{2}\hbar a_-\chi_-\right) d\tau$$

$$= \frac{1}{2}\hbar \cos^2\frac{1}{2}\theta - \frac{1}{2}\hbar \sin^2\frac{1}{2}\theta$$

$$= \frac{1}{2}\hbar \cos\theta.$$

The resulting expectation values are the projections on the coordinate axes of a vector of length $\frac{1}{2}\hbar$ oriented with longitudinal angle φ and colatitudinal angle θ. Therefore, χ itself describes a spin state in which the angular momentum $\frac{1}{2}\hbar$ appears as Figure 8.4 indicates.

8.10. Relating State Functions for Different Magnetic Quantum Numbers

Because the properties of angular-momentum shift operators are derived from the commutation rules, these properties apply wherever the rules hold. But, the effect of the step-up or step-down operator on a pertinent eigenfunction is a constant times a similarly normalized eigenfunction with the magnetic quantum number changed by 1. Here we will determine this constant by considering the action of J^2 on the original eigenfunction and considering the Hermitian character of J_x and J_y.

Equation (8.32) tells us that shift operator $J_\pm$ acts on a state function for rotation, spin, or composite behavior to produce a number times the state function for one unit more or less angular momentum around the z axis. Thus, we have

$$J_\pm Y_{J,M} = C^\pm_{J,M} Y_{J,M\pm 1} \tag{8.89}$$

where the subscripts on state function Y are the quantum numbers for the non-magnetic angular energy and the angular momentum about the axis of rotation.

Let us multiply (8.89) by $Y^*_{J,M\pm1}$, integrate over the pertinent coordinates, and assume that the function is normalized to 1:

$$\int Y^*_{J,M\pm1} J_\pm Y_{J,M}\, \mathrm{d}^2\Omega = C^\pm_{J,M} \int Y^*_{J,M\pm1} Y_{J,M\pm1}\, \mathrm{d}^2\Omega$$

$$= C^\pm_{J,M}. \tag{8.90}$$

Choosing the positive sign leaves

$$C^+_{J,M} = \int Y^*_{J,M+1} J_+ Y_{J,M}\, \mathrm{d}^2\Omega, \tag{8.91}$$

while choosing the negative sign and replacing M with $M + 1$ yields

$$C^-_{J,M+1} = \int Y^*_{J,M} J_- Y_{J,M+1}\, \mathrm{d}^2\Omega$$

$$= \left[\int Y^*_{J,M+1} J_+ Y_{J,M}\, \mathrm{d}^2\Omega\right]^*. \tag{8.92}$$

The last step is based on the Hermiticity of J_x and J_y. (See Examples 8.6 and 8.7.)

On comparison, we see that

$$C^+_{J,M} = C^{-*}_{J,M+1}. \tag{8.93}$$

Because the relative phases of the different eigenfunctions are unobservable, we can choose

$$C^+_{J,M} = C^-_{J,M+1} \tag{8.94}$$

with no loss of generality.

Since Y is a simultaneous eigenfunction of (8.21) and (8.22), it is also an eigenfunction of $J_x^2 + J_y^2 + J_z^2$. Setting the eigenvalue of this operator equal to $J(J+1)\hbar$, with the nature of J to be determined shortly, yields

$$[(J_x^2 + J_y^2) + J_z^2]\, Y = J_- J_+ Y + \hbar J_z Y + J_z^2 Y$$

$$= C^-_{J,M+1}\, C^+_{J,M}\, Y + \hbar^2 M Y + \hbar^2 M^2 Y$$

$$= \hbar^2 J(J + 1) Y. \tag{8.95}$$

The first equality has introduced (8.33); the second equality has introduced (8.89) and (8.6).

From the last equality, we obtain

$$C^-_{J,M+1}\, C^+_{J,M} = \hbar^2 [J(J + 1) - M(M + 1)]. \tag{8.96}$$

Introducing the relationship between the phases implied by (8.94) leads to

$$C^{+}_{J,M} = C^{-}_{J,M+1} = \hbar\sqrt{J(J+1) - M(M+1)}\,. \tag{8.97}$$

Equations (8.89) then become

$$J_{+}Y_{J,M} = \hbar\sqrt{J(J+1) - M(M+1)}\;Y_{J,M+1}, \tag{8.98}$$

$$J_{-}Y_{J,M} = \hbar\sqrt{J(J+1) - (-M)(-M+1)}\,Y_{J,M-1}. \tag{8.99}$$

If a system obeying the commutation rules can exist in a state for which M is J, it can also exist with M equal to $-J$. From (8.98) and (8.99), J_{+} acting on $Y_{J,M=J}$ for the system is zero and J_{-} acting on $Y_{J,M=-J}$ is zero. As a consequence, the sequence of M's cannot be carried beyond J or $-J$ and state functions exist only with

$$-J \leqslant M \leqslant J. \tag{8.100}$$

For a rotating system, Y is a spherical harmonic with a dependence on θ given by (7.84) and (7.85). According to the argument following Equations (7.65), (7.66), the limits on m are $\pm l$:

$$-l \leqslant m \leqslant l. \tag{8.101}$$

Since m corresponds to M in the general theory, l corresponds to the general J. For spin, the limits on m_s are $\pm s$,

$$-s \leqslant m_s \leqslant s, \tag{8.102}$$

and s corresponds to J. Since the magnetic quantum number has limits for any rotating or spinning system with given nonmagnetic angular energy, it must also have limits for composite systems.

Example 8.6. Show that operators J_x, J_y, and J_z for rotational motion are Hermitian.

Let Y_1 and Y_2 be the angular functions for two rotational states while J_z is operator (8.5) for the system. Then since

$$\begin{aligned}\int_0^{2\pi} Y_1^* \frac{\hbar}{i}\frac{\partial}{\partial\varphi} Y_2\,\mathrm{d}\varphi &= \int Y_1^* \frac{\hbar}{i}\,\mathrm{d}Y_2 \\ &= Y_1^* \frac{\hbar}{i} Y_2 \Big|_0^{2\pi} - \int Y_2 \frac{\hbar}{i}\,\mathrm{d}Y_1^* \\ &= 0 - \int Y_2 \frac{\hbar}{i}\frac{\partial}{\partial\varphi} Y_1^*\,\mathrm{d}\varphi \\ &= \left[\int_0^{2\pi} Y_2^* \frac{\hbar}{i}\frac{\partial}{\partial\varphi} Y_1\,\mathrm{d}\varphi\right]^*,\end{aligned}$$

we have

$$\int_0^{2\pi}\int_0^{\pi} Y_1^* \frac{\hbar}{i}\frac{\partial}{\partial\varphi} Y_2 \sin\theta \, d\theta \, d\varphi = \int Y_1^* J_z Y_2 \, d^2\Omega$$

$$= \left[\int Y_2^* J_z Y_1 \, d^2\Omega\right]^*$$

$$= \int (J_z Y_1)^* Y_2 \, d^2\Omega.$$

The x, y, and z axes are geometrically equivalent. Consequently, rotating the given physical system so x is replaced by y, y by z, and z by x alters the above equation to

$$\int Y_1^* J_x Y_2 \, d^2\Omega = \left[\int Y_2^* J_x Y_1 \, d^2\Omega\right]^*$$

$$= \int (J_x Y_1)^* Y_2 \, d^2\Omega,$$

while replacing x by z, y by x, and z by y leads to

$$\int Y_1^* J_y Y_2 \, d^2\Omega = \left[\int Y_2^* J_y Y_1 \, d^2\Omega\right]^*$$

$$= \int (J_y Y_1)^* Y_2 \, d^2\Omega.$$

Note that the only restrictions on Y_1 and Y_2 are that they be well-behaved functions of x/r, y/r, z/r (and thus, of θ and φ).

From (8.73), the same formula applies to spin eigenfunctions. As a consequence, the formula applies to the composite eigenfunctions also.

Example 8.7. Use the Hermitian property of J_x and J_y to transform

$$\int Y_1^* J_- Y_2 \, d^2\Omega.$$

Express J_- as $J_x - iJ_y$, introduce the pertinent results from Example 8.6 into each part, and recombine:

$$\int Y_1^* (J_x - iJ_y) Y_2 \, d^2\Omega = \int Y_1^* J_x Y_2 \, d^2\Omega - i \int Y_1^* J_y Y_2 \, d^2\Omega$$

$$= \left[\int Y_2^* J_x Y_1 \, d^2\Omega \right]^* - i \left[\int Y_2^* J_y Y_1 \, d^2\Omega \right]^*$$

$$= \left[\int Y_2^* (J_x + iJ_y) Y_1 \, d^2\Omega \right]^* .$$

Thus,

$$\int Y_1^* J_- Y_2 \, d^2\Omega = \left[\int Y_2^* J_+ Y_1 \, d^2\Omega \right]^* .$$

8.11. Combining Angular Momenta

In a confining field, the orbital motions and the spins of particles may fit together in various ways to yield a given composite result. When the field to which a constituent particle is subject is approximately spherical, each possible state of the particle is determined by an orbital J and a spin J, together with the possible components along a given axis. A set of such particles involves an orbital quantum number and a spin quantum number for each particle, together with the possible single-axis components.

In the complete system, interactions between the orbital and spin motions may be relatively weak; so we first need to combine the orbital motions and the spin motions by themselves. If these interactions are relatively strong, we may need to combine the orbital motion with the spin motion of each particle, before combining the motions of the particles, in establishing the composite states.

In determining how the constituent states combine, we look at the no-interaction limit. There, the axiom that independent probabilities combine multiplicatively requires the eigenfunctions corresponding to a given set of quantum numbers to combine multiplicatively. Whenever more than one combination of quantum numbers is allowed, the different pure states mix following the superposition principle. The overall state function must be normalized to 1, as are the constituent eigenfunctions.

The result is a series of the type

$$Y_{J,M} = \sum_k c_k Y_{J_1,(M_1)_k} Y_{J_2,(M_2)_k} \cdots \tag{8.103}$$

in which $J_j, (M_j)_k$ is the kth set of quantum numbers for the jth angular momentum (rotation or spin), while J and M constitute the quantum numbers for the composite system.

The maximum angular momenta of the different rotations and spins combine vectorially to produce the various possible total angular momenta. Thus we have

$$\left| J_1 \pm J_2 \pm \ldots \pm J_n \right|_{\min} \leqslant J \leqslant \left| J_1 + J_2 + \ldots + J_n \right| , \tag{8.104}$$

where the number on the left is the minimum absolute value obtainable from the indicated combinations. Since the components along the z axis combine additively, we also have

$$M = M_1 + M_2 + \ldots + M_n \tag{8.105}$$

for each k.

When only two constituent motions are involved, we write

$$Y_{J_1, J_2, J, M} = \sum_{M_1} \sum_{M_2} C_{M_1, M_2} (J_1, J_2, J) \, Y_{J_1, M_1} Y_{J_2, M_2} , \tag{8.106}$$

where the summations are over M_1 and M_2 with

$$-J_1 \leqslant M_1 \leqslant J_1 , \tag{8.107}$$

$$-J_2 \leqslant M_2 \leqslant J_2 . \tag{8.108}$$

Numbers $C_{M_1, M_2} (J_1, J_2, J)$ are called *Clebsch-Gordon coefficients*.

To determine these coefficients for a given composite state, we construct the pertinent equation, (8.106), and let J_+ or J_- act on it repeatedly until all nonzero terms can be collected into two terms. The corresponding coefficients are obtained. These are substituted back into preceding equations and the additional coefficients are calculated.

Example 8.8. How is the system with $J = \frac{3}{2}$, $M = \frac{1}{2}$ constructed from a particle with orbital J equal to 1 and spin J equal to $\frac{1}{2}$?

The possible magnetic quantum numbers for $J_1 = 1$ and $J_2 = \frac{1}{2}$ are

$$M_1 = -1, 0, 1 \qquad \text{and} \qquad M_2 = -\frac{1}{2}, \frac{1}{2} .$$

From (8.105), the only combinations that make $M = \frac{1}{2}$ are

$$M_1 = 0, \qquad M_2 = \frac{1}{2},$$

and

$$M_1 = 1, \qquad M_2 = -\frac{1}{2} .$$

Consequently, Equation (8.106) becomes

$$Y_{3/2,1/2}(1,2) = c_1 Y_{1,0}(1) Y_{1/2,1/2}(2) + c_2 Y_{1,1}(1) Y_{1/2,-1/2}(2)$$

where 1 and 2 in parentheses refer to coordinates for the orbital and spin motions.

Applying formula (8.98) to each eigenfunction in turn yields

$$J_+ Y_{3/2,1/2} = \hbar\sqrt{3}\, Y_{3/2,3/2},$$

$$J_+ Y_{1,0} = \hbar\sqrt{2}\, Y_{1,1}, \qquad J_+ Y_{1,1} = 0,$$

$$J_+ Y_{1/2,1/2} = 0, \qquad J_+ Y_{1/2,-1/2} = \hbar Y_{1/2,1/2}.$$

Since J_+ is a differentiating operator, we also have

$$J_+ YZ = (J_+ Y)Z + Y(J_+ Z).$$

Therefore, action of J_+ on the superposition constructed using (8.106) yields

$$\hbar\sqrt{3}\, Y_{3/2,3/2} = c_1 \hbar\sqrt{2}\, Y_{1,1} Y_{1/2,1/2} + c_2 \hbar Y_{1,1} Y_{1/2,1/2}$$

$$= (\sqrt{2}\, c_1 + c_2)\hbar Y_{1,1} Y_{1/2,1/2}.$$

Because $Y_{3/2,3/2}$ and $Y_{1,1} Y_{1/2,1/2}$ are both normalized to the same value, their coefficients must be equal:

$$\sqrt{3} = \sqrt{2}\, c_1 + c_2.$$

The normalization argument applied to the equation before action of J_+ yields

$$c_1^2 + c_2^2 = 1.$$

These two conditions on c_1 and c_2 lead to

$$c_1 = \sqrt{\frac{2}{3}},$$

$$c_2 = \sqrt{\frac{1}{3}}.$$

Discussion Questions

8.1. For what kinds of systems may shift operators be formulated?

8.2. How can reorientation of a particle be a significant operation? How does such reorientation differ from reorientation of an array of particles?

8.3. In what sense are the operators for spin and for composite angular momentum similar to the operator for simple rotational angular momentum?
8.4. How are expressions for operators J_x, J_y, and J_z found?
8.5. Explain the nature of operators for which the rotational standing wave is an eigenfunction.
8.6. How do the commutation rules for J_x, J_y, and J_z arise from the expressions for these operators?
8.7. Explain why $J_\pm$ shifts the M but does not alter the J of an eigenfunction. Why does $J_x^2 + J_y^2$ not alter either M or J?
8.8. How can the cross product of a vector operator with itself be different from zero?
8.9. How does the lack of handedness in space limit the magnetic quantum numbers?
8.10. Recall the complete Lorentz force law and how it was established.
8.11. Why is the divergence of **B** zero?
8.12. In the two-dimensional magnetic field, why is B_y equal to a function of z times y? Why is the radial component B_r equal to $B_y \sin \varphi$? Why is the mean $-B_r$ equal to $\frac{1}{2} r(\partial B_z/\partial z)$?
8.13. What is the magnetic moment of a current loop?
8.14. Describe the Stern-Gerlach experiment.
8.15. Why is a velocity filter needed in the given set-up? How does such a filter operate?
8.16. What happens if the particles in the beam of a Stern-Gerlach system are charged?
8.17. Why does a continuous variation in a coordinate imply that the number of independent eigenfunctions involving the coordinate is infinite?
8.18. How are (a) spin operators, (b) spin eigenfunctions defined?
8.19. What is the statistical weight of eigenfunction χ_i in the mixture $\Sigma\, a_j \chi_j$? Explain.
8.20. By what reasoning do we obtain the expectation values for S_x and S_y?
8.21. When is a particle in a single-spin state? Explain.
8.22. How does the identity $\sin^2 \alpha + \cos^2 \alpha = 1$ help us to parameterize the relationship $a^2 + b^2 = 1$?
8.23. Show that a given free electron is in a pure spin state with respect to some axis.
8.24. Explain how the state function for a given rotation, spin, or composite behavior is related to functions for one unit higher or lower angular momentum.
8.25. Why can M be $-J$ if it can equal J?
8.26. Why does the sequence of eigenfunctions terminate at $M = \pm J$?
8.27. Explain how rotations and spins combine.

Problems

8.1. Simplify the operator

$$\frac{\hbar}{i}\left(x\frac{\partial}{\partial x} - \frac{\partial}{\partial x}x\right).$$

8.2. Show that a function of r behaves like a constant when it is acted on by J_x and J_y.

8.3. What standing-wave functions arise when (a) operator J_z and (b) operator J_z^2 act on $x(x^2 - 3y^2)R(r)$? From these results, construct the eigenfunctions of J_z that contribute to the given function.

8.4. Operate repeatedly on $(3z^2 - r^2)R(r)$ with J_+ and explain the results.

8.5. Show when J_+J_- acting on Y yields an eigenfunction of J_z.

8.6. A beam of helium atoms from a gas reservoir impinges on a velocity filter that consists of two similar rotating disks mounted 5.0 cm apart on an axle parallel to the beam. Each disk has 12 holes arranged equidistantly on a circumference 2.0 cm from its center. If the holes are aligned when the disks are at rest, how fast should the axle turn so that the filter will pass helium atoms traveling at 1.0×10^5 cm s^{-1}?

8.7. Show that when $S_\pm \chi_j$ exists, it is an eigenfunction of S_z with the eigenvalue $(m_s \pm 1)\hbar$.

8.8. Use properties of operators to calculate how two particles with angular momentum quantum numbers $J_1 = \frac{1}{2}$ and $J_2 = \frac{1}{2}$ combine to form a $J = 0, M = 0$ state.

8.9. Determine how the angular momentum states $J_1 = 1, M_1 = 1$ and $J_2 = 1, M_2 = -1$ combine.

8.10. If $AB - BA = 1$, what is $AB^2 - B^2A$?

8.11. Simplify the operator $J_xJ_y^2 - J_y^2J_x$.

8.12. Express J_x and J_y in spherical coordinates.

8.13. Operate on $2ixyzR(r)$ with J_z and J_z^2. Then construct the eigenfunctions for J_z that contribute to this standing-wave function.

8.14. Explain what happens when $J_\pm$ acts on $xR(r)$.

8.15. Show that rotating the spin angular momentum of an electron by one turn about an axis perpendicular to the spin axis reverses the sign of wave function χ.

8.16. Employ properties of the pertinent operators to show that

$$c_{J,M-1}^{+}c_{J,M}^{-} = \hbar^2[J(J+1) - (-M)(-M+1)].$$

8.17. Use properties of operators to calculate the coefficients in the equation

$$Y_{1,1}(1)\,Y_{1/2,-1/2}(2) = c_1 Y_{3/2,1/2}(1,2) + c_2 Y_{1/2,1/2}(1,2).$$

8.18. Determine how two particles with angular momentum quantum numbers $J_1 = \frac{3}{2}$ and $J_2 = \frac{3}{2}$ combine to form a $J = 1, M = 1$ state.

References

Books

Edmonds, A. R.: 1957, *Angular Momentum in Quantum Mechanics*, Princeton University Press, Princeton, N.J., pp. 10–52.

El Baz, E., and Castel, B.: 1972, *Graphical Methods of Spin Algebras in Atomic, Nuclear, and Particle Physics*, Marcel Dekker, New York, pp. 1–46.

Ikenberry, E.: 1962, *Quantum Mechanics for Mathematicians and Physicists*, Oxford University Press, New York, pp. 162–179.

Park, D.: 1974, *Introduction to the Quantum Theory*, 2nd edn, McGraw-Hill, New York, pp. 135–206.
Perkins, D. H.: 1972, *Introduction to High Energy Physics*, Addison-Wesley, Reading, Mass., pp. 318–333.
Rose, M. E.: 1957, *Elementary Theory of Angular Momentum*, Wiley, New York, pp. 15–47.

Articles

Andrews, M.: 1979, 'Vector Operators and Spherical Harmonics in Quantum Mechanics', *Am. J. Phys.* **47**, 274–277.
Gray, C. G.: 1979, 'Remark on Integral Orbital Angular Momentum', *Am. J. Phys.* **37**, 559–560.
Hoshino, Y.: 1978, 'An Elementary Application of Exterior Differential Forms in Quantum Mechanics', *Am. J. Phys.* **46**, 1148–1150.
Merzbacher, E.: 1962, 'Single Valuedness of Wave Functions', *Am. J. Phys.* **30**, 237–247.
Smorodinskii, Ya, A., and Shelepin, L. A.: 1972, 'Clebsch-Gordon Coefficients, Viewed from Different Sides', *Soviet Phys. Uspekhi* **15**, 1–24.
Wolf, A. A.: 1969, 'Rotation Operators', *Am. J. Phys.* **37**, 531–536.
Zaharia, M.: 1972, 'On the Quantum Theory of the Symmetric Top', *Am. J. Phys.* **40**, 844–849.

Chapter 9

Propagation, Spreading, and Scattering

9.1. The Current Associated with Propagation

The state of each submicroscopic particle in a system is described, as we have noted before, by a function Ψ. The observable probability-density distribution of the particle over space at each time t and the observable changes associated with increasing time are derivable from this function. Analogous to the classic mechanical current found in a moving fluid is an apparent current of streaming probability density.

The fundamental equation governing changes is the *time-dependent Schrödinger equation*

$$-\frac{\hbar}{i}\frac{\partial\Psi}{\partial t} = H\Psi = -\frac{\hbar^2}{2\mu}\nabla^2\Psi + V\Psi \tag{9.1}$$

established in Section 6.7. Let us multiply (9.1) by Ψ^* from the left,

$$-\frac{\hbar}{i}\Psi^*\frac{\partial\Psi}{\partial t} = -\frac{\hbar^2}{2\mu}\Psi^*\nabla^2\Psi + V\Psi^*\Psi. \tag{9.2}$$

Let us also take the complex conjugate of (9.1) and multiply the result by Ψ from the right,

$$+\frac{\hbar}{i}\frac{\partial\Psi^*}{\partial t}\Psi = -\frac{\hbar^2}{2\mu}(\nabla^2\Psi^*)\Psi + V\Psi^*\Psi. \tag{9.3}$$

Subtracting (9.3) from (9.2) yields

$$-\frac{\hbar}{i}\frac{\partial}{\partial t}(\Psi^*\Psi) = -\frac{\hbar^2}{2\mu}[\Psi^*\nabla^2\Psi - (\nabla^2\Psi^*)\Psi] \tag{9.4}$$

or

$$\frac{\partial}{\partial t}(\Psi^*\Psi) + \frac{\hbar}{2\mu i}\nabla\cdot[\Psi^*\nabla\Psi - (\nabla\Psi^*)\Psi] = 0. \tag{9.5}$$

Now, the *divergence* of a vector function is the negative of the time rate of change in the mass density of the fluid for which the function is the momentum density. In symbols,

$$\frac{\partial \rho}{\partial t} + \nabla \cdot \rho \mathbf{v} = 0. \tag{9.6}$$

On comparing (9.5) with (9.6), we see that when

$$\Psi^* \Psi = \rho, \tag{9.7}$$

the probability density, we have

$$\frac{\hbar}{2\mu i} [\Psi^* \nabla \Psi - (\nabla \Psi^*) \Psi] = \rho \mathbf{v} = \mathbf{J}, \tag{9.8}$$

the corresponding *current density*. Equation (9.5) is called the quantum mechanical *equation of continuity.*

Example 9.1. Apply Equation (9.8) to a beam of particles translating in the direction of increasing x.

For movement only in the x direction, formula (1.19) reduces to

$$\Psi = A\, e^{ikx}\, e^{-i\omega t}.$$

Then

$$\Psi^* = A^*\, e^{-ikx}\, e^{i\omega t}$$

and

$$\nabla \Psi = \hat{\mathbf{x}} \frac{\partial}{\partial x} A\, e^{ikx}\, e^{-i\omega t} = ikA\, e^{ikx}\, e^{-i\omega t}\, \hat{\mathbf{x}},$$

$$\nabla \Psi^* = -ikA^*\, e^{-ikx}\, e^{i\omega t}\, \hat{\mathbf{x}}.$$

Substituting these expressions into (9.8) yields

$$\rho \mathbf{v} = \frac{\hbar}{2\mu i} (A^* ikA\, \hat{\mathbf{x}} + ikA^* \hat{\mathbf{x}} A) = \frac{\hbar k}{\mu} A^* A\, \hat{\mathbf{x}} = \frac{\mu \mathbf{v}}{\mu} A^* A$$

$$= A^* A\, \mathbf{v},$$

whence

$$\rho = A^* A,$$

as (9.7) implies.

9.2. An Evolution Operator

When a system is conservative – that is, when the Hamiltonian does not depend on time explicitly – an operator that displaces a prevailing Ψ in time can be readily constructed from the time-dependent Schrödinger equation.

Consider a particle whose behavior is described by a known function $\Psi(\mathbf{r}, 0)$ at time $t = 0$. Furthermore, suppose that the energy of the particle is conserved, that its Hamiltonian H is independent of t. Then the first equation in (9.1) can be formally integrated to

$$\Psi(\mathbf{r}, t) = \exp\left(-\frac{it}{\hbar} H\right)\Psi(\mathbf{r}, 0). \tag{9.9}$$

Since the exponential acts to change $\Psi(\mathbf{r}, 0)$ to $\Psi(\mathbf{r}, t)$, it is called the *evolution operator* for the state function.

For simplicity, let us limit our subsequent discussion to one dimension. We consider a particle moving freely in the x direction, with $V = 0$. The state function is then $\Psi(x, t)$ and formula (6.88) reduces to

$$H = -\frac{\hbar^2}{2\mu}\frac{\partial^2}{\partial x^2}. \tag{9.10}$$

Substituting (9.10) into the exponential in (9.9) and expanding leads to

$$\exp\left(\frac{i\hbar t}{2\mu}\frac{\partial^2}{\partial x^2}\right) = \Sigma\,\frac{1}{k!}\left(\frac{i\hbar t}{2\mu}\right)^k \frac{\partial^{2k}}{\partial x^{2k}}. \tag{9.11}$$

In determining the action of this operator on a Gaussian wave in free space, we will make use of two identities. By carrying out the indicated differentiations, we find that

$$\frac{\partial^2}{\partial x^2}\,\frac{\mathrm{e}^{-x^2/4s^2}}{s} = \frac{\partial}{\partial(s^2)}\,\frac{\mathrm{e}^{-x^2/4s^2}}{s}. \tag{9.12}$$

Furthermore, using the rule for expanding an exponential, then identifying the expansion as a Taylor expansion, leads to

$$\left[\exp\alpha\frac{\partial}{\partial(s^2)}\right] f(s^2) = \Sigma\,\frac{\alpha^k}{k!}\,\frac{\partial^k}{\partial(s^2)^k}\,f(s^2)$$

$$= f(s^2 + \alpha). \tag{9.13}$$

9.3. Spreading of a Gaussian Wave Packet

A submicroscopic particle may be confined in its movements (a) by the potential field to which it is subjected and (b) by the mixed state in which it finds itself. Even when the pure states permit extensive movements, these may superpose to confine the particle, or restrict it, to a very small region.

The resulting wave packet may be constructed so that initially it contracts, momentarily remains constant, or expands. When the potential is constant over space, the rate of a contraction tends to decrease with time and turn into an expansion; and a packet momentarily constant in shape immediately begins to spread. Furthermore, the rate of spreading increases with time.

As a simple example, consider a one-dimensional wave packet initially concentrated about the origin of the x axis, in a null potential field. Also, suppose the particle density has the normalized Gaussian form

$$\rho = \frac{1}{(2\pi)^{\frac{1}{2}} s} \, e^{-x^2/2s^2} \tag{9.14}$$

when $t = 0$. (A result in Example 3.5 yields the normalization coefficient.)

Let us choose the phase so the corresponding ψ is real. Then ψ is the square root of (9.14) and

$$\Psi(x, 0) = \frac{1}{(2\pi)^{\frac{1}{4}} s^{\frac{1}{2}}} \, e^{-x^2/4s^2}. \tag{9.15}$$

According to (9.9) and (9.10), we obtain $\Psi(x, t)$ in the $V = 0$ field by letting operator (9.11) act on (9.15):

$$\begin{aligned}
\Psi(x, t) &= \exp\left(\frac{i\hbar t}{2\mu} \frac{\partial^2}{\partial x^2}\right) \frac{1}{(2\pi)^{\frac{1}{4}} s^{\frac{1}{2}}} \, e^{-x^2/4s^2} \\
&= \frac{s^{\frac{1}{2}}}{(2\pi)^{\frac{1}{4}}} \exp\left(\frac{i\hbar t}{2\mu} \frac{\partial^2}{\partial x^2}\right) \frac{e^{-x^2/4s^2}}{s} \\
&= \frac{s^{\frac{1}{2}}}{(2\pi)^{\frac{1}{4}}} \exp\left(\frac{i\hbar t}{2\mu} \frac{\partial}{\partial (s^2)}\right) \frac{e^{-x^2/4s^2}}{s} \\
&= \frac{s^{\frac{1}{2}}}{(2\pi)^{\frac{1}{4}}} \frac{1}{(s^2 + i\hbar t/2\mu)^{\frac{1}{2}}} \exp\left[-\frac{x^2}{4(s^2 + i\hbar t/2\mu)}\right],
\end{aligned} \tag{9.16}$$

whence

$$\rho = \frac{1}{(2\pi)^{\frac{1}{2}}} \frac{s}{(s^4 + \hbar^2 t^2/4\mu^2)^{\frac{1}{2}}} \exp\left[-\frac{s^2 x^2}{2(s^4 + \hbar^2 t^2/4\mu^2)}\right]. \tag{9.17}$$

The wave packet remains Gaussian. Furthermore, the variance of its probability density,

$$\sigma^2 = s^2 \left(1 + \frac{\hbar^2 t^2}{4\mu^2 s^4}\right), \tag{9.18}$$

increases quadratically as t increases (or decreases) from zero. The effect is symmetric about the point $t = 0$,

$$\rho(x, -t) = \rho(x, t). \tag{9.19}$$

With (9.18), formula (9.17) becomes

$$\rho = \frac{1}{(2\pi)^{\frac{1}{2}}\,\sigma}\, e^{-x^2/2\sigma^2}. \tag{9.20}$$

Similarly, we may consider a freely moving one-dimensional packet that is Gaussian, with a fixed shape at time $t = 0$. The center of mass is the expectation value for the coordinate. When $t = 0$, we suppose that

$$\langle x \rangle = x_0 \tag{9.21}$$

and

$$\langle p \rangle = p_0 = \hbar k_0. \tag{9.22}$$

The velocity of the center of mass is constant at

$$v = \frac{p_0}{\mu} = \frac{\hbar k_0}{\mu}. \tag{9.23}$$

Hence this center appears where

$$x = x_0 + \frac{\hbar k_0}{\mu}\, t \tag{9.24}$$

and the Galilean transformation that puts the center of mass at the origin of an inertial frame is

$$x' = x - x_0 - \frac{\hbar k_0}{\mu}\, t. \tag{9.25}$$

Coordinate x' now behaves as coordinate x did in (9.20). For the freely moving one-dimensional packet, (9.20) is replaced by

$$\rho = \frac{1}{(2\pi)^{\frac{1}{2}}\,\sigma} \exp\left\{-\frac{[x - x_0 - (hk_0/\mu)t]^2}{2\sigma^2}\right\}. \tag{9.26}$$

9.4. Existence of Scattering Centers in Materials

From the manner in which rays of alpha particles were scattered by various targets, Ernest Rutherford inferred in 1911 that most of the mass of a material is concentrated in relatively small positively charged portions. Each of these concentrations of mass was presumed to form a nucleus for a different atom. The negative particles called electrons were assumed to fill the space around the nuclei because some of them could be readily removed from the material.

Subsequent workers have steadily accumulated more information about nuclei and about the constituents of nuclei from additional scattering experiments. In the simplest systems, a macroscopically homogeneous beam is allowed to strike a macroscopically homogeneous target and the projectile particles are deflected elastically by the microscopic, or submicroscopic, nonhomogeneities. The intensities appearing at various angles are then determined.

The energy of a projectile particle depends on the magnitude and extent of its acceleration. The intensity in a beam depends on the rate at which particles are supplied to the pertinent cross-sectional area. The energy fixes the wavevector, the intensity the normalization, of the function for the de Broglie wave.

Deflections out of an incident beam are caused by nonhomogeneities in the target material. In general, each nonhomogeneity extends over only a small volume, which at great distances appears as a *scattering center*. From each center, the scattered particles move radially. Neighboring rays of the corresponding wavefunction are diverging straight lines. Since the movement is free, the wavevector does not change along a ray.

However, the scattering center determines the fraction deflected at each possible angle and the phase shift introduced in the corresponding deflection. The wavelets from different scattering centers interfere to form a diffraction pattern, but this effect will not be considered in this chapter.

9.5. Cross Sections Presented by Scattering Centers

Let us consider a macroscopically homogeneous beam falling on a macroscopically homogeneous target as Figure 9.1 depicts. Let us also suppose that small nonhomogeneities in the target scatter particles from the beam without change in energy. Figure 9.2 indicates how the scattering regions may be distributed. For simplicity, we assume that the nonhomogeneities are all alike, so each one scatters incident particles in a certain direction with the same probability.

Far from the target, where the observations are to be made, each scatterer appears approximately at the same point. Let us place the origin at this point, orient the z axis so it points in the direction of motion of the incident particles, put the x axis in the vertical plane, and the y axis in the horizontal plane, as Figure 9.3 illustrates. Let the solid angle that the detector subtends about colatitude θ and aximuthal angle φ, at distance r, be Ω.

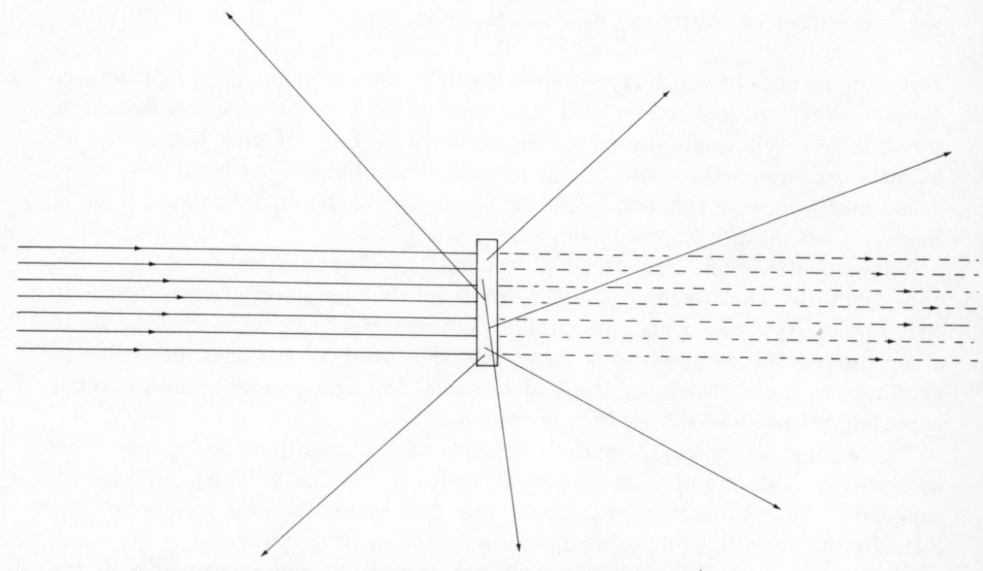

Fig. 9.1. Scattering of particles from a beam by material in a target.

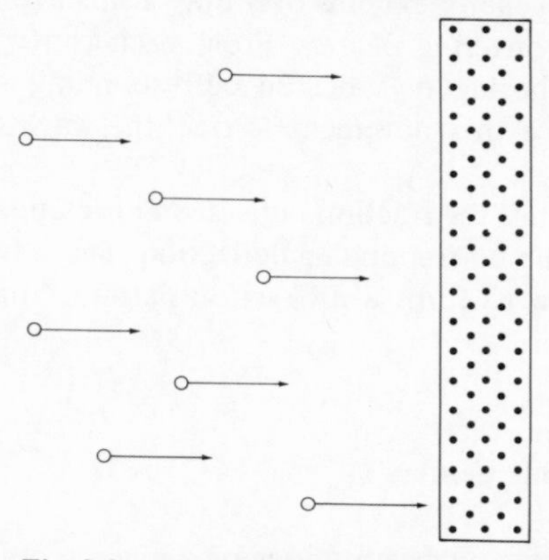

Fig. 9.2. Distribution of scattering centers in a target.

A wave function is usually defined so the square of its absolute value at a point equals the probability density at the point. If $\Psi_s(r, \theta, \varphi, t)$ represents the wave scattered by a single center from the incident wave

$$\Psi_i = A\, e^{ikz}\, e^{-i\omega t}, \tag{9.27}$$

the probability of finding a particle scattered from the center into the element of solid angle $d\Omega$ and thickness dr at distance r from the center is

$$\Psi_s^* \Psi_s r^2\, d\Omega\, dr. \tag{9.28}$$

See Figure 9.4.

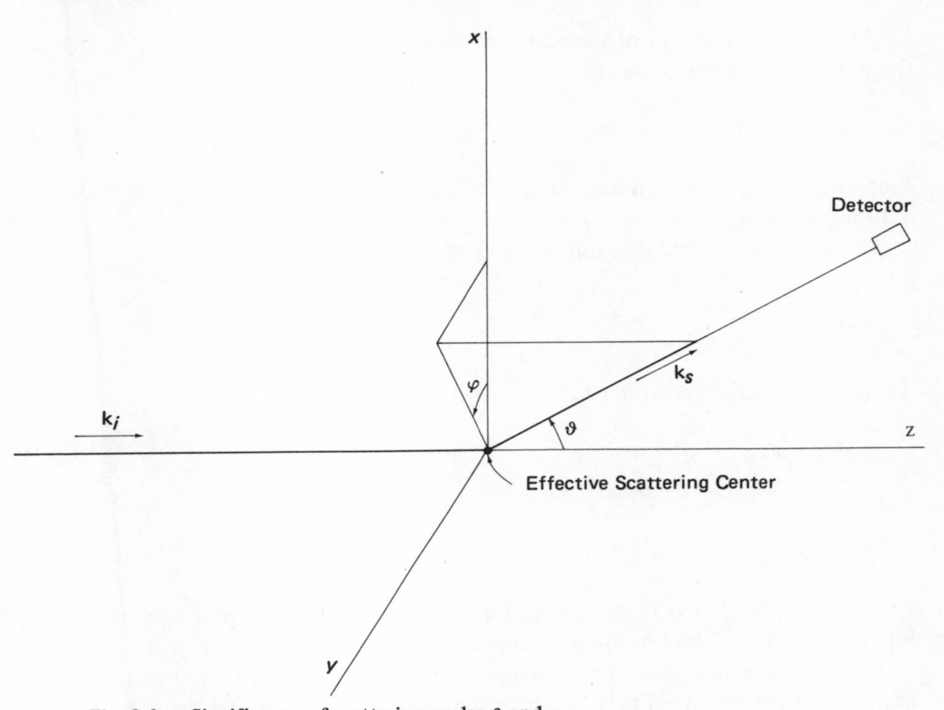

Fig. 9.3. Significance of scattering angles θ and φ.

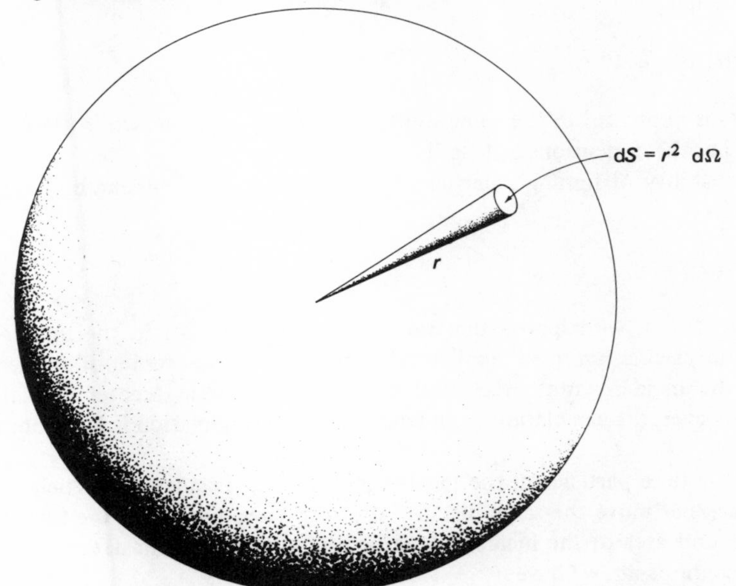

Fig. 9.4. Cross-sectional area a distance r from the center of a cone enclosing solid angle $d\Omega$.

If k_s is the wavevector of the scattered wavelet and ω_s the corresponding angular frequency, the wavelet must fit

$$B\, e^{ik_s r}\, e^{-i\omega_s t} \tag{9.29}$$

over small changes in r. Indeed, recall the asymptotic solution obtained in Example 4.3 for positive E.

To determine how the magnitude of B varies with r, we let

$$B = \frac{Af}{r} \tag{9.30}$$

where A is the coefficient in (9.27). Then

$$\Psi_s r = Af\, e^{ik_s r}\, e^{-i\omega_s t} \tag{9.31}$$

and

$$\Psi_s^* \Psi_s r^2 = A^* f^* A f = |A|^2 |f|^2 . \tag{9.32}$$

For probability (9.28) to be independent of distance r, but proportional to the intensity in the incident beam, magnitude $|f|$ must not depend on r or A.

For (9.29) to hold over small changes in r, as we have assumed, function f needs also to be independent of t. It can, however, vary with k, θ, φ, and the nature of the scattering center:

$$f = f(k, \theta, \varphi). \tag{9.33}$$

Because Ψ_s is expressed in the same units as Ψ_i and A is expressed in these units, function f has the dimensions of length.

The probability of finding a particle in unit cross section and length of incident beam $dz = dr$ is

$$|A|^2\, dr \tag{9.34}$$

As we have noted, we suppose that no kinetic energy is lost in the interaction. Strictly, the calculation then applies to properties in the center-of-momentum frame of the incident and target particles. If the scattering center is relatively massive, however, the calculation is applicable as an approximation in the laboratory frame.

During the time particles in the incident beam move distance dr, particles in the scattered wavelet move the same distance dr, on the average. So if the fraction of particles in unit area of the incident beam scattered into element $d\Omega$ by the single scattering center is $d\sigma$, we have

$$d\sigma\, |A|^2\, dr = |A|^2 |f|^2\, d\Omega\, dr \tag{9.35}$$

or

$$d\sigma = |f|^2 \, d\Omega. \tag{9.36}$$

Since f has the dimensions of length, $d\sigma$ has the dimensions of area.

The fraction scattered through all angles from the unit cross-sectional area is

$$\sigma \equiv \int_0^{4\pi} \frac{d\sigma}{d\Omega} \, d\Omega = \int_0^{4\pi} |f|^2 \, d\Omega. \tag{9.37}$$

Since it appears to be the area removed from unit area of the beam by the scattering center, σ is called the *cross section* of the center for the scattering process. The apparent area removed and deflected into angle $d\Omega$ is $d\sigma$; expression $d\sigma/d\Omega$ is consequently called the *differential cross section* of the center for the incident beam.

When the concentration of scattering centers in the target is c, the number of scattering centers encountered in distance dz is $c \, dz$ per unit area. But σ is the fraction of unit area of the beam scattered by a single center. Therefore,

$$\sigma c \, dz \tag{9.38}$$

represents the total fraction scattered from thickness dz of the target.

A convenient unit for cross sections is the *barn*, defined as 10^{-28} m^2. A convenient unit for distances within nuclei is the *fermi*, which is 10^{-15} m.

Example 9.2. How does the intensity of a beam vary with the distance it has traveled through a homogeneous target?

The fraction scattered by the target between z and $z + dz$ is

$$\frac{I(z) - I(z + dz)}{I(z)} \equiv -\frac{dI}{I}$$

if $I(z)$ is the intensity of the beam at z. Setting this expression equal to (9.38) yields

$$\frac{dI}{I} = -\sigma c \, dz,$$

whence

$$\ln \frac{I}{I_0} = -\sigma c z$$

or

$$I = I_0 \, e^{-\sigma c z}.$$

Example 9.3. A beam of thermal neutrons is reduced to 0.0100 per cent of its initial intensity by 0.083 cm of cadmium. Calculate the corresponding cross section.

Solve the integrated expression in Example 9.2 for the cross section:

$$\sigma = \frac{-2.303 \log (I/I_0)}{cz}.$$

The ratio I/I_0 and distance z are given. To obtain the concentration c, take the density of cadmium, divide it by the weight per mole, and multiply by the number of atoms in a mole (Avogadro's number):

$$c = \frac{8.642 \text{ g cm}^{-3}}{112.40 \text{ g mole}^{-1}} 6.02 \times 10^{23} \text{ atoms mole}^{-1}$$

$$= 4.63 \times 10^{22} \text{ atoms cm}^{-3}.$$

Then

$$\sigma = \frac{2.303 \times 4.00}{(4.63 \times 10^{22} \text{ atoms cm}^{-3})(0.083 \text{ cm})} = 2.4 \times 10^{-21} \text{ cm}^2,$$

whence

$$\sigma = 2400 \text{ barns}.$$

9.6. Pure Outgoing, Incoming, and Standing Spherical Waves

Away from a scattering center, where the potential energy of a moving particle is constant, the particle's momentum and the corresponding propagation vector do not vary. Furthermore, by de Broglie's equation, these exhibit the same direction. So a wavelet scattered elastically along **r** is sinusoidal with a **k** pointing in the same direction.

But according to Equation (9.30), the amplitude of such a wavelet is

$$\frac{Af}{r}. \tag{9.39}$$

Since f is a function of wavevector k and spherical coordinates θ, φ, the spatial factor for the wavelet is

$$\psi_s = A \frac{f(k, \theta, \varphi)}{r} e^{ikr}; \tag{9.40}$$

and in the path of the incident beam, we have

$$\psi = \psi_i + \psi_s$$
$$= A\left[e^{ikz} + \frac{f(k, \theta, \varphi)}{r}\, e^{ikr}\right]. \tag{9.41}$$

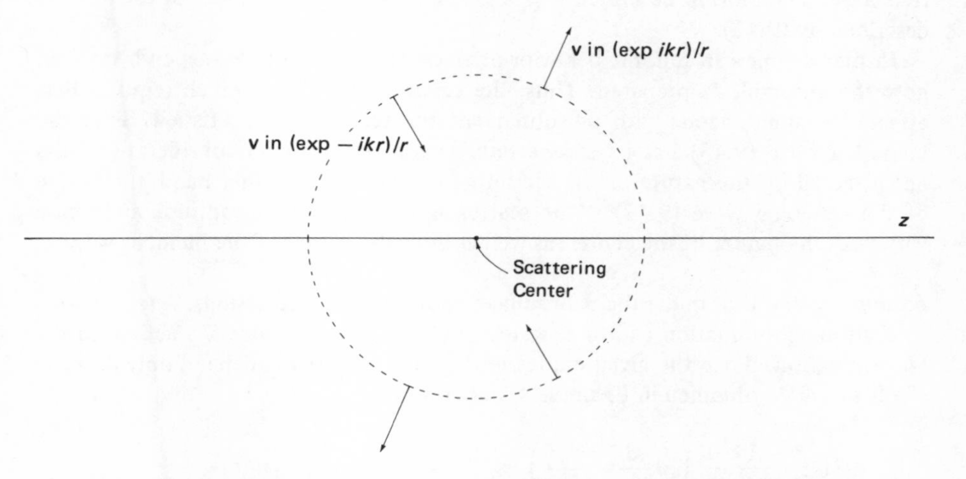

Fig. 9.5. Velocities of particles in the outgoing and the incoming unit waves.

If the amplitude of (9.40) on all the surface $r = 1$ were unity, the wave would have the form

$$G_+(r) = \frac{e^{ikr}}{r}. \tag{9.42}$$

As a consequence, expression (9.42) represents a unit-outgoing wave. Since reversing the sign in the exponent reverses the direction of travel of the wave, the corresponding unit-incoming wave is

$$G_-(r) = \frac{e^{-ikr}}{r}. \tag{9.43}$$

(See Figure 9.5.) Superposing the incoming wave on the outgoing wave and dividing by 2 yields the standing wave

$$G_1(r) = \frac{\cos kr}{r}. \tag{9.44}$$

Subtracting the incoming wave from the outgoing wave and dividing by $2i$ yields the independent standing wave

$$G_2(r) = \frac{\sin kr}{r}. \tag{9.45}$$

Outside the scattering region, there is no variation in potential to produce reflection. The motion described by (9.43) is therefore independent of the motion described by (9.42).

Insofar as one can tell, the behavior of an ensemble of particles depends only on how the ensemble is prepared. Thus, the *causality principle*, which requires that effects be simultaneous with or subsequent to their causes, is satisfied. Since the incoming wave (9.43) has its source outside the region under consideration, it is not affected by the nature of the scattering center. On the other hand, the source of the outgoing wave (9.42) is the scattering center. So its magnitude and phase vary with the nature of the center, as well as with the nature of the incident wave.

Example 9.4. Determine the Schrödinger equation that expressions $G_\pm(r)$ satisfy.

Schrödinger equation (4.30) describes the effect of operator ∇^2 acting on the wave function. Since the given expressions (9.42) and (9.43) depend only on r, let the form of ∇^2 obtained in Example 4.1 act on them:

$$\nabla^2 G_\pm = \frac{1}{r^2}\frac{\mathrm{d}}{\mathrm{d}r}\left(r^2\frac{\mathrm{d}}{\mathrm{d}r}\frac{\mathrm{e}^{\pm ikr}}{r}\right) = \frac{1}{r^2}\frac{\mathrm{d}}{\mathrm{d}r}\left[(\pm ikr - 1)\,\mathrm{e}^{\pm ikr}\right]$$

$$= \frac{1}{r^2}\left[(\pm ikr - 1)(\pm ik) \pm ik\right]\mathrm{e}^{\pm ikr}$$

$$= -k^2\frac{\mathrm{e}^{\pm ikr}}{r} = -k^2 G_\pm .$$

The given expressions satisfy (4.30) with k constant.

9.7. Possible Effects of Centers on the Partial Waves

Let us now consider how scattering centers can alter appropriate components of a homogeneous planar incident wave. For simplicity, we will assume that each scattering system is circularly symmetric around the z axis. Thus, the effects of spin are to be neglected. Factor f then depends only on k and θ.

When the scattering centers have no influence, the spatial factor in the wave function is

$$\psi_i = A\,\mathrm{e}^{ikz}. \tag{9.46}$$

According to Equations (7.97), (7.101), and (7.104), this can be obtained by the superposition

$$\psi_i = A \sum_l \frac{2l+1}{2ikr} \left[e^{ikr} - (-1)^l e^{-ikr}\right] P_l(\cos\theta). \tag{9.47}$$

Outside the scattering regions, there is no variation of potential which could mix the partial waves. So each term on the right of (9.47) is physically independent. Furthermore, the incoming waves cannot be influenced by a scattering center before they reach the center. However, the center can change both the phase and amplitude of each outgoing wave.

When the alteration in phase of the lth outgoing wave is $2\delta_l$ and the factor by which its amplitude is reduced is η_l, sum (9.47) becomes

$$\psi_{\text{total}} = A \sum_l \frac{2l+1}{2ikr} \left[\eta_l e^{2i\delta_l} e^{ikr} - (-1)^l e^{-ikr}\right] P_l(\cos\theta). \tag{9.48}$$

By convention, δ_l is called the lth *phase shift*, while η_l is called the *amplitude reduction factor*. The difference caused by the scattering center is the scattered wave

$$\begin{aligned}
\psi_s &= \psi_{\text{total}} - \psi_i \\
&= A \frac{e^{ikr}}{kr} \sum_l (2l+1) \left(\frac{\eta_l e^{2i\delta_l} - 1}{2i}\right) P_l(\cos\theta) \\
&= A \frac{e^{ikr}}{r} f.
\end{aligned} \tag{9.49}$$

In the last step, the definition of f from (9.30) has been introduced. We see that

$$f = \sum_l \frac{2l+1}{k} \left(\frac{\eta_l e^{2i\delta_l} - 1}{2i}\right) P_l(\cos\theta). \tag{9.50}$$

There are $2l + 1$ independent pure partial waves with quantum number l. From (9.50), the amplitude of each of these is determined by the factor

$$f_l = \frac{\eta_l e^{2i\delta_l} - 1}{2ik}. \tag{9.51}$$

Now, a measure of the effect of η_l and δ_l that is independent of k is the product

$$kf_l = \frac{\eta_l e^{2i\delta_l} - 1}{2i}. \tag{9.52}$$

We may replace the reciprocal wavevector by the wavelength divided by 2π. Indeed from Example 1.2,

$$\lambda\!\!\!{}^{-} \equiv \frac{\lambda}{2\pi} = \frac{1}{k}. \tag{9.53}$$

Since we have taken the wavevector to be the same after as before scattering, our formulas describe results in the center-of-momentum frame. When the scattering center is relatively massive, this frame does not differ appreciably from the laboratory frame.

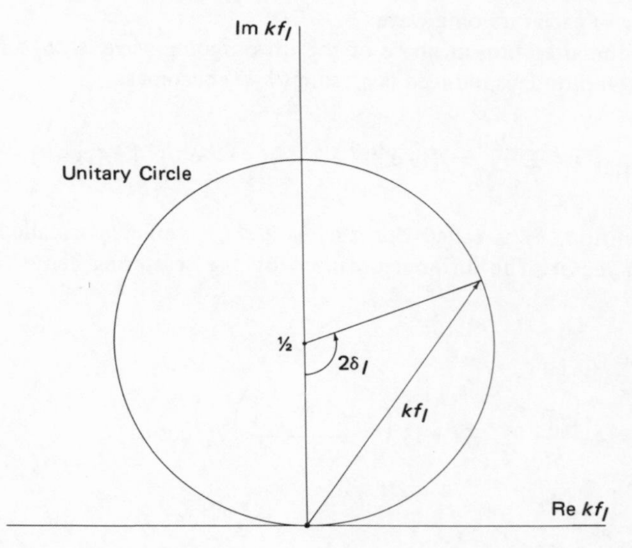

Fig. 9.6. Product kf_l when $\eta_l = 1$.

Example 9.5. How does the wavevector–scattering amplitude product vary as the energy of the bombarding particles increases?

From Equation (9.52) we obtain

$$kf_l = \frac{\eta_l \, e^{2i\delta_l} - 1}{2i} = \frac{1}{2}(-i)\eta_l \, e^{2i\delta_l} + \frac{1}{2} i$$

$$= \frac{1}{2} \eta_l \, e^{i(2\delta_l - \pi/2)} + \frac{1}{2} i.$$

A plot of this product appears in Figure 9.6.

When no reaction occurs at the scattering center, η_l equals 1. When the particle energy is small enough, the wavelength in the incident beam is very large with respect to the effective radius of a scattering center and the phase shift vanishes. Increasing the energy decreases the wavelength and increases the magnitude of the phase shift. The end point of kf_l then traces out the unitary circle of Figure 9.6. Decreasing η_l, by introducing reaction, causes the end point of vector kf_l to move inside the unitary circle.

9.8. Contributions to Various Cross Sections

According to (9.36), the derivative of a cross section with respect to solid angle is the corresponding $|f|^2$. Integrating $|f|^2$ over the complete solid angle about a scattering center therefore yields the cross section for scattering by any angle.

The expression we have obtained for factor f contains a $P_l(\cos\theta)$ in each term. To calculate the corresponding σ, we need the integral of the product of any two Legendre polynomials. But from the last equation in (7.74), the overall formula (7.77), and (7.92), we obtain

$$\int_0^{4\pi} P_{l_1} P_{l_2}\, d\Omega = \frac{4\pi}{2l_1 + 1}\, \delta_{l_1 l_2} \tag{9.54}$$

where

$$\delta_{l_1 l_2} = 1 \qquad \text{when} \quad l_1 = l_2 \tag{9.55}$$

and

$$\delta_{l_1 l_2} = 0 \qquad \text{when} \quad l_1 \neq l_2 . \tag{9.56}$$

Substituting (9.50) and (9.53) into (9.37) now yields the elastic scattering cross section

$$\sigma_{\text{el}} = 4\pi\lambda\!\!\!^{-2} \sum_l (2l+1) \left| \frac{\eta_l\, e^{2i\delta_l} - 1}{2i} \right|^2 \tag{9.57}$$

When there is no absorption of the incoming wavelets, each η_l is 1 and

$$\sigma_{\text{el}} = 4\pi\lambda\!\!\!^{-2} \sum_l (2l+1) \sin^2 \delta_l . \tag{9.58}$$

The *probability* of *reaction* in the time a typical particle travels unit distance is given by the probability that a particle is coming in towards the center minus the probability that it is leaving in that distance:

$$\int_0^{4\pi} \left(|\psi_{\text{in}}|^2 - |\psi_{\text{out}}|^2 \right) r^2\, d\Omega . \tag{9.59}$$

Wave function ψ_{in} is the second sum on the right of (9.48), ψ_{out} the first sum. Introducing these expressions and carrying out the integration yields

$$|A|^2 \frac{\pi}{k^2} \sum_l (2l+1)(1 - |\eta_l|^2). \tag{9.60}$$

Dividing (9.60) by $|A|^2$ gives us the cross section for reaction:

$$\sigma_{re} = \pi \lambda\!\!\!^{-2} \sum_l (2l+1)(1 - |\eta_l|^2). \tag{9.61}$$

The total cross section is the sum of (9.57) and (9.61). Since $\eta_l = \eta_l^*$, we obtain

$$\sigma_{tot} = 2\pi \lambda\!\!\!^{-2} \sum_l (2l+1)(1 - \eta_l \cos 2\delta_l). \tag{9.62}$$

In the path of the incident beam, angle θ is 0 and $P_l(\cos\theta)$ is 1. Solving for the imaginary part of (9.50) for this angle yields

$$\operatorname{Im} f(k, 0, \varphi) = \frac{1}{2i}\,[f(k, 0, \varphi) - f^*(k, 0, \varphi)]$$

$$= \frac{1}{2k} \sum_l (2l+1)(1 - \eta_l \cos 2\delta_l). \tag{9.63}$$

Combining (9.62) and (9.63) yields

$$\sigma_{tot} = \frac{4\pi}{k} \operatorname{Im} f(k, 0, \varphi), \tag{9.64}$$

whence

$$\operatorname{Im} f(k, 0, \varphi) = \frac{k}{4\pi}\,\sigma_{tot}. \tag{9.65}$$

Relationship (9.65) is known as the *optical theorem.*

9.9. A Classical Model for Scattering Centers

In the Newtonian approximation, a freely moving particle approaches a target nucleus along a straight line, as Figure 9.7 shows. The distance b of this line from the parallel straight line passing through the center of the nucleus determines whether the projectile strikes the nucleus or not.

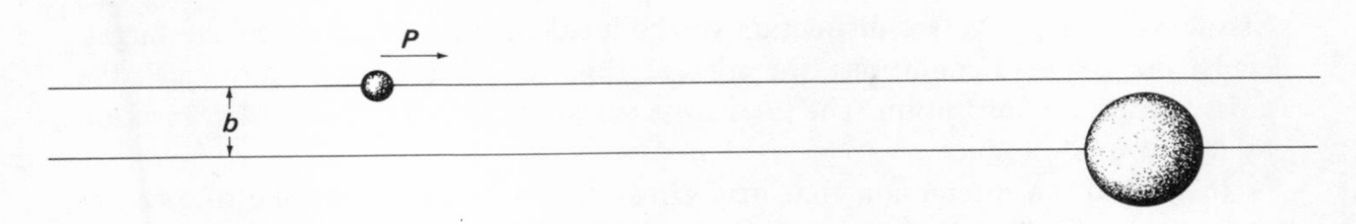

Fig. 9.7. Projectile particle moving with linear momentum p along a straight line distance b from a parallel line through the center of a target particle.

In the simplest quantum description, one thing is added, quantization of the angular momentum. Employing (2.27), (2.32) in their pertinent forms,

$$pb = l\hbar, \tag{9.66}$$

and introducing (1.30) for the linear momentum leads to

$$b = \frac{l\hbar}{\hbar k} = l\lambda\!\!\!^{-}. \tag{9.67}$$

In the *black sphere* model, a target nucleus acts as a totally absorbing sphere with a definite radius. All projectiles whose centers come within distance R of its center are assumed to react while the others do not. The maximum b for absorption is then R and the corresponding l is

$$l_{max} = \frac{R}{\lambda\!\!\!^{-}}. \tag{9.68}$$

Furthermore, the amplitude reduction factor η_l is 0 when $l \leqslant l_{max}$, 1 when $l > l_{max}$, and the phase shift δ_l is 0 when $l > l_{max}$. Equations (9.57) and (9.61) then yield

$$\sigma_{el} = \pi\lambda\!\!\!^{-2} \sum_0^{l_{max}} (2l + 1), \tag{9.69}$$

$$\sigma_{re} = \pi\lambda\!\!\!^{-2} \sum_0^{l_{max}} (2l + 1), \tag{9.70}$$

Introducing the identity proved in Example 9.6, (9.68),

$$\sum_0^{l_{max}} (2l + 1) = (l_{max} + 1)^2 \simeq l_{max}^2 = \frac{R^2}{\lambda\!\!\!^{-2}}, \tag{9.71}$$

and substituting the result into (9.69) and (9.70) leads to πR^2 for both σ_{el} and σ_{re}. The total cross section is the sum

$$\sigma_{tot} = \sigma_{el} + \sigma_{re} = \pi R^2 + \pi R^2 = 2\pi R^2. \tag{9.72}$$

Elastic scattering involves diffraction of the incident wave function by the target. Under our assumed conditions, the cross section due to this diffraction equals the cross section for absorption. The total cross section is therefore twice what reaction by itself would produce.

In general, an interaction that acts effectively only within *range a* of a center appreciably affects only those partial waves with

$$l \leqslant \frac{a}{\lambda\!\!\!^{-}}, \tag{9.73}$$

following formula (9.67). Furthermore, as $l \to 0$, we would have

$$\eta_l \to 0. \tag{9.74}$$

On the other hand, the effects on amplitude and phase become small as l increases beyond $a/\lambda\!\!\!^{-}$ and vanish when $l >> a/\lambda\!\!\!^{-}$.

Example 9.6. By mathematical induction, establish the first equality in (9.71),

$$\sum_{0}^{l_{\max}} (2l + 1) = (l_{\max} + 1)^2 .$$

When $l_{\max} = 0$, both sides of the formula equal 1. When $l_{\max} = 1$, the left side is

$$1 + (2 + 1) = 4,$$

while the right side is

$$(1 + 1)^2 = 4.$$

Suppose that the relationship has been proved for the summation to a:

$$\sum_{0}^{a} (2l + 1) = (a + 1)^2 .$$

For the summation to $a + 1$, add the next term $2(a + 1) + 1$ to the sum and factor the result:

$$\sum_{0}^{a+1} (2l + 1) = (a + 1)^2 + 2(a + 1) + 1 = [(a + 1) + 1]^2 .$$

The general form is the same. But the form is valid when a is 0 and 1. Therefore, it has to be valid for a equal to 2, 3, 4, ...

9.10. Nuclear Radii

According to the nuclear model introduced by Rutherford, an atom consists of a small positive nucleus of definite size surrounded by a relatively large region through which the orbital electrons move.

For a projectile to be effectively free of a nucleus when it is not in contact with the nucleus, the projectile has to be uncharged. For the projectile to interact strongly with the nucleus when it is in contact with it, the projectile has to be a meson or a baryon. Furthermore, it must not move too slowly or absorption will occur at a distance; and it must not move too fast, in the range where the nucleus becomes transparent.

One does have to allow for the radius of the projectile. Indeed, if both incident particle and target nucleus are spherical, a projectile whose center strikes within distance R of the center of the nucleus, where

$$R = R_1 + R_2, \tag{9.75}$$

hits the nucleus. Here R_1 is the radius of the nucleus while R_2 is the radius of the projectile. So from (9.72), the total cross section is

$$\sigma = 2\pi R^2 = 2\pi(R_1 + R_2)^2. \tag{9.76}$$

Cross sections obtained with moderate energy neutrons indicate that the nuclear radius varies as the cube root of the *mass number* A_j, the integer closest to the atomic weight,

$$R_j = R_0 A_j^{1/3}, \tag{9.77}$$

to a good approximation. Furthermore,

$$R_0 \simeq 1.4 \text{ fermi}. \tag{9.78}$$

Relationship (9.77) implies that the volume of a nucleus is

$$V_j = \frac{4}{3}\pi R_j^3 = \frac{4}{3}\pi R_0^3 A_j. \tag{9.79}$$

A person can explain this result by assuming that the jth nucleus contains A_j particles, each occupying the volume $\frac{4}{3}\pi R_0^3$.

When both projectile and target are charged, the particles interact at a distance through the Coulomb potential. The barrier that the incoming particle must penetrate has the height

$$V = \frac{Z_1 Z_2 e^2}{4\pi\epsilon_0(R_1 + R_2)}, \tag{9.80}$$

where e is the charge on a proton, Z_1 the number of charges on the nucleus, Z_2 the number of charges on the projectile, R_1 the radius of the nucleus, R_2 the radius of the projectile, and ϵ_0 the permittivity of space.

Example 9.7. Estimate the Coulomb barrier for an alpha particle reacting with a nitrogen nucleus of mass number 14.

From (9.77) and (9.78) the pertinent radii are

$$R_1 = 1.4 \times 10^{-15}\ (14)^{1/3}\ \text{m} = 3.3_7 \times 10^{-15}\ \text{m},$$

$$R_2 = 1.4 \times 10^{-15}\ (4)^{1/3}\ \text{m} = 2.2_2 \times 10^{-15}\ \text{m}.$$

Substituting these into (9.80) yields

$$V = c^2 \times 10^{-7}\ \frac{Z_1 Z_2 e^2}{R_1 + R_2}$$

$$= (2.9979 \times 10^8)^2\ (10^{-7}\ \text{C}^{-2}\ \text{J m})\ \frac{(7)\ (2)\ (1.6022 \times 10^{-19}\ \text{C})^2}{(5.5_9 \times 10^{-15}\ \text{m})\ (1.6022 \times 10^{-13}\ \text{J MeV}^{-1})}$$

$$= 3.6_1\ \text{MeV}.$$

9.11. Resonance

The scattering cross section for a given angular-momentum quantum number l passes through a maximum at the center-of-momentum wavevector k that makes the magnitude of the expression between the vertical lines in (9.57) a maximum. The projectile and the target particles are then said to *resonate*. The mass of the resonant state is the total center-of-momentum mass of the colliding particles.

When the overall process is essentially elastic,

$$\eta = 1, \tag{9.81}$$

expression (9.52) reduces to

$$kf = \frac{e^{i\delta}\ (e^{i\delta} - e^{-i\delta})}{2i} = \frac{\sin\delta}{e^{-i\delta}} = \frac{\sin\delta}{\cos\delta - i\sin\delta}$$

$$= \frac{1}{\cot\delta - i} \tag{9.82}$$

for the given l.

The phase shift δ varies with wavevector k, which depends on the center-of-momentum energy E. At the resonance energy E_R, the magnitude of the denominator in the final expression of (9.82) is as small as possible. Then $\cot\delta$ is 0 and δ is $\pi/2$. But an expansion of $\cot\delta$ is

$$\cot\delta = \cot\delta_R + (E - E_R)\left(\frac{d}{dE}\cot\delta\right)_{E = E_R} + \ldots$$

$$\simeq -(E - E_R)\frac{2}{\Gamma} \tag{9.83}$$

if we define

$$\left(\frac{\mathrm{d}}{\mathrm{d}E}\cot\delta\right)_{E=E_R} = -\frac{2}{\Gamma}. \tag{9.84}$$

Therefore, Equation (9.82) can be rewritten in the form

$$kf \simeq \frac{1}{-(E-E_R)\,2/\Gamma - i} = -\frac{\frac{1}{2}\Gamma}{(E-E_R)+\frac{1}{2}i\Gamma} \tag{9.85}$$

and the lth term in (9.57) reduces to the *Breit–Wigner formula*

$$\sigma_{\mathrm{el}} = 4\pi\lambda^2\,(2l+1)\,\frac{\frac{1}{4}\Gamma^2}{(E-E_R)^2+\frac{1}{4}\Gamma^2}. \tag{9.86}$$

At the two energies where

$$E - E_R = \pm\frac{1}{2}\Gamma, \tag{9.87}$$

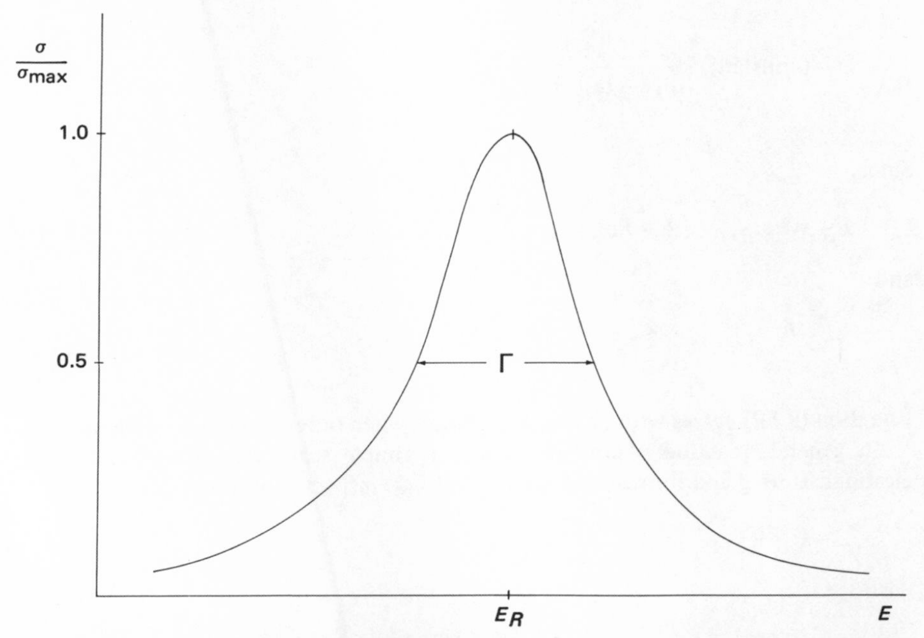

Fig. 9.8. Variation in the reduced cross section $\sigma/\sigma_{\mathrm{max}}$ near a resonance, for a given l.

the cross section is one-half its maximum. Consequently, parameter Γ equals the difference between these two energies; Γ is the *width* of the *peak* at the height where σ_{el} has fallen to one-half its maximum. A typical plot appears in Figure 9.8.

When both projectile and target particle do not exhibit spin, l equals J, the total angular-momentum quantum number for the resonant state, and $2l + 1$ in formula (9.86) is $2J + 1$. When a spinless projectile falls on a nucleus or nucleon with a half-unit of spin, the factor $2J + 1$ still applies. But since the spin of the resonant state is a certain value, only one-half the target spin states can contribute. Then (9.86) becomes

$$\sigma_{el} = \frac{\pi \lambda\!\!\!^{-2}}{2} \frac{(2J + 1)\Gamma^2}{(E - E_R)^2 + \frac{1}{4}\Gamma^2}. \tag{9.88}$$

9.12. The Resonant State with Reaction

When the projectile particle is close to the target particle, the two interact, forming an unstable state dependent on quantum numbers l and J. The break-up of such a state is governed by (5.144) and (6.30).

Now, the cross section for a given energy E and angular frequency ω is proportional to the density of the contribution of the corresponding state to the mixed state; that is, proportional to (6.33). Thus,

$$\sigma = (\text{constant}) \frac{A^*A}{(\omega_R - \omega)^2 + 1/(4\tau^2)}. \tag{9.89}$$

Since

$$E_R = \hbar\omega_R, \qquad E = \hbar\omega, \tag{9.90}$$

and

$$\Gamma = \frac{\hbar}{\tau}. \tag{9.91}$$

Equation (9.89) agrees with (9.88), and (9.86), when only elastic scattering occurs.

In general, reaction is present, as well as simple scattering. If the widths for elastic scattering and for reaction are Γ_{el} and Γ_{re}, respectively, we have

$$\Gamma = \Gamma_{el} + \Gamma_{re}. \tag{9.92}$$

Also

$$\sigma = \sigma_{el} + \sigma_{re}. \tag{9.93}$$

The appropriate generalizations of (9.88) are

$$\sigma_{\mathrm{el}} = \frac{\pi\lambda\!\!\!^{-2}}{2} \frac{(2J+1)\,\Gamma_{\mathrm{el}}^2}{(E-E_R)^2 + \frac{1}{4}\Gamma^2} \tag{9.94}$$

and

$$\sigma_{\mathrm{re}} = \frac{\pi\lambda\!\!\!^{-2}}{2} \frac{(2J+1)\,\Gamma_{\mathrm{el}}\,\Gamma_{\mathrm{re}}}{(E-E_R)^2 + \frac{1}{4}\Gamma^2}\,. \tag{9.95}$$

9.13. Singularity in the Partial Wave Amplitude Ratio at a Physical State

When parameter k is real, the complete wave function (9.48) represents the scattering of an incident planar wave by a point or small region. But mathematically, we can let k become complex or imaginary and consider a component partial wave with given l. The imaginary k corresponds to a negative energy E, an energy associated with a bound state.

But for the partial wave function to be normalizable when k is complex or imaginary, the amplitude of the term that increases with r must vanish. At the energy where such vanishing occurs, there is a singularity in the ratio of the amplitude of the decreasing exponential term to the amplitude of the increasing term. Since a complex function is determined by its properties at singularities, this behavior turns out to be fundamental.

The lth contribution to wave function (9.48) has the form

$$\psi_l = [A(k)\,\mathrm{e}^{-ikr} + B(k)\,\mathrm{e}^{ikr}]\,\frac{P_l(\cos\theta)}{r} \tag{9.96}$$

in which $A(k)$ and $B(k)$ are the expressions multiplying the pertinent exponential times $P_l(\cos\theta)/r$. Wavevector k is related to energy E as in (4.67). With V zero, we have

$$k = \frac{(2\mu E)^{\frac{1}{2}}}{\hbar}. \tag{9.97}$$

In a mathematical sense, we may consider E to be complex and plotted in a plane. Two revolutions of the E vector then correspond to one revolution of the k vector about the origin. Wavevector k as a function of E appears on two Riemannian sheets. The necessary cut may run from 0 to ∞, as Figure 9.9 shows. The continuum of states corresponding to positive real energies lies along this cut; but each bound state, for which E is negative, appears on the negative real axis.

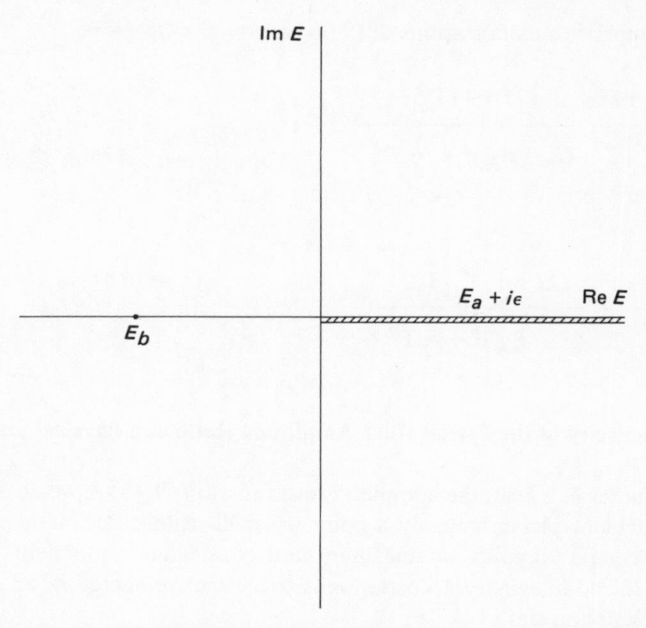

Fig. 9.9. Cut and pole for amplitude ratio $B(k)/A(k)$ in the complex energy plane. Physical states appear at the singularities.

There is no difficulty in replacing k with $i\kappa$, as in (1.134):

$$k = i\kappa \tag{9.98}$$

with

$$\kappa = \frac{(-2\mu E)^{\frac{1}{2}}}{\hbar}. \tag{9.99}$$

Then (9.96) becomes

$$\psi_l = [A(k)\,\mathrm{e}^{\kappa r} + B(k)\,\mathrm{e}^{-\kappa r}]\,\frac{P_l(\cos\theta)}{r}. \tag{9.100}$$

In the negative energy region, κ is real and the first term in (9.100) is not normalizable. For ψ_l itself to be normalizable and a bound state to appear, amplitude $A(k)$ must vanish. The constant term in a power series expansion of this amplitude is therefore zero and

$$A(k) = a_1(E_b - E) + a_2(E_b - E)^2 + \ldots$$
$$\simeq a_1(E_b - E) \tag{9.101}$$

if E_b is the bound-state energy.

The ratio of the second amplitude to the first one is therefore

$$S(k) = \frac{B(k)}{A(k)} \simeq \frac{B(k_b)}{a_1(E_b - E)} = \frac{\lambda}{E_b - E} \tag{9.102}$$

when E is near E_b. A pole with residue λ is thus associated with the bound state. This state may appear as a physical particle.

Discussion Questions

9.1. How is the probability density $\Psi^*\Psi$ analogous to the density of a fluid?

9.2. Why does the equation for the time rate of change of $\Psi^*\Psi$ not contain the potential V?

9.3. How is the dependence of the current density $\Psi^*\Psi\mathbf{v}$ on Ψ obtained?

9.4. Integrate the time-dependent Schrödinger equation to obtain a form for the evolution operator.

9.5. In what ways may a particle be confined?

9.6. When does a wave packet (a) contract, (b) expand?

9.7. Construct the evolution operator for particles moving freely parallel to the x axis.

9.8. Apply this operator to a free Gaussian wave and show that such a state function remains Gaussian with time.

9.9. What effect does movement of the center of mass of the Gaussian wave have on the packet?

9.10. Show that

$$\frac{\partial^2}{\partial x^2} \frac{e^{-x^2/4s^2}}{s} = \frac{\partial}{\partial (s^2)} \frac{e^{-x^2/4s^2}}{s}.$$

9.11. What is a Galilean transformation?

9.12. What is a scattering center? How is the existence of nuclei established?

9.13. Explain what kind of incident beam the function $A\, e^{ikz}\, e^{-i\omega t}$ describes.

9.14. Why does the scattered wave have the form

$$Af(k, \theta, \varphi) \frac{e^{ikr}}{r} e^{-i\omega t}?$$

9.15. What is the fraction of particles in unit area of beam scattered into element $d\Omega$ by a scattering center?

9.16. Why is $d\sigma/d\Omega$ equal to $|f|^2$? How is the cross section for scattering by any angle obtained from f?

9.17. How does the absorption law $I = I_0\, e^{-\sigma c z}$ arise?

9.18. Explain the expressions for (a) the unit-outgoing wave, (b) the unit-incoming wave, (c) the unit-standing waves. Why are waves (a) and (b) independent of each other?

9.19. What is the causality principle? Why is each incoming wavelet unaffected by the nature of the scattering center? How does the center affect the outgoing wavelets?

9.20. Explain what is represented by the sum

$$A \sum_l \frac{2l+1}{2ikr} [e^{ikr} - (-1)^l e^{-ikr}] P_l(\cos\theta).$$

9.21. How are (a) the phase shift δ_l and (b) the amplitude reduction factor η_l introduced into the sum in Question 9.20?

9.22. How is factor f related to this sum?

9.23. Explain how the wavevector-scattering amplitude product varies with energy. What is the unitary circle?

9.24. How is the elastic scattering cross section related to δ_l and η_l?

9.25. How does consideration of the probability of reaction lead to an expression for the reaction cross section?

9.26. When is the reaction cross section equal to the elastic scattering cross section? Why is the cross section of a black disk equal to twice its cross-sectional area?

9.27. What is the range of a center? Why should $\eta_l \to 0$ as $l \to 0$? Under what conditions do the effects on amplitude and phase become small?

9.28. How can σ become as large as the result in Example 9.3?

9.29. Why can nuclear radii not be determined using (a) very slowly moving projectiles, (b) fast moving projectiles?

9.30. Why is nuclear volume proportional to the mass number?

9.31. Under what conditions does a projectile particle resonate with a target particle? Explain.

9.32. How is the phase shift related to the width of the resonance peak? How is the cross section related to this width?

9.33. How is a resonant state similar to a radioactive nucleus?

9.34. Explain what pole may be associated with an observed physical particle.

9.35. How should ψ_l in (9.96) be modified when quantum number m is not 0? Would this modification alter the subsequent argument establishing a pole at a bound state?

Problems

9.1. Show that if μ is the mass of a particle and p the operator for its momentum, then

$$\frac{\partial}{\partial t}(\mu\Psi^*\Psi) + \nabla \cdot \mathrm{Re}\,\Psi^* p \Psi = 0.$$

9.2. Show how formula (9.9) applies when the system is in a single energy state.

9.3. How thick must a layer of zinc be to reduce a beam of thermal neutrons to 2.00 per cent of its initial intensity? The cross section of zinc for such neutrons is 1.06 barns.

9.4. A target of magnesium 7.80 cm thick absorbed 1.96 per cent of a beam of thermal neutrons. What is the average cross section of a magnesium atom in the target?

9.5. Determine the Schrödinger equation that

$$G_{\pm}(r)\, Y_l^m(\theta, \varphi)$$

satisfies sufficiently far from the origin.

9.6. Show that $1 + 2ikf_l \equiv S_l$ is the operator that introduces the effect of the scattering center on the lth outgoing wave. Show that $S_l^* S_l = 1$ when $\eta_l = 1$.

9.7. What is the Coulomb barrier for the reaction

$${}^{24}_{12}\mathrm{Mg} + {}^{2}_{1}\mathrm{H} \rightarrow {}^{1}_{1}\mathrm{H} + {}^{25}_{12}\mathrm{Mg}?$$

9.8. At what energies near a resonance is $\sigma/\sigma_{\max}$ equal to 0.900?

9.9. Determine the mass current density in the outgoing wave

$$\Psi = A\frac{f}{r} e^{ikr}\, e^{-i\omega t}.$$

9.10. Determine the spatial and temporal variations in the probability density associated with a free wave packet for which the wavevector representation is $\phi = e^{-au^2/2}$, if ω is a quadratic function of u and u is $k - \bar{k}$.

9.11. How much is the intensity of a beam of thermal neutrons reduced on passing through 0.250 cm nickel and 2.00 cm iron? The pertinent cross section of Ni is 4.50 barns, that of Fe 2.43 barns.

9.12. Express the contribution of the partial wave with quantum number l to exp ikz as a function of θ times a trigonometric function of kr divided by kr.

9.13. Show what Schrödinger equation is satisfied by the expression

$$\frac{1}{r}(A \cos kr + B \sin kr).$$

9.14. What are the phase shifts for $l = 0, 1, 2, \ldots$ when each outgoing wave is displaced distance a toward the center by an attractive potential? What is the corresponding scattering amplitude-wavevector product kf_l?

9.15. If resonance occurs for a given l when k is 10^{13} cm^{-1}, by how much does the potential shift the corresponding partial wave in toward the center?

9.16. Determine the normal distribution curve that passes through the Breit–Wigner curve at the points where $\sigma/\sigma_{\max}$ is 1 and 0.500.

References

Books

Burkhardt, H.: 1969, *Dispersion Relation Dynamics*, American Elsevier, New York, pp. 80–92.

Merzbacher, E.: 1970, *Quantum Mechanics*, 2nd edn, Wiley, New York, pp. 215–250.

Newton, R. G.: 1966, *Scattering Theory of Waves and Particles*, McGraw-Hill, New York, pp. 3–150.

Perkins, D. H.: 1972, *Introduction to High Energy Physics*, Addison-Wesley, Reading, Mass., pp. 265–287.

Rapp, D.: 1971, *Quantum Mechanics*, Holt, Rinehart and Winston, New York, pp. 519–551.

Articles

Balasubramanian, S.: 1980, 'Comment on "Use of the Time Evolution Operator"', *Am. J. Phys.* **48**, 985–986.

Berrondo, M., and Garcia-Calderon, G.: 1982, 'Resonances in Repulsive Singular Potentials as an Eigenvalue Problem', *Eur. J. Phys.* **3**, 34–38.

Berry, M. V., and Balazs, N. L.: 1979, 'Nonspreading Wave Packets', *Am. J. Phys.* **47**, 264–267.

Bramhall, M. H., and Casper, B. M.: 1970, 'Reflections on a Wave Packet Approach to Quantum Mechanical Barrier Penetration', *Am. J. Phys.* **38**, 1136–1145.

Chalk, J.: 1981, 'Elementary Approach to Separable Potentials of Exponential Form', *Am. J. Phys.* **49**, 1069–1071.

Clark, C. W.: 1979, 'Coulomb Phase Shift', *Am. J. Phys.* **47**, 683–684.

Corinaldesi, E., and Rafeli, F.: 1978, 'Aharonov–Bohm Scattering by a Thin Impenetrable Solenoid', *Am. J. Phys.* **46**, 1185–1187.

Delville, A., Jasselette, P., and Vandermeulen, J.: 1978, 'Elementary Quantum Mechanics in a High-Energy Process', *Am. J. Phys.* **46**, 907–909.

Downs, B. W., and Ram, B.: 1978, 'Hard-Core Potentials in Quantum Mechanics', *Am. J. Phys.* **46**, 164–168.

Farina, J. E. G.: 1977, 'Classical and Quantum Spreading of Position Probability', *Am. J. Phys.* **45**, 1200–1202.

Fernow, R. C.: 1976, 'Expansions of Spin-1/2 Expectation Values', *Am. J. Phys.* **44**, 560–563.

Gijzeman, O. L. J., and Naqvi, K. R.: 1979, 'Use of the Time Evolution Operator', *Am. J. Phys.* **47**, 384–385.

Greenman, J. V.: 1972, 'Non-Dispersive Mirror Wavepackets', *Am. J. Phys.* **40**, 1193–1201.

Grübl, G. J., and Leubner, C.: 1980, 'Current Misconception Concerning the Time Evolution Operator', *Am. J. Phys.* **48**, 484–486.

Hobbie, R. K.. 1962, 'A Simplified Treatment of the Quantum-Mechanical Scattering Problem Using Wave Packets', *Am. J. Phys.* **30**, 857–864.

Holstein, B. R., and Swift, A. R.: 1972, 'Spreading Wave Packets – A Cautionary Note', *Am. J. Phys.* **40**, 829–832.

Ignatovich, V. K.: 1978, 'Nonspreading Wave Packets in Quantum Mechanics', *Found. Phys.* **8**, 565–571.

Kayser, B.: 1974, 'Classical Limit of Scattering in a $1/r^2$ Potential', *Am. J. Phys.* **42**, 960–964.

Keller, J. B.: 1972, 'Quantum Mechanical Cross Sections for Small Wavelengths', *Am. J. Phys.* **40**, 1035–1036.

Kujawski, E.: 1971, 'Additivity of Phase Shifts for Scattering in One Dimension', *Am. J. Phys.* **39**, 1248–1254.

Lapidus, I. R.: 1973, 'Scattering of Charged Particles by a Localized Magnetic Field', *Am. J. Phys.* **41**, 822–823.

Lapidus, I. R.: 1982, 'Quantum-Mechanical Scattering in Two Dimensions', *Am. J. Phys.* **50**, 45–47.

Lee, H.-W.: 1982, 'Spreading of a Free Wave Packet', *Am. J. Phys.* **50**, 438–440.

Lieber, M.: 1973, 'The Complete Asymptotic Expansion of the Continuum Schrödinger Wave Function', *Am. J. Phys.* **41**, 497–502.

Meijer, P. H. E., and Repace, J. L.: 1975, 'Phase Shifts of the Three-Dimensional Spherically Symmetric Square Wave Potential', *Am. J. Phys.* **43**, 428–433.

Segre, C. U., and Sullivan, J. D.: 1976, 'Bound-State Wave Packets', *Am. J. Phys.* **44**, 729–732.

Singh, I., and Whitaker, M. A. B.: 1982, 'Role of the Observer in Quantum Mechanics and the Zeno Paradox', *Am. J. Phys.* **50**, 882–887.

Synge, J. L.: 1972, 'A Special Class of Solutions of the Schrödinger Equation for a Free Particle', *Found. Phys.* **2**, 35–40.
Woodsum, H. C., and Brownstein, K. R.: 1977, 'Tumbling Motion of Free Three-Dimensional Wave Packets', *Am. J. Phys.* **45**, 667–670.

Chapter 10

Investigating Multiparticle Systems

10.1. Quantum Mechanical Particles and Quasi Particles

In classical physics, each element of mass, each particle, is supposed to move as a coherent body along a smooth curve through space and time. It is distinguishable from each of its neighbors, whether or not it interacts with any of them.

But in modern physics, we relinquish this naïve picture. A submicroscopic particle, or quasi particle, makes itself known by what it does at a certain point in space and time, indeed, by an identifiable event such as a momentum transfer in a collision. When the entity survives, it may produce a sequence of events, as the vapor trail in a cloud chamber. When the entity can exist by itself in a vacuum, it is called a real *particle*. When it appears merely as a mode in a field or in material, the entity is called a *quasi particle*.

The potentiality for appearing at a given point in the space and time imposed on the submicroscopic world is measured by the probability density ρ. This density is related to the spatial function describing the state of the particle by the equation

$$\rho = \Psi^*\Psi. \tag{10.1}$$

Alternatively, one can employ a density over momentum. This equals $\Phi^*\Phi$, where Φ is the appropriate Fourier transform of Ψ, as we have seen.

The state described by either function may be a mixed state with respect to the procedure used. Such a mixed state is representable as a superposition of states that are pure with respect to the observation; and each pure state is an eigenstate with respect to the operator for the observation. The most important eigenequation is the energy equation, for which the operator is the Hamiltonian operator.

The generalization to a multiparticle system is not quite obvious, since particles of the same kind may not be distinguishable. Indeed, the different molecules in a macroscopic amount of a pure substance are indistinguishable whenever they mix freely. Different atoms of the same species occupying positions that permute are indistinguishable. Different electrons in an atom or molecule mix freely and so are not distinguishable. Different protons in a nucleus similarly are not distinguishable; and different neutrons in a nucleus are not distinguishable. Photons carrying the same energy in a given field are not distinguishable, and a like remark applies to phonons, excitons, rotons, and so on.

When a particle, or quasi particle, is assigned coordinates in the imposed space and time, it is made distinguishable. Yet such assignment is necessary in a mathematical discussion. Fortunately, one can later introduce the property of indistinguishability wherever it is needed.

10.2. The Variation Theorem

From our earlier discussions, we know that a multiparticle system is governed by a wave function Ψ which varies with the coordinates of each particle. These coordinates are assigned subject to the caveat just noted.

When the system is in a definite energy state E_j, the wave function satisfies a Schrödinger equation of the form

$$H\Psi_j = E_j\Psi_j. \tag{10.2}$$

With the radius vector locating the kth particle being

$$\mathbf{r}_k = x_k\hat{\mathbf{x}} + y_k\hat{\mathbf{y}} + z_k\hat{\mathbf{z}}, \tag{10.3}$$

the corresponding Laplacian operator is

$$\nabla_k^2 = \frac{\partial^2}{\partial x_k^2} + \frac{\partial^2}{\partial y_k^2} + \frac{\partial^2}{\partial z_k^2} \tag{10.4}$$

and the Hamiltonian operator is

$$H = \sum_{k=1}^{N} -\frac{\hbar^2}{2\mu k}\nabla_k^2 + V(\mathbf{r}_1, \mathbf{r}_2, \ldots, \mathbf{r}_N), \tag{10.5}$$

as long as the vector potential and the corresponding magnetic effects can be neglected.

In general, there are an infinite number of eigenfunctions that are well behaved and that fit the boundary conditions. These superpose linearly to produce any possible state. Furthermore, with the infinite flexibility available, an arbitrary well-behaved function meeting the boundary conditions can be represented in this way:

$$\Psi = \Sigma\, c_j\Psi_j. \tag{10.6}$$

If we multiply (10.2) by Ψ_j^*, integrate over all coordinates, and solve for the energy, we obtain

$$E_j = \frac{\int_R \Psi_j^* H\Psi_j\, \mathrm{d}^{3N}\mathbf{r}}{\int_R \Psi_j^* \Psi_j\, \mathrm{d}^{3N}\mathbf{r}}. \tag{10.7}$$

When the jth function is normalized to 1, this reduces to

$$E_j = \int_R \Psi_j^* H \Psi_j \, d^{3N} \mathbf{r}. \tag{10.8}$$

With an orthogonal set of normalized eigenfunctions, the equation

$$\sum_{j=1}^{\infty} c_j^* c_j = 1 \tag{10.9}$$

then ensures the like normalization of function Ψ.

Let us construct

$$E = \int_R \Psi^* H \Psi \, d^{3N} \mathbf{r}$$

$$= \int_R \Sigma \, c_j^* \Psi_j^* H \, \Sigma \, c_k \Psi_k \, d^{3N} \mathbf{r}$$

$$= \Sigma \, c_j^* c_j E_j. \tag{10.10}$$

There is a lower limit on the energy of the system. Furthermore, we may order the energies so

$$E_j \geqslant E_{j-1}. \tag{10.11}$$

Then (10.10) and (10.9) tell us that

$$E_1 \leqslant \int_R \Psi^* H \Psi \, d^{3N} \mathbf{r}. \tag{10.12}$$

Consequently, if one substitutes an approximate Ψ that varies with one or more parameters into (10.12), the best values for the parameters are those that make the integral on the right a minimum. This result is a form of the *variation theorem*. When trial function Ψ is not normalized, one employs

$$E_1 \leqslant \frac{\int_R \Psi^* H \Psi \, d^{3N} \mathbf{r}}{\int_R \Psi^* \Psi \, d^{3N} \mathbf{r}}. \tag{10.13}$$

If one subtracts out the contribution of the first eigenfunction to Ψ, then minimizing the right side of (10.13) yields a value equal to or greater than E_2. Furthermore, if one subtracts out the contributions of the first $j-1$ eigenfunctions, minimizing the right side of (10.13) yields a value equal to or greater than E_j. As a consequence, the first n minima of the original ratio of integrals yields an approximation to $E_1, E_2, \ldots, E_n$.

We have

$$E_j \leqslant \text{Min} \frac{\int_R \Psi^* H \Psi \, d^{3N} \mathbf{r}}{\int_R \Psi^* \Psi \, d^{3N} \mathbf{r}}. \tag{10.14}$$

When the Hamiltonian does not contain the time variable, the preceding arguments can be carried through with the temporal factor canceled. In place of (10.14), we obtain

$$E_j \leqslant \text{Min}\,\frac{\int_R \psi^* H\psi\, \mathrm{d}^{3N}\mathbf{r}}{\int_R \psi^*\psi\, \mathrm{d}^{3N}\mathbf{r}}. \tag{10.15}$$

For convenience, the ratio of integrals may be designated E.

Example 10.1. If a particle is confined between $x = 0$ and $x = 1$ at $V = 0$, what exponent b in the approximate wave function

$$\psi = A(x - x^b)$$

yields the best ground state energy?

The variation theorem in the form of (10.15) enables one to choose the best value for b and obtain the corresponding energy. First, construct the Hamiltonian operator with μ the mass of the particle and with V equal to zero:

$$H = -\frac{\hbar^2}{2\mu}\frac{\mathrm{d}^2}{\mathrm{d}x^2} = -\frac{h^2}{8\mu\pi^2}\frac{\mathrm{d}^2}{\mathrm{d}x^2} = -B\frac{\mathrm{d}^2}{\mathrm{d}x^2}.$$

Let this Hamiltonian act on the given function:

$$H\psi = -B\frac{\mathrm{d}^2}{\mathrm{d}x^2}A(x - x^b) = ABb(b-1)x^{b-2}.$$

Then

$$\psi^*\psi = A^2(x^2 - 2x^{b+1} + x^{2b})$$

and

$$\psi^* H\psi = A^2Bb(b-1)(x^{b-1} - x^{2b-2}).$$

Substituting into the integrals in (10.15) yields

$$\int_R \psi^* H\psi\,\mathrm{d}x = A^2Bb(b-1)\int_0^1 (x^{b-1} - x^{2b-2})\,\mathrm{d}x$$

$$= A^2Bb(b-1)\left(\frac{x^b}{b} - \frac{x^{2b-1}}{2b-1}\right)\Big|_0^1$$

$$= A^2B\,\frac{(b-1)\,[(2b-1)-b]}{2b-1} = A^2B\,\frac{(b-1)(b-1)}{2b-1}$$

and

$$\int_R \psi^* \psi \, dx = A^2 \int_0^1 (x^2 - 2x^{b+1} + x^{2b}) \, dx$$

$$= A^2 \left(\frac{x^3}{3} - \frac{2x^{b+2}}{b+2} + \frac{x^{2b+1}}{2b+1} \right) \Bigg|_0^1$$

$$= A^2 \frac{2(b^2 - 2b + 1)}{3(b+2)(2b+1)},$$

whence

$$E = \frac{\int_R \psi^* H \psi \, dx}{\int_R \psi^* \psi \, dx} = B \frac{3}{2} \frac{2b^2 + 5b + 2}{2b - 1}.$$

To find the minimum, differentiate and set the result equal to zero:

$$\frac{dE}{db} = B \frac{3}{2} \frac{4b^2 - 4b - 9}{(2b-1)^2} = 0.$$

Solve this equation for the positive b,

$$b = 2.0811,$$

and substitute the result into the equation for E:

$$E = \frac{h^2}{8\mu\pi^2} \, 1.5 \, \frac{(5.1623)(4.0811)}{3.1623} = \frac{h^2}{8\mu} \, 1.0125.$$

From (1.125), the exact energy for the system is

$$E = \frac{h^2}{8\mu}.$$

The variational function

$$A(x - x^b)$$

has yielded an energy that is 1.25 per cent high.

10.3. Separation of Particle Variables

Systems containing more than a few particles generally yield intractable Schrödinger equations. However, approximation methods exist which yield much information when certain conditions are met.

Thus, a system may consist of weakly interacting, relatively simple, composite particles, or single particles. Recall, for instance, the situation in a dilute gas. In any case, the Laplacians for the composites, or singles, separate from each other. And, the potentials separate in the approximation that the interactions between the units are feeble.

If H_j is the part of the Hamiltonian operator involving only coordinates of the jth composite or single, when these weak interactions are neglected, we have

$$H \simeq H_1 + H_2 + \ldots + H_n. \tag{10.16}$$

From Section 1.7, the rule for combining probabilities is met when the constituent state functions are multiplied together. Letting $\Psi^{(j)}$ be the state function for the jth unit, we set

$$\Psi = \Psi^{(1)} \Psi^{(2)} \ldots \Psi^{(n)}. \tag{10.17}$$

Then letting the eigenvalue of H_j be ϵ_j when its eigenfunction is $\Psi^{(j)}$ leads to

$$\begin{aligned} H\Psi &= (H_1 + H_2 + \ldots + H_n)\Psi^{(1)} \Psi^{(2)} \ldots \Psi^{(n)} \\ &= \Psi^{(2)} \ldots \Psi^{(n)} H_1 \Psi^{(1)} + \Psi^{(1)} \Psi^{(3)} \ldots \Psi^{(n)} H_2 \Psi^{(2)} + \ldots + \\ &\quad + \Psi^{(1)} \Psi^{(2)} \ldots \Psi^{(n-1)} H_n \Psi^{(n)} \\ &= \Psi^{(2)} \ldots \Psi^{(n)} \epsilon_1 \Psi^{(1)} + \Psi^{(1)} \Psi^{(3)} \ldots \Psi^{(n)} \epsilon_2 \Psi^{(2)} + \ldots + \\ &\quad + \Psi^{(1)} \Psi^{(2)} \ldots \Psi^{(n-1)} \epsilon_n \Psi^{(n)} \\ &= (\epsilon_1 + \epsilon_2 + \ldots + \epsilon_n)\Psi^{(1)} \Psi^{(2)} \ldots \Psi^{(n)} \\ &= E\Psi. \end{aligned} \tag{10.18}$$

The energy of the system equals the sum of the energies of the individual particles:

$$E = \epsilon_1 + \epsilon_2 + \ldots + \epsilon_n. \tag{10.19}$$

In many systems, no breakdown into nearly independent, materially distinct, composites exists. Yet, in these, one may be able to find interlocking composites, modes of cooperative movements, which are nearly independent and which lead to reductions of type (10.16). The quantum entities resulting are called quasi particles, as we noted in Section 10.1.

When the particles interact strongly with each other, the average effect on any one particle may be estimated and the resulting energy and eigenfunction calculated. This may be done for each particle in turn and the resulting potential acting on each particle obtained. If this is different from the assumed potential, adjustments are made and the process repeated. The cycles are continued until a check is found; that is, until the interaction field is *self-consistent*.

This procedure was developed by Douglas R. Hartree in 1928 and refined by Vladimir A. Fock in 1930. They considered various atomic structures. In these and in molecules, the self-consistent function for a single electron is called an *orbital*.

The kth orbital satisfies the pseudo-eigenvalue equation

$$Fu_k = \epsilon_k u_k, \tag{10.20}$$

in which F is the Hartree–Fock operator and ϵ_k is the orbital energy. This operator does not contain the internuclear potential (when it exists). But in it, the average interelectronic potential separates:

$$F = -\frac{\hbar^2}{2\mu}\nabla_k^2 - \sum_i \frac{Z_i e^2}{4\pi\epsilon_0 r_{ik}} + V^{\text{eff}}. \tag{10.21}$$

Here r_{ik} is the distance from the kth electron to the ith nucleus while V^{eff} represents the effective potential of the electron in the kth orbital due to the field of the $n - 1$ other electrons.

If one adds all the orbital energies, the interelectron repulsion energy G is counted twice. Therefore, this energy has to be subtracted from the sum to get the total energy of the system. We write

$$E = \sum_{k=1}^{n} \epsilon_k - G. \tag{10.22}$$

Example 10.2. Construct an approximation to the ground-state wave function for a helium-like atom from the appropriate hydrogen-like orbitals.

In the ground state the two electrons in a helium-like atom are in their 1s orbitals. When the effects of interelectron interaction are neglected, the radial factor from Table 4.2 applies. Combining it with the angular factor from Table 7.2 for each electron in turn yields

$$\psi^{(1)} = \frac{1}{\sqrt{\pi}}\left(\frac{Z}{a_0}\right)^{3/2} e^{-Zr_1/a_0},$$

$$\psi^{(2)} = \frac{1}{\sqrt{\pi}}\left(\frac{Z}{a_0}\right)^{3/2} e^{-Zr_2/a_0}.$$

Multiplying these as in (10.17) then leads to

$$\psi = \frac{1}{\pi}\left(\frac{Z}{a_0}\right)^{3} e^{-Z(r_1 + r_2)/a_0}.$$

Example 10.3. Estimate the interelectron repulsion energy G for the wave function in Example 10.2.

The interelectron potential operator is

$$V_{12} = \frac{e^2}{4\pi\epsilon_0 r_{12}},$$

where r_{12} equals the distance between the first electron and the second electron. From (5.39), the corresponding expectation value is

$$G = \int_R \psi^* V_{12} \psi \, \mathrm{d}^6\mathbf{r}$$

$$= \iint_R \frac{1}{\pi}\left(\frac{Z}{a_0}\right)^3 \mathrm{e}^{-Z(r_1 + r_2)/a_0} \frac{e^2}{4\pi\epsilon_0 r_{12}} \frac{1}{\pi}\left(\frac{Z}{a_0}\right)^3 \mathrm{e}^{-Z(r_1 + r_2)/a_0} \, \mathrm{d}^3\mathbf{r}_1 \, \mathrm{d}^3\mathbf{r}_2$$

$$= \frac{1}{\pi^2}\left(\frac{Z}{a_0}\right)^6 \frac{e^2}{4\pi\epsilon_0} \iint_R \frac{\mathrm{e}^{-2Zr_1/a_0} \, \mathrm{e}^{-2Zr_2/a_0}}{r_{12}} \, \mathrm{d}^3\mathbf{r}_1 \, \mathrm{d}^3\mathbf{r}_2 .$$

Carrying out the indicated integration yields

$$G = \frac{5}{8} \frac{e^2}{4\pi\epsilon_0 a_0} Z.$$

Since the energy of a $1s$ orbital in a hydrogen-like atom is

$$\epsilon_{1s} = -\frac{1}{2} \frac{e^2}{4\pi\epsilon_0 a_0} Z^2,$$

the approximate energy of the helium-like atom is

$$E = 2\epsilon_{1s} + G$$

$$= -\frac{e^2}{4\pi\epsilon_0 a_0}\left(Z^2 - \frac{5}{8} Z\right)$$

$$= 2E_{1s}(\mathrm{H})\left(Z^2 - \frac{5}{8} Z\right),$$

where $E_{1s}(\mathrm{H})$ is the energy of the ground state of hydrogen; that is, 13.606 eV.

For He itself, we obtain

$$E = -27.212\left(4 - \frac{10}{8}\right) \text{eV}$$

$$= -74.83 \text{ eV}.$$

The observed energy is -78.98 eV.

Example 10.4. How may the six-dimensional integration in Example 10.3 be carried out? How might a Hartree-Fock calculation for the helium-like atom proceed?

The integral after the constants have been factored out may be interpreted as the repulsion energy, in units such that $e/(4\pi\epsilon_0)^{1/2} = q$, of a charge cloud of density $\exp(-2Z'r_1)$ on one of density $\exp(-2Z'r_2)$, where $Z' = Z/a_0$. Since the first cloud is spherically symmetric, Gauss's law tells us that the electrostatic potential due to a spherical shell of this cloud of radius r_1 and thickness dr_1 is

$$4\pi r_1^2\, e^{-2Z'r_1}\, dr_1 \frac{1}{r_1} = 4\pi r_1\, e^{-2Z'r_1}\, dr_1$$

when $r < r_1$ and

$$4\pi r_1^2\, e^{-2Z'r_1}\, dr_1 \frac{1}{r}$$

when $r > r_1$. The net electrostatic potential at radius r is

$$\phi(r) = \frac{4\pi}{r}\int_0^r e^{-2Z'r_1} r_1^2\, dr_1 + 4\pi\int_r^\infty e^{-2Z'r_1} r_1\, dr_1$$

$$= \frac{\pi}{Z'^3}\frac{1}{r}\left[1 - e^{-2Z'r}(1 + Z'r)\right].$$

Integrating $\phi(r)$ over the density of the second cloud yields

$$\frac{\pi}{Z'^3}\int_0^\infty \frac{1}{r}\left(e^{-2Z'r} - e^{-4Z'r} - Z'r\, e^{-4Z'r}\right) 4\pi r^2\, dr$$

$$= \frac{4\pi^2}{Z'^3}\left(\frac{1}{4Z'^2} - \frac{1}{16Z'^2} - \frac{1}{32Z'^2}\right) = \frac{5\pi^2}{8Z'^5} = \frac{5}{8}\pi^2\left(\frac{a_0}{Z}\right)^5.$$

Substituting this result into the formula for G gives us

$$G = \frac{1}{\pi^2}\left(\frac{Z}{a_0}\right)^6 \frac{e^2}{4\pi\epsilon_0}\frac{5}{8}\pi^2\left(\frac{a_0}{Z}\right)^5 = \frac{5}{8}\frac{e^2}{4\pi\epsilon_0 a_0} Z,$$

the expression used in Example 10.3.

One could solve the Hartree–Fock equation numerically with V^{eff} the potential of the electron in the field of the other electron in a hydrogen-like 1*s* orbital. In this way, an altered orbital would be obtained. One can construct an orbital intermediate between this and the 1*s* orbital to use in the next calculation. One would repeat the iterations until a check is found. The resulting orbital energies and interelectronic repulsion energy are calculated and substituted into (10.22).

10.4. Indistinguishability of Identical Particles in a Quantal System

Even though classical bodies may be made exactly alike in principle, they always remain distinct. One can be picked out and followed through its gyrations and collisions because it has a definite trajectory. But in quantum mechanics, the movements of a particle are described by a wave function. No definite trajectory exists. As a result, a particle can be kept distinct from other particles of the same species only if its wave function is zero wherever the wave function of each of the other identical particles is not zero.

Whenever a given particle moves through the same region as other particles having the same attributes, distinct paths do not exist to distinguish it from the others. No person can then assign the given particle to any of the constituent states. Every identical particle in the set must appear distributed among the pertinent orbitals.

No physically determinable property can depend on how the exchangeable particles are labeled. In particular, expectation values must remain unaltered in any allowed permutation of the particles. The labeling is necessary in the mathematical description, however.

There is no unique way to construct a wave function that meets these conditions. The wave function for the system may be either symmetric or antisymmetric with respect to the interchange of any two like particles. Or, it might exhibit a definite mixed symmetry.

Example 10.5. In what different independent ways can two particles be placed in the orbitals $u_1(\mathbf{r}_j)$ and $u_2(\mathbf{r}_j)$?

When the particles are distinguishable, each can be put into either orbital. To satisfy the rule for combining probabilities, we combine the orbitals multiplicatively, getting

$$u_1(\mathbf{r}_1)\,u_1(\mathbf{r}_2),$$

$$u_2(\mathbf{r}_1)\,u_2(\mathbf{r}_2),$$

$$u_1(\mathbf{r}_1)\,u_2(\mathbf{r}_2),$$

$$u_1(\mathbf{r}_2)\,u_2(\mathbf{r}_1).$$

But when the particles are indistinguishable, each product containing a permutation of the particles from the order in the first expression must also appear with the same statistical weight. Since exchanging $\mathbf{r}_1$ and $\mathbf{r}_2$ in the first two products does not change either, both are suitable as they stand. However, exchanging $\mathbf{r}_1$ and $\mathbf{r}_2$ in the last two products interchanges the expressions. So the last two possibilities must be superpositions of these.

One may get orthogonal, independent, forms by choosing the sum and the difference. When $u_1(\mathbf{r}_j)$ and $u_2(\mathbf{r}_j)$ are normalized and orthogonal, the normalized forms are

$$\frac{1}{\sqrt{2}}\left[u_1(\mathbf{r}_1)\,u_2(\mathbf{r}_2)+u_1(\mathbf{r}_2)\,u_2(\mathbf{r}_1)\right]$$

and

$$\frac{1}{\sqrt{2}}\left[u_1(\mathbf{r}_1)\,u_2(\mathbf{r}_2)-u_1(\mathbf{r}_2)\,u_2(\mathbf{r}_1)\right].$$

The first two products, and the next to last superposition, are symmetric, while the last is antisymmetric, with respect to interchange of the two particles.

It is convenient to replace $\mathbf{r}_j$ by merely the integer j. Then the three symmetric combinations are

$$u_1(1)\,u_2(2),$$

$$\frac{1}{\sqrt{2}}\left[u_1(1)\,u_2(2)+u_1(2)\,u_2(1)\right],$$

$$u_1(2)\,u_2(1),$$

while the antisymmetric combination is

$$\frac{1}{\sqrt{2}}\left[u_1(1)\,u_2(2)-u_1(2)\,u_2(1)\right].$$

Example 10.6. In what different mutually exclusive ways can two particles be placed in the spin functions $\chi_+(\tau_j)=\alpha(\tau_j)$, for $m_s=+\frac{1}{2}$, and $\chi_-(\tau_j)=\beta(\tau_j)$, for $m_s=-\frac{1}{2}$?

As in Example 10.4, when the particles are distinguishable, we have the independent forms

$$\alpha(\tau_1)\alpha(\tau_2),$$

$$\beta(\tau_1)\beta(\tau_2),$$

$$\alpha(\tau_1)\beta(\tau_2),$$

$$\alpha(\tau_2)\beta(\tau_1).$$

When the particles are indistinguishable, one must superpose on a given form the different forms which are obtained by permuting the particles in the given form, each with the same weight. From the first two in the list here, no different forms are obtained. But from the third form, one obtains the fourth; and from the fourth form, one obtains the third.

Two suitable normalized orthogonal combinations obtained from these latter two include the symmetric combination

$$\frac{1}{\sqrt{2}}\,[\alpha(\tau_1)\beta(\tau_2)+\alpha(\tau_2)\beta(\tau_1)]$$

and the antisymmetric combination

$$\frac{1}{\sqrt{2}}\,[\alpha(\tau_1)\beta(\tau_2)-\alpha(\tau_2)\beta(\tau_1)].$$

It is convenient to replace the spin coordinate of the jth particle τ_j by integer j. Then the symmetric combinations are

$$\alpha(1)\,\alpha(2), \qquad m_s = 1,$$

$$\frac{1}{\sqrt{2}}\,[\alpha(1)\,\beta(2)+\alpha(2)\,\beta(1)], \qquad m_s = 0,$$

$$\beta(1)\,\beta(2), \qquad m_s = -1,$$

while the antisymmetric combination is

$$\frac{1}{\sqrt{2}}\,[\alpha(1)\,\beta(2)-\alpha(2)\beta(1)], \qquad m_s = 0.$$

10.5. Fundamental Combinatory Rules

The product in (10.17) puts the first particle in the first state function, the second particle in the second state function, . . . , and the nth particle in the nth state function. To the extent that the functions are distinguishable, it labels the particles, making them distinguishable.

But if the particles are really indistinguishable, as we have just argued, this form with distinguishable functions is not suitable. Instead, a superposition of such forms, or a set of superpositions, must then be employed. The result must reflect the prevailing permutational symmetry.

Let us consider a system of n identical particles in a definite energy state E. Now, coordinates have to be assigned the particles in constructing the Hamiltonian operator H. However, the resulting mathematical expression has to remain unaffected by any permutation of the identical particles.

Thus, if P_k is an operator that effects a certain permutation of these particles, we must have

$$P_k H = H. \tag{10.23}$$

The Schrödinger equation for the state is

$$H\Psi = E\Psi. \tag{10.24}$$

Letting permutation P_k act on (10.24) yields

$$P_k(H\Psi) = HP_k\Psi = EP_k\Psi. \tag{10.25}$$

From the last equality, expression $P_k\Psi$ is an eigenfunction of H with the same eigenvalue E as Ψ itself exhibits. When the energy level is not degenerate, the new eigenfunction is a constant times the old one:

$$P_k\Psi = c\Psi. \tag{10.26}$$

Operator P_k may merely interchange the ith and jth particles; we then write

$$P_k = (ij) \tag{10.27}$$

and

$$(ij)\Psi = c\Psi. \tag{10.28}$$

But repetition of this operation restores the initial order; so

$$\Psi = (ij)^2\Psi = c(ij)\Psi = c^2\Psi \tag{10.29}$$

and

$$c^2 = 1, \tag{10.30}$$

whence

$$c = \pm 1. \tag{10.31}$$

The positive sign implies that interchange of the two identical particles leaves Ψ unchanged; the wave function is then said to be *symmetric* with respect to the interchange. Systems of such particles obey the statistics developed by Satyandra N. Bose and Albert Einstein. The particles themselves are called *bosons*.

The negative sign in (10.31) implies that interchange of the two identical particles changes the sign of Ψ. The wave function is then said to be *antisymmetric* with respect to the interchange. Systems of such particles obey the statistics developed by Enrico Fermi and Paul A. M. Dirac. The particles themselves are called *fermions*.

To explain experimental data, one has to treat fundamental particles, such as electrons, neutrinos, protons, and neutrons, as fermions. The field particles, photons and gravitons, are bosons.

Example 10.7. Show how the orbital part of ψ may be either symmetric or antisymmetric when ψ itself is antisymmetric, as in describing an electron system.

In an atom or molecule, the spins of the electrons are independent of the orbital motions to the approximation that magnetic effects can be neglected. Then the wave function factors into an orbital part and a spin part:

$$\psi = u(\mathbf{r}_1, \mathbf{r}_2, \ldots, \mathbf{r}_n)\chi(\tau_1, \tau_2, \ldots, \tau_n).$$

Let exchange operator (ij) act on this equation:

$$(ij)\psi = [(ij)u]\ [(ij)\chi].$$

We suppose that i and j may be any two different integers from 1 to, and including, n.

When u is symmetric with respect to the interchange of the ith particle with the jth particle, operator (ij) does not alter u:

$$(ij)u = u.$$

But then, we may have

$$(ij)\chi = -\chi;$$

that is, the spin factor can be antisymmetric with respect to i and j. Substituting in then yields

$$(ij)\psi = [u]\ [-\chi] = -u\chi = -\psi,$$

and ψ is an antisymmetric wave function.

When u is antisymmetric with respect to the ith and jth particles, we have

$$(ij)u = -u.$$

We now choose the spin factor so that

$$(ij)\chi = \chi.$$

Then

$$(ij)\psi = [-u]\ [\chi] = -u\chi = -\psi,$$

and the wave function is still antisymmetric.

10.6. Symmetrizing and Antisymmetrizing Operations

A function that varies with the coordinates of the identical particles in a system can be made symmetric or antisymmetric with respect to exchange of some or all of these particles by the application of suitable operators.

Let the identical particles be initially labeled by the numbers 1, 2, . . . , N and let $\mathbf{r}_j$ be the radius vector drawn from the origin to the jth particle. Also, consider any suitable function of these coordinates

$$\psi_1(\mathbf{r}_1, \mathbf{r}_2, \ldots, \mathbf{r}_N). \tag{10.32}$$

Some permutations of particles may not affect (10.32) at all, while others may alter it. Let us identify the permutations that do change (10.32) and label the operator that effects the kth permutation by the symbol P_k.

Now, one may merely add the results and renormalize. Thus, one may construct

$$\psi = B \sum_k P_k \psi_1. \tag{10.33}$$

Since interchanging particles in the result ψ merely rearranges terms in the sum, the operator

$$\mathscr{S} = B \sum_k P_k \tag{10.34}$$

is a symmetrizing operator.

When the different permutations yield orthogonal terms, coefficient B equals $1/\sqrt{n!}$, where $n!$ is the number of terms. Then

$$\mathscr{S} = \frac{1}{\sqrt{n!}} \sum_k P_k \tag{10.35}$$

is the *symmetrizing operator*.

In general, operation P_k may be effected by a series of p exchanges, such as (ij). If the number of interchanges needed is odd, the *parity* of the permutation is odd; if the number of interchanges in P_k is even the *parity* of the permutation is even.

One may let the sign of a contribution be negative when the parity is odd, positive when the parity is even. Thus, we construct

$$\psi = C \sum_k (-1)^p P_k \psi_1 \tag{10.36}$$

with C chosen to normalize the result. Each of the exchanges in any P_k permutation applied to this ψ changes a positive term to negative and a negative term to positive; so it reverses the sign of the wave function. Consequently, the expression

$$\mathscr{A} = C \sum_k (-1)^p P_k \tag{10.37}$$

is an antisymmetrizing operator for the interchanges making up the P_k's.

When the different terms in the sum are orthogonal, C is $1/\sqrt{n!}$, where $n!$ equals the number of terms. Then

$$\mathscr{A} = \frac{1}{\sqrt{n!}} \sum_k (-1)^p P_k \tag{10.38}$$

is the *antisymmetrizing operator*.

When ψ_1 is expressed as a product of single-particle functions, formula (10.36) can be rewritten in determinantal form. Thus, when

$$\psi_1 = u_1(1)\chi_1(1)\, u_2(2)\chi_2(2) \ldots u_N(N)\chi_N(N) \tag{10.39}$$

and the different single-particle functions are orthogonal, we have

$$\psi = \frac{1}{\sqrt{N!}} \begin{vmatrix} u_1(1)\chi_1(1) & u_1(2)\chi_1(2) & \ldots & u_1(N)\chi_1(N) \\ u_2(1)\chi_2(1) & u_2(2)\chi_2(2) & \ldots & u_2(N)\chi_2(N) \\ \vdots & \vdots & & \vdots \\ u_N(1)\chi_N(1) & u_N(2)\chi_N(2) & \ldots & u_N(N)\chi_N(N) \end{vmatrix} \tag{10.40}$$

This form is called a *Slater determinant* after John C. Slater, who introduced it in 1929.

Often, a given configuration yields more than one Slater determinant. Then, each allowable form may contribute equally to the complete wave function.

Example 10.8. Construct a Slater determinant for the ground state of lithium.

In the lowest energy state of lithium, two electrons with opposite spins are in the $1s$ orbital and one electron is in the $2s$ orbital. For the first electron, with $m_s = \frac{1}{2}$, in the $1s$ orbital, we write

$$u_1(1)\chi_1(1) = 1s(1)\alpha(1) = 1s(1);$$

for the second electron, with $m_s = -\frac{1}{2}$, in the $1s$ orbital, we write

$$u_2(2)\chi_2(2) = 1s(2)\beta(2) = 1\bar{s}(2);$$

and for the third electron, with $m_s = \frac{1}{2}$, in the $2s$ orbital, we write

$$u_3(3)\chi_3(3) = 2s(3)\alpha(3) = 2s(3).$$

In the final forms, a bar over the letter for the orbital function indicates β spin, while absence of the bar indicates α spin.

The Slater determinant is now

$$\psi = \frac{1}{\sqrt{6}} \begin{vmatrix} 1s(1) & 1s(2) & 1s(3) \\ 1\bar{s}(1) & 1\bar{s}(2) & 1\bar{s}(3) \\ 2s(1) & 2s(2) & 2s(3) \end{vmatrix}$$

No other Slater determinant for the $1s^2 2s$ configuration of Li with a net $m_s = \frac{1}{2}$ can be constructed.

Example 10.9. Construct Slater determinants for the $1s2s$ configuration of He with no net spin.

Two determinants with no net spin can be set up, specifically

$$\frac{1}{\sqrt{2}} \begin{vmatrix} 1s(1) & 1s(2) \\ 2\bar{s}(1) & 2\bar{s}(2) \end{vmatrix} \quad \text{and} \quad \frac{1}{\sqrt{2}} \begin{vmatrix} 1\bar{s}(1) & 1\bar{s}(2) \\ 2s(1) & 2s(2) \end{vmatrix}$$

Two orthogonal, independent, forms containing equal amounts of each can be constructed by taking $(1/\sqrt{2})$ times the sum and $(1/\sqrt{2})$ times the difference; thus

$$\psi_{\text{I}} = \frac{1}{2} \left(\begin{vmatrix} 1s(1) & 1s(2) \\ 2\bar{s}(1) & 2\bar{s}(2) \end{vmatrix} + \begin{vmatrix} 1\bar{s}(1) & 1\bar{s}(2) \\ 2s(1) & 2s(2) \end{vmatrix} \right)$$

and

$$\psi_{\text{II}} = \frac{1}{2} \left(\begin{vmatrix} 1s(1) & 1s(2) \\ 2\bar{s}(1) & 2\bar{s}(2) \end{vmatrix} - \begin{vmatrix} 1\bar{s}(1) & 1\bar{s}(2) \\ 2s(1) & 2s(2) \end{vmatrix} \right)$$

These form the allowed wave functions for the configuration.

10.7. Basis for the Pauli Exclusion Principle

We have seen how the indistinguishability of member particles and the possibility of nondegeneracy lead to restrictions on the nature of the state function for a multiparticle system. Nevertheless when ψ is symmetric with respect to any exchange of identical particles, linear dependence can exist among the constituent single-particle functions. But when ψ is antisymmetric, such linear dependence cannot exist; for, if we introduce such dependence into a Slater determinant, the determinant vanishes.

Indeed, consider formula (10.40). Suppose the relationship

$$u_1(j)\chi_1(j) + c_2u_2(j)\chi_2(j) + \ldots + c_Nu_N(j)\chi_N(j) = 0 \tag{10.41}$$

exists. Also, multiply the kth row, where $k = 2, 3, \ldots, N$, in turn, by c_k and add the results to the first row of the determinant. In the end, one obtains a row of zeros; so the determinant vanishes.

If the linear dependence does not involve the functions in the first $l - 1$ rows, but does involve $u_l(j)\chi_l(j)$, then we replace (10.41) with

$$u_l(j)\chi_l(j) + c_{l+1}u_{l+1}(j)\chi_{l+1}(j) + \ldots + c_Nu_N(j)\chi_N(j) = 0. \tag{10.42}$$

We multiply the kth row, where $k = l + 1, l + 2, \ldots, N$, in turn, by c_k and add the results to the lth row. At the end, we obtain a row of zeros at the lth level. Again, the determinant and the state function constructed from it vanish.

On the other hand, introducing either dependence into (10.33) does not make the symmetric ψ vanish.

When the various single-particle functions in (10.39) are either the same except for possible phase shifts or completely independent, orthogonal, then linear dependences (10.41) or (10.42) exist only among the functions describing the same states and each such state is multiply occupied. We see that such multipresence is allowed only when ψ is symmetric; that is, when the particles are bosons. Two or more fermions cannot occupy the same state; that is, have the same set of quantum numbers.

This *exclusion principle* is named after Wolfgang Pauli, who first stated it in 1925. A general form of the rule is that linear dependences among the constituent single-particle functions are not allowed when the particles are fermions.

10.8. Relating the Energy of a Plural Paired-Particle System to One- and Two-Particle Effects

The state function for a multiparticle system can be compounded from single-particle orbitals and spin functions, as we have seen. Any physical property that rests on the behavior of individual particles and on two-particle interactions depends on one- and two-particle integrals. Thus, the expectation value for the energy reduces to a sum of terms involving one and two orbitals in the integrands. For an N-particle system, a total of $(N!)^2$ terms arise. However, these can be classified into a few types and useful formulas extracted.

For simplicity, we will consider a multiparticle system in which the particles are paired in each occupied orbital. Thus, the formulas will apply to common molecules and to singlet atoms.

In particular, let us consider a four-electron system for which the Slater determinant is

$$\psi_4 = \frac{1}{\sqrt{4!}} \begin{vmatrix} \phi_1(1) & \phi_1(2) & \phi_1(3) & \phi_1(4) \\ \bar{\phi}_1(1) & \bar{\phi}_1(2) & \bar{\phi}_1(3) & \bar{\phi}_1(4) \\ \phi_2(1) & \phi_2(2) & \phi_2(3) & \phi_2(4) \\ \bar{\phi}_2(1) & \bar{\phi}_2(2) & \bar{\phi}_2(3) & \bar{\phi}_2(4) \end{vmatrix}, \tag{10.43}$$

where

$$\phi_j(k) = u_j(k)\alpha(k) \tag{10.44}$$

and

$$\bar{\phi}_j(k) = u_j(k)\beta(k). \tag{10.45}$$

We can rewrite (10.43) in the form

$$\psi_4 = (4!)^{-1/2} \sum_k (-1)^p P_k \phi_1(1)\bar{\phi}_1(2)\phi_2(3)\bar{\phi}_2(4). \tag{10.46}$$

The results will be generalized to apply to the $2n$-electron system for which

$$\psi_{2n} = [(2n)!]^{-1/2} \sum_k (-1)^p P_k \phi_1(1)\bar{\phi}_1(2)\phi_2(3)\bar{\phi}_2(4)\ldots$$
$$\ldots\phi_n(2n-1)\bar{\phi}_n(2n). \tag{10.47}$$

The operator for the system energy is Hamiltonian operator (10.5), neglecting magnetic effects. When an internuclear potential exists, conditions at fixed internuclear distances will be considered, and the contributions from this source will be subtracted out.

The expectation value for the energy assumes the form

$$E = \int \psi^* H \psi \, \mathrm{d}^{6n}\mathbf{r}$$
$$= \frac{1}{4!}\int \sum_j (-1)^{p_j} P_j \phi_1^*(1)\bar{\phi}_1^*(2)\phi_2^*(3)\bar{\phi}_2^*(4) \times$$
$$\times\ H \sum_k (-1)^{p_k} P_k \phi_1(1)\bar{\phi}_1(2)\phi_2(3)\bar{\phi}_2(4)\, \mathrm{d}^{12}\mathbf{r}. \tag{10.48}$$

With respect to the first summation, a single term appears as

$$\frac{1}{4!}\int (-1)^{p_j} P_j \phi_1^*(1)\bar{\phi}_1^*(2)\phi_2^*(3)\bar{\phi}_2^*(4) \times$$
$$\times\ H \sum_k (-1)^{p_k} P_k \phi_1(1)\bar{\phi}_1(2)\phi_2(3)\bar{\phi}_2(4)\, \mathrm{d}^{12}\mathbf{r}. \tag{10.49}$$

Now, let P_j^{-1} be the inverse of P_j and let $(-1)^{p_j} P_j^{-1}$ act on the factor in front of H, let P_j^{-1} act on H, and let $(-1)^{p_j} P_j^{-1}$ act on ψ. Since $(-1)^{p_j} (-1)^{p_j}$ equals 1, this is equivalent to permuting the particles in the integrand. But the particles are equivalent by assumption. So the value of the integral is not changed thereby.

When this process is imposed on each term of the first summation in (10.48) and the results added, the simpler form

$$E = \int \phi_1^*(1)\,\bar{\phi}_1^*(2)\,\phi_2^*(3)\,\bar{\phi}_2^*(4)\,H \sum_k (-1)^{p_k} P_k\,\phi_1(1)\,\bar{\phi}_1(2)\,\phi_2(3)\,\bar{\phi}_2(4)\,\mathrm{d}^{12}\mathbf{r}. \quad (10.50)$$

is obtained.

For the $2n$-electron system, we similarly have

$$E = \int \phi_1^*(1)\,\bar{\phi}_1^*(2) \ldots \phi_n^*(2n-1)\,\bar{\phi}_n^*(2n) \times$$
$$\times H \sum_l (-1)^{p_l} P_l\,\phi_1(1)\,\bar{\phi}_1(2) \ldots \phi_n(2n-1)\,\bar{\phi}_n(2n)\,\mathrm{d}^{6n}\mathbf{r}. \quad (10.51)$$

The Hamiltonian can be separated into one-electron operators and two-electron operators. The part dealing with the kinetic energy of the jth electron and its interaction with the nuclei is labeled H_j^{core}. The part determining the interaction between the jth and the kth electrons is of the form $e^2/4\pi\epsilon_0 r_{jk}$. The complete operator has the form

$$H = \sum_{j=1}^{2n} H_j^{\mathrm{core}} + \sum_{j=1}^{2n-1} \sum_{k=j+1}^{2n} \frac{e^2}{4\pi\epsilon_0 r_{jk}}. \quad (10.52)$$

When (10.52) is substituted into (10.50), the integrals containing the one-electron operators separate from those containing the two-electron operators. For the four-electron system, the former set includes

$$\int \phi_1^*(1)\,\bar{\phi}_1^*(2)\,\phi_2^*(3)\,\bar{\phi}_2^*(4)\,(H_1^{\mathrm{core}} + H_2^{\mathrm{core}} + H_3^{\mathrm{core}} + H_4^{\mathrm{core}}) \times$$
$$\times \sum_l (-1)^{p_l} P_l\,\phi_1(1)\,\bar{\phi}_1(2)\,\phi_2(3)\,\bar{\phi}_2(4)\,\mathrm{d}^{12}\mathbf{r}. \quad (10.53)$$

Four terms arise from each permutation P_l. When P_l is the identity, the first term has the form

$$\int \phi_1^*(1)\,\bar{\phi}_1^*(2)\,\phi_2^*(3)\,\bar{\phi}_2^*(4)\,H_1^{\mathrm{core}}\,\phi_1(1)\,\bar{\phi}_1(2)\,\phi_2(3)\,\bar{\phi}_2(4)\,\mathrm{d}^{12}\mathbf{r}$$

$$= \int \phi_1^*(1) H_1^{\mathrm{core}}\,\phi_1(1)\,\mathrm{d}^3\mathbf{r}_1 \int \bar{\phi}_1^*(2)\,\bar{\phi}_1(2)\,\mathrm{d}^3\mathbf{r}_2 \times$$

$$\times \int \phi_2^*(3)\,\phi_2(3)\,\mathrm{d}^3\mathbf{r}_3 \int \bar{\phi}_2^*(4)\,\bar{\phi}_2(4)\,\mathrm{d}^3\mathbf{r}_4$$

$$= \int \phi_1^*(1) H_1^{\mathrm{core}}\,\phi_1(1)\,\mathrm{d}^3\mathbf{r}_1 \equiv H_{11}. \quad (10.54)$$

In the second equality, we have made use of the normalization of the functions. In the last equality, the one-electron integral for the expectation value of H_1^{core} has been labeled H_{11}. The subscripts on H identify the two orbitals in the integral. The other three terms reduce to

$$\int \bar{\phi}_1^*(2) H_2^{\text{core}} \bar{\phi}_1(2)\, d^3\mathbf{r}_2 \equiv H_{11}, \tag{10.55}$$

$$\int \phi_2^*(3) H_3^{\text{core}} \phi_2(3)\, d^3\mathbf{r}_3 \equiv H_{22}, \tag{10.56}$$

$$\int \bar{\phi}_2^*(4) H_4^{\text{core}} \bar{\phi}_2(4)\, d^3\mathbf{r}_4 \equiv H_{22}. \tag{10.57}$$

Whenever P_l is not the identity, the corresponding four terms are all zero, as long as ϕ_1 and ϕ_2 are orthogonal, as we assume. For ϕ_1 and $\bar{\phi}_1$ are orthogonal because of their spin factors. Also, ϕ_2 and $\bar{\phi}_2$ are orthogonal for the same reason. To see what happens, consider a typical example:

$$\begin{aligned}
&\int \phi_1^*(1)\bar{\phi}_1^*(2)\phi_2^*(3)\bar{\phi}_2^*(4) H_1^{\text{core}} \phi_1(2)\bar{\phi}_1(1)\phi_2(3)\bar{\phi}_2(4)\, d^{12}\mathbf{r} \\
&\quad = \int \phi_1^*(1) H_1^{\text{core}} \bar{\phi}_1(1)\, d^3\mathbf{r}_1 \int \bar{\phi}_1^*(2)\phi_1(2)\, d^3\mathbf{r}_2 \times \\
&\quad\quad \times \int \phi_2^*(3)\phi_2(3)\, d^3\mathbf{r}_3 \int \bar{\phi}_2^*(4)\bar{\phi}_2(4)\, d^3\mathbf{r}_4 \\
&\quad = 0.
\end{aligned} \tag{10.58}$$

Since $\bar{\phi}_1$ is orthogonal to ϕ_1, the second integral is zero and the product is zero.

The general contribution to the expectation value of H from $\sum_j H_j^{\text{core}}$ is therefore

$$2 \sum_{j=1}^{n} H_{jj}. \tag{10.59}$$

For the four-electron system, the two-electron operators yield the integral

$$\begin{aligned}
&\frac{e^2}{4\pi\epsilon_0} \int \phi_1^*(1)\bar{\phi}_1^*(2)\phi_2^*(3)\bar{\phi}_2^*(4) \left(\frac{1}{r_{12}} + \frac{1}{r_{13}} + \frac{1}{r_{14}} + \frac{1}{r_{23}} + \frac{1}{r_{24}} + \frac{1}{r_{34}} \right) \times \\
&\quad \times \sum_l (-1)^{p_l} P_l \phi_1(1)\bar{\phi}_1(2)\phi_2(3)\bar{\phi}_2(4)\, d^{12}\mathbf{r}.
\end{aligned} \tag{10.60}$$

Now, six terms arise for each permutation P_l. When P_l is the identity, the first appears as

$$\frac{e^2}{4\pi\epsilon_0}\int \phi_1^*(1)\bar{\phi}_1^*(2)\phi_2^*(3)\bar{\phi}_2^*(4)\,\frac{1}{r_{12}}\,\phi_1(1)\bar{\phi}_1(2)\phi_2(3)\bar{\phi}_2(4)\,\mathrm{d}^{12}\mathbf{r}$$

$$= \frac{e^2}{4\pi\epsilon_0}\int \phi_1^*(1)\bar{\phi}_1^*(2)\,\frac{1}{r_{12}}\,\phi_1(1)\bar{\phi}_1(2)\,\mathrm{d}^3\mathbf{r}_1\,\mathrm{d}^3\mathbf{r}_2\,\times$$

$$\times\int \phi_2^*(3)\phi_2(3)\,\mathrm{d}^3\mathbf{r}_3\int \bar{\phi}_2^*(4)\bar{\phi}_2(4)\,\mathrm{d}^3\mathbf{r}_4$$

$$= \int \phi_1^*(1)\phi_1(1)\,\frac{e^2}{4\pi\epsilon_0 r_{12}}\,\bar{\phi}_1^*(2)\bar{\phi}_1(2)\,\mathrm{d}^3\mathbf{r}_1\,\mathrm{d}^3\mathbf{r}_2 \equiv J_{11}. \tag{10.61}$$

In the second equality, we have made use of the normalization of the functions, again. In the last equality, the *coulombic* integral has been labeled J_{11}. The subscripts on J identify the two orbitals involved in the six-dimensional integral.

Thus, for all the terms in (10.60) with P_l the identity, we obtain

$$J_{11} + J_{12} + J_{12} + J_{12} + J_{12} + J_{22} = \sum_j \big(J_{jj} + \sum_{k\neq j} 2J_{jk}\big). \tag{10.62}$$

When

$$P_l = (jk) \quad \text{and the operator is} \quad \frac{e^2}{4\pi\epsilon_0 r_{jk}}, \tag{10.63}$$

exponent p_l is 1 and the contribution to the integral is negative. And when this permutation exchanges electrons of like spin, it also differs from zero. Thus, when the permutation is (13), we have

$$-\frac{e^2}{4\pi\epsilon_0}\int \phi_1^*(1)\bar{\phi}_1^*(2)\phi_2^*(3)\bar{\phi}_2^*(4)\,\frac{1}{r_{13}}\,\phi_1(3)\bar{\phi}_1(2)\phi_2(1)\bar{\phi}_2(4)\,\mathrm{d}^{12}\mathbf{r}$$

$$= -\frac{e^2}{4\pi\epsilon_0}\int \phi_1^*(1)\phi_2^*(3)\,\frac{1}{r_{13}}\,\phi_1(3)\phi_2(1)\,\mathrm{d}^3\mathbf{r}_1\,\mathrm{d}^3\mathbf{r}_3\,\times$$

$$\times\int \bar{\phi}_1^*(2)\bar{\phi}_1(2)\,\mathrm{d}^3\mathbf{r}_2\int \bar{\phi}_2^*(4)\bar{\phi}_2(4)\,\mathrm{d}^3\mathbf{r}_4$$

$$= -\int \phi_1^*(1)\phi_2(1)\,\frac{e^2}{4\pi\epsilon_0 r_{13}}\,\phi_2^*(3)\phi_1(3)\,\mathrm{d}^3\mathbf{r}_1\,\mathrm{d}^3\mathbf{r}_3 = -K_{12}. \tag{10.64}$$

In the second equality, we have made use of the normalization of the functions. In the last equality, the *single-exchange* integral has been labeled K_{12}. The subscripts on K identify the two orbitals involved.

When the permutation is (24), one also gets $-K_{12}$. The other four (jk)'s exchange electrons with opposing spins. Since the spin functions factor out and they are orthogonal, the corresponding integrals are zero.

When the operator is $e^2/4\pi\epsilon_0 r_{jk}$, any permutation that exchanges an electron not in the jk pair leads to a factor consisting of the integral of the product of two different functions from the set

$$\phi_1(n), \quad \phi_2(n), \quad \bar{\phi}_1(n), \quad \bar{\phi}_2(n). \tag{10.65}$$

As long as these functions are orthogonal, as we have assumed, this factor equals zero. So, such permutations do not contribute to the result.

From the definitions, we know that

$$K_{jk} = K_{kj}, \qquad K_{jj} = J_{jj}, \qquad J_{jk} = J_{kj}. \tag{10.66}$$

Consequently, the nonzero terms we have found can be expanded as follows:

$$\begin{aligned} &J_{11} + 4J_{12} + J_{22} - 2K_{12} \\ &\quad = 2J_{11} + 2J_{12} + 2J_{21} + 2J_{22} - K_{11} - K_{12} - K_{21} - K_{22} \\ &\quad = \sum_{j=1}^{2} \sum_{k=1}^{2} (2J_{jk} - K_{jk}). \end{aligned} \tag{10.67}$$

For the energy of the state, we obtain

$$E = \sum_{j=1}^{2} 2H_{jj} + \sum_{j=1}^{2} \sum_{k=1}^{2} (2J_{jk} - K_{jk}). \tag{10.68}$$

Generalizing to the $2n$-electron closed-shell structure yields the formula

$$E = \int \psi_{2n}^* H \psi_{2n} \, d^{6n}\mathbf{r} = \sum_{j=1}^{n} 2H_{jj} + \sum_{j=1}^{n} \sum_{k=1}^{n} (2J_{jk} - K_{jk}). \tag{10.69}$$

Discussion Questions

10.1. What is (a) a particle, (b) a quasi particle?

10.2. When are the particles in a system (a) distinguishable, (b) indistinguishable?

10.3. How is the general wave function for a given system constructed?

10.4. Why can an arbitrary, single-valued, smooth function satisfying the boundary conditions be represented as a superposition of the eigenfunctions for an operator governing the system?

10.5. What does the variation theorem tell us about the parameters on which a suitable function depends?
10.6. How can one approximate the lowest n energies $E_1, E_2, \ldots, E_n$ of a system?
10.7. Under what conditions does the Hamiltonian operator break down into relatively simple separate operators?
10.8. How may such a breakdown be effected when the particles interact strongly with each other, as in an atom, a molecule, or a crystal?
10.9. How is the total energy related to the sum of the orbital energies? Why does one have to subtract out the interparticle energy from $\Sigma\, \epsilon_k$?
10.10. Describe the Hartree-Fock procedure for treating a multiparticle system.
10.11. Why is the indistinguishability of identical particles in a multiparticle system considered a quantum mechanical effect?
10.12. Describe (a) the permutation operator P_k, (b) the exchange operator (ij).
10.13. In what sense is $(ij) = \pm 1$? Explain how each possibility arises.
10.14. What are (a) fermions, (b) bosons?
10.15. Why may a wave function be considered the product of an orbital factor, a temporal factor, and a spin factor?
10.16. When can the orbital factor be (a) symmetric, (b) antisymmetric, as in Example 10.5?
10.17. What is the parity of a permutation?
10.18. How can an operator make a function (a) symmetric, (b) antisymmetric with respect to the identical particles on whose coordinates the function depends?
10.19. What is a Slater determinant?
10.20. How can a given configuration yield more than one Slater determinant? How do we obtain the contribution of each of these to the wave function?
10.21. What basis is there for the Pauli exclusion principle?
10.22. How is the state function constructed from single-particle orbitals and spin functions?
10.23. A state function for the N-fermion paired-particle system is constructed from single-particle functions. How many terms appear in the resulting summation?
10.24. When the function from 10.23 is employed, how many terms appear in the expectation value for energy? How many of these are equal?
10.25. How does one represent the contribution to E from interactions with the cores?
10.26. How does one obtain the contribution to E from the interactions among like particles?

Problems

10.1. Calculate the expectation value for energy of a free particle confined between $x = 0$ and $x = 1$ with $\psi = A(x - x^2)$. How does this expectation value change when the last exponent is changed from 2 to 2.0811?

10.2. A deuteron is modeled by a particle of reduced mass moving a varying distance r from the center of potential. If the interaction between a proton and a neutron yields the potential $V = -A\,e^{-r/a}$, what constant α makes $\psi = c\,e^{-(\alpha r)/2a}$ the best approximation to the wave function for the ground state?

10.3. Use the result from Problem 10.2 to calculate the ground-state energy of a deuteron when the depth of the well A is 32 MeV and the range a is 2.2×10^{-15} m.

10.4. Construct the Schrödinger equation for a system of two electrons and a nucleus of charge Ze. Show that when the interelectron interaction is neglected, there is a solution of the form $\psi = \psi^{(1)}\psi^{(2)}$, where $\psi^{(1)}$ and $\psi^{(2)}$ are hydrogen-like solutions for the first and second electron, respectively.

10.5. Apply (a) the symmetrizing operator, (b) the antisymmetrizing operator to the spin function $\alpha(1)\,\beta(2)$.

10.6. Show how ψ_{I} and ψ_{II} of Example 10.9 factor into an orbital function and a spin function. Note the permutation symmetry of each factor.

10.7. Show that for a system of N identical particles the completely antisymmetric wave function is orthogonal to the completely symmetric one. (*Hint*: Let an arbitrary exchange (ij) act on the integrand of the orthogonality integral.)

10.8. Using a given set of single-particle orbitals and spin functions, one may construct for the N-particle system a symmetric wave function $\psi^{(S)}$ and an antisymmetric wave function $\psi^{(A)}$. Suppose these could be superposed to form

$$\psi = \lambda\psi^{(S)} + \mu\psi^{(A)}$$

and determine how the probability density ρ for the particles varies with λ and μ.

10.9. Again consider the superposition in Problem 10.8. If $\psi^{(S)}$, $\psi^{(A)}$, and ψ can be simultaneously real, what would λ or μ equal?

10.10. Calculate the expectation value for the energy of a harmonic oscillator in the state for which $\psi = A\,e^{-\alpha x^2/2}$. Using the variational method, determine the magnitude of α which yields the best approximation to the ground-state energy.

10.11. A one-node function orthogonal to the one in Problem 10.10 for the harmonic oscillator is $\psi = Ax\,e^{-\alpha x^2/2}$. Using the variational method, determine the form for α and the best approximation to the energy for the first excited level.

10.12. What is the formula for the expectation value of the energy of a hydrogen-like atom when it is in the state described by $\psi = A\,e^{-\alpha r}$. Apply the variational method to find the appropriate value of α in terms of the conventional parameters. Then deduce the corresponding energy.

10.13. Write down the possible forms for the wave function of a system of three identical bosons in given one-particle states.

10.14. Consider the helium atom and apply (a) the symmetrizing operator, (b) the antisymmetrizing operator to the candidate functions $1s(1)\,2\bar{s}(2)$ and $1\bar{s}(1)\,2s(2)$.

10.15. Consider a system of two identical particles, each with the possible spin quantum numbers $s_z = -s, -s+1, \ldots, s$. Determine how many independent symmetric spin states and antisymmetric spin states occur and calculate the ratio between these numbers.

10.16. A system consists of two identical particles rotating with a definite J. Note that reflection through the origin corresponds to interchanging the particles. Determine what J values are not allowed when the particles are (a) bosons with no spin, (b) fermions with one-half unit of spin opposed (antiparallel).

10.17. Use a result from Problem 10.16 to determine the possibility of the decay

$$^{8}_{4}\mathrm{Be} \longrightarrow {}^{4}_{2}\mathrm{He} + {}^{4}_{2}\mathrm{He}$$

when the beryllium nucleus is in a state with its total angular momentum quantum number equal to 1.

10.18. Show that if ψ_1 and ψ_2 are Slater determinants which differ in only one row, then one can form a single Slater determinant ψ for which

$$\psi = \cos\alpha\,\psi_1 + \sin\alpha\,\psi_2,$$

but when they differ in two or more rows, such a result does not generally follow.

References

Books

Constantinescu, F., and Magyari, E. (translated by Grecu, V. V., and Spiers, J. A.): 1971, *Problems in Quantum Mechanics*, Pergamon Press, Oxford, pp. 180–181, 204.
Lowe, J. P.: 1978, *Quantum Chemistry*, Academic Press, New York, pp. 509–519.
Pauncz, R.: 1967, *Alternant Molecular Orbital Method*, Saunders, Philadelphia, Penn., pp. 1–19.

Articles

Banyard, K. E.: 1970, 'Electron Correlation in Atoms and Molecules', *J. Chem. Educ.* **47**, 668–671.
Harriss, D. K., and Rioux, F.: 1980, 'A Simple Hartree SCF Calculation on a One-Dimensional Model of the He Atom', *J. Chem. Educ.* **57**, 491–493.
Latham, W. P., Jr, and Kobe, D. H.: 1973, 'Comparison of the Hartree-Fock and Maximum Overlap Orbitals for a Simple Model', *Am. J. Phys.* **41**, 1258–1266.
Mackintosh, A. R., and Mackintosh, P. E.: 1981, 'Atomic Structure with a Programmable Calculator', *Eur. J. Phys.* **2**, 3–9.
Mirman, R.: 1973, 'Experimental Meaning of the Concept of Identical Particles', *Nuovo Cimento* **18B**, 110–122.
Pilar, F. L.: 1978, '4s is Always above 3d!', *J. Chem. Educ.* **55**, 2–6.

Answers to Problems

Chapter 1

1.2. 1.0 m. **1.3.** $(A/2)\,e^{iax}\,e^{-ibt} + (A/2)\,e^{-iax}\,e^{-ibt}$. **1.4.** $7.94 \times 10^{10}\ m^{-1}$, 0.792 Å.
1.5. (a) $(3.16\ Å^{-1}\ s^{-\frac{1}{2}})\,e^{(6.28\ Å^{-1})ix}\,e^{-(2.29 \times 10^{17}\ s^{-1})it}$;
(b) $(1.17 \times 10^{-8}\ Å^{-3/2})\,e^{(6.28\ Å^{-1})ix}\,e^{-(2.29 \times 10^{17}\ s^{-1})it}$. **1.6.** $R(r, \ldots)\,e^{ikq}\,e^{-i\omega t}$.
1.7. 1.53 eV. **1.8.** $4\pi V k^2$, $\frac{4\pi V}{h^2}\,(2m^3)^{\frac{1}{2}}\,E^{\frac{1}{2}}$,
1.9. (a) 222; (b) 123, 231, 312, 213, 321, 132. **1.10.** 0.325 Å.
1.12. $A\,e^{(2.20 \times 10^{11}\ cm^{-1})ix}$. **1.13.** $\partial f/\partial x$, $\partial f/\partial t$. **1.15.** 11 600 K.
1.16. $\sqrt{2/a}\,\cos 5\pi\,(x/a)$. **1.17.** 333, 115, 151, 511.
1.18. $(A/2)\,e^{ax}\,e^{-ibt} + (A/2)\,e^{-ax}\,e^{-ibt}$. **1.19.** $|B| = |A|$, $B = \frac{k - i\kappa}{k + i\kappa} A$, $C = \frac{2k}{k + i\kappa} A$.
1.20. $-\frac{me^4}{37.9\,\pi^2 \epsilon_0^2 \hbar^2}$.

Chapter 2

2.1. M. **2.2.** $e^{i3\varphi}$. **2.3.** $J(J+1)\hbar^2/3, \ldots, 0$. **2.4.** $10.64\ cm^{-1}$.
2.5. $0.228\ cm^{-1}$, $0.457\ cm^{-1}$, $0.685\ cm^{-1}$. **2.6.** $0.356\ cm^{-1}$, $0.178\ cm^{-1}$.
2.8. $1.543C$, $1.500C$. **2.9.** $(\cos 2\alpha)\,\psi_{d_{xy}} + (\sin 2\alpha)\,\psi_{d_{x^2-y^2}}$.
2.10. $Ey(y^2 - 3x^2)R_{n3}(r)/r^3$. **2.11.** (a) 0.00027429 u, (b) 0.000548286 u.
2.12. $\sqrt{2}\,\cos 4\varphi$. **2.13.** 3. **2.14.** $J(J+1)$ $7.47\ cm^{-1}$, $14.94\ cm^{-1}$, $29.89\ cm^{-1}, \ldots$.
2.15. $0.62\ cm^{-1}$, $1.24\ cm^{-1}$, $1.86\ cm^{-1}, \ldots$. **2.16.** $27.35\ cm^{-1}$, $14.58\ cm^{-1}$, $9.51\ cm^{-1}$.
2.17. 0, $24.09\ cm^{-1}$, $36.87\ cm^{-1}$, $41.93\ cm^{-1}, \ldots$.
2.18. $(\sqrt{3}/2)\,\psi_{d_{x^2-y^2}} - \frac{1}{2}\,\psi_{d_{3z^2-r^2}}$. **2.19.** $Dz\,(x^2 - y^2)/r^3$. **2.20.** $2B = \sqrt{15}$.

Chapter 3

3.1. 0.8427. **3.2.** $\sqrt{2}\left(\frac{a}{\pi}\right)^{\frac{1}{4}} w\,e^{-w^2/2}$.
3.3. $A\,e^{\pm ax^2/2}$, $B\,e^{\pm bx^2/2}$, where $a = (f\mu)^{\frac{1}{2}}/\hbar$, $b = (g\mu)^{\frac{1}{2}}/\hbar$.
3.4. $10.33\ cm^{-1}$, $2886\ cm^{-1}$. **3.5.** 5.1360×10^5 dyne cm^{-1}, 5.3056×10^5 dyne cm^{-1}.
3.6. $4395.2\ cm^{-1}$. **3.7.** $38\,729.9\ cm^{-1}$. **3.8.** $10.31\ cm^{-1}$, $2886\ cm^{-1}$.
3.11. 0.1116. **3.12.** $\pm 1/\sqrt{2}$. **3.13.** $A\,e^{\pm w^3/3}$, $a^3 = \mu f/\hbar^2$, $w^4 = a^2 x^4$.
3.14. $8.08_3\ cm^{-1}$, $2555\ cm^{-1}$. **3.15.** 18.56×10^5 dyne cm^{-1}, 15.48×10^5 dyne cm^{-1}.
3.16. $378.4\ cm^{-1}$. **3.17.** $1482.0\ cm^{-1}$, $4367.9\ cm^{-1}$. **3.18.** $41\,795\ cm^{-1}$.

Chapter 4

4.1. $\frac{\partial^2\psi}{\partial x^2} + \frac{\partial^2\psi}{\partial y^2} + \frac{\partial^2\psi}{\partial z^2} + \frac{2\mu E}{\hbar^2}\psi + \frac{\mu Z e^2}{2\pi\epsilon_0\hbar^2}\frac{1}{r}\psi = 0.$

4.2. $a = \frac{\mu Z e^2}{4\pi\epsilon_0\hbar^2}, E = -\frac{\mu Z^2 e^4}{32\pi^2\epsilon_0^2\hbar^2}$. **4.3.** $\frac{1}{r}\frac{\partial}{\partial r}\left(r\frac{\partial\psi}{\partial r}\right) + \frac{1}{r^2}\frac{\partial^2\psi}{\partial\varphi^2} + \frac{\partial^2\psi}{\partial z^2}.$

4.4. $V = V_1(r) + (1/r^2)V_2(\varphi) + V_3(z)$. **4.6.** $(1/\pi a^3)^{\frac{1}{2}}$. **4.7.** 0.238. **4.8.** a.
4.9. $1S - 3P$, $2S - 4P$ or $2P - 4S$ or $2P - 4D$. **4.10.** 7.01 keV.

4.11. $a = \frac{\mu Z e^2}{8\pi\epsilon_0\hbar^2}, E = -\frac{\mu Z^2 e^4}{32\pi^2\epsilon_0^2\hbar^2}\frac{1}{4}$. **4.12.** $c = \frac{\mu Z e^2}{8\pi\epsilon_0\hbar^2}, E = -\frac{\mu Z^2 e^4}{32\pi^2\epsilon_0^2\hbar^2}\frac{1}{4}, b = -c.$

4.13. $(1/8\pi a^3)^{\frac{1}{2}}$. **4.14.** 0.188. **4.15.** 5.236a. **4.16.** $5 \to 4$.

4.17. (a) 82 258.19 cm^{-1}, (b) 740 668.9 cm^{-1}. **4.18.** $\frac{\sqrt{\xi+\eta}}{2\sqrt{\xi}}, \frac{\sqrt{\xi+\eta}}{2\sqrt{\eta}}, \sqrt{\xi\eta}$.

4.19. $\frac{4}{\xi+\eta}\frac{\partial}{\partial\xi}\left(\xi\frac{\partial\psi}{\partial\xi}\right) + \frac{4}{\xi+\eta}\frac{\partial}{\partial\eta}\left(\eta\frac{\partial\psi}{\partial\eta}\right) + \frac{1}{\xi\eta}\frac{\partial^2\psi}{\partial\varphi^2} + \frac{2\mu}{\hbar^2}(E - V)\psi = 0.$

Chapter 5

5.2. $V(r) = V^*(r)$. **5.3.** Limits on φ: 0 and 2π. **5.4.** Eigenvalues: p_x and p_x^3.
5.5. $(3/2)a$. **5.6.** $-e^2/4\pi\epsilon_0 a$. **5.7.** $\delta(x)$. **5.10.** 0.945.
5.12. (a) When boundary conditions periodic in p_x prevail, (b) when momenta p_x are confined within limits. **5.13.** The constant, ∞. **5.14.** For $P(A), P(a_j)$. **5.15.** 5a.

5.16. $-\frac{e^2}{4\pi\epsilon_0 a}\frac{1}{4}$. **5.17.** $\delta(x)$. **5.20.** 1.31×10^{-12}.

Chapter 6

6.1. $\frac{8Aa^2}{\pi^3}\left(\sin\frac{\pi x}{a} + \frac{1}{27}\sin\frac{3\pi x}{a} + \frac{1}{125}\sin\frac{5\pi x}{a} + \ldots\right).$

6.2. $\frac{960}{(\pi n_j)^6}$ for odd n_j, 0 otherwise; $1.013E_1$. **6.3.** 2.53×10^{-8} eV.

6.4. $2A\cos\left(\frac{\Delta k}{2}x - \frac{\Delta\omega}{2}t\right)e^{i(kx-\omega t)}, \frac{\Delta\omega}{\Delta k}$. **6.5.** $\frac{1}{2}w$, 24.18 cm s^{-1}.

6.7. $e^{i(\bar{k}x-\omega t)}$. **6.10.** $\left[\frac{1}{2\mu}\Sigma\left(\frac{\hbar}{i}\frac{\partial}{\partial x_j} - qA_j - \frac{\partial S}{\partial x_j}\right)^2 + V - \frac{\partial S}{\partial t}\right]\Psi_1 = -\frac{\hbar}{i}\frac{\partial\Psi_1}{\partial t}.$

6.11. (a) $2\left(\sin x - \frac{1}{2}\sin 2x + \frac{1}{3}\sin 3x - \ldots\right)$,

(b) $\frac{\pi^2}{3} - 4\left(\cos x - \frac{1}{4}\cos 2x + \frac{1}{9}\cos 3x - \ldots\right).$

6.12. (a) $A\left[\frac{\pi^2}{6} - \left(\cos 2\varphi + \frac{1}{4}\cos 4\varphi + \frac{1}{9}\cos 6\varphi + \ldots\right)\right]$,

(b) $\frac{8A}{\pi}\left(\sin\varphi + \frac{1}{27}\sin 3\varphi + \frac{1}{125}\sin 5\varphi + \ldots\right)$. **6.13.** 5.49×10^{-24} s. **6.14.** $\frac{v}{2} + \frac{V}{\mu v}$.

6.15. $\frac{3}{2}w$, 101.4 cm s^{-1}. **6.17.** $\frac{1}{(2\pi)^{3/2}}e^{-i\bar{\mathbf{r}}\cdot\mathbf{k}}$, the same.

6.19. When V does not depend on φ. **6.20.** $\Psi' = \Psi\exp(i/\hbar)q\,\Delta f$.

Chapter 7

7.1. $a + b + c = 0$. **7.2.** Y_{xy}. **7.3.** $Nr^3 \dfrac{\partial^2}{\partial z^2}\left(\dfrac{1}{r}\right)$. **7.5.** $e^{-im\alpha}$. **7.6.** $Y' = (-1)^l Y$.

7.8. All have $l = 3$; first pair, $m = \pm 1$; second pair, $m = \pm 2$; thid pair, $m = \pm 3$.

7.9. $\dfrac{1}{\sqrt{3}} Y_1^0 + \sqrt{\dfrac{2}{3}} Y_1^1$. **7.11.** $3a + g + h = 0, 3b + d + j = 0, 3c + e + f = 0$.

7.12. $Y_{x^2 - y^2}$. **7.13.** $-Nr^4 \dfrac{\partial^3}{\partial x\, \partial y\, \partial z}\left(\dfrac{1}{r}\right)$. **7.14.** $-\sqrt{\dfrac{15}{8\pi}}\, e^{i\varphi} \cos\theta \sin\theta$.

7.15. $N \dfrac{z^2 - y^2}{r^2}$. **7.16.** $\psi' = (-1)^l \psi,\ Y' = (-1)^l Y$.

7.17. $-\sqrt{\dfrac{3}{8}}\, Y_{x(5z^2 - r^2)} + \sqrt{\dfrac{5}{8}}\, Y_{x(x^2 - 3y^2)}$.

7.18. $-\sqrt{\dfrac{5}{8}}\, Y_{x(5z^2 - r^2)} - \sqrt{\dfrac{3}{8}}\, Y_{x(x^2 - 3y^2)}$. **7.19.** Spherically symmetric.

Chapter 8

8.1. $-\dfrac{\hbar}{i}$. **8.3.** $\dfrac{1}{\sqrt{2}} (x \pm iy)^3 R(r)$. **8.4.** Given function is $Y_{2,0}R$.

8.5. When $J_z Y = M\hbar Y$. **8.6.** 1.67×10^3 rev s^{-1}.

8.8. $\dfrac{1}{\sqrt{2}} Y_{\frac{1}{2},\frac{1}{2}}(1)\, Y_{\frac{1}{2},-\frac{1}{2}}(2) - \dfrac{1}{\sqrt{2}} Y_{\frac{1}{2},-\frac{1}{2}}(1)\, Y_{\frac{1}{2},\frac{1}{2}}(2)$.

8.9. $\dfrac{1}{\sqrt{3}} Y_{0,0}(1,2) + \dfrac{1}{\sqrt{2}} Y_{1,0}(1,2) + \dfrac{1}{\sqrt{6}} Y_{2,0}(1,2)$. **8.10.** $2B$.

8.11. $2i\hbar J_y J_z + \hbar^2 J_x$. **8.12.** $J_x = \dfrac{\hbar}{i}\left(-\sin\varphi \dfrac{\partial}{\partial\theta} - \cot\theta \cos\varphi \dfrac{\partial}{\partial\varphi}\right)$;

$J_y = \dfrac{\hbar}{i}\left(\cos\varphi \dfrac{\partial}{\partial\theta} - \cot\theta \sin\varphi \dfrac{\partial}{\partial\varphi}\right)$. **8.13.** $\pm \dfrac{1}{\sqrt{2}} (x \pm iy)^2 zR(r)$.

8.14. $J_\pm$ acting on each component of the standing wave yields 0 or $cY_{1,0}$.

8.17. $c_1 = \dfrac{1}{\sqrt{3}},\ c_2 = \dfrac{\sqrt{2}}{\sqrt{3}}$. **8.18.** $\sqrt{\dfrac{3}{10}}\, Y_{3/2,3/2}(1)\, Y_{3/2,-1/2}(2)$ –

$- \dfrac{2}{\sqrt{10}} Y_{3/2,1/2}(1)\, Y_{3/2,1/2}(2) + \sqrt{\dfrac{3}{10}}\, Y_{3/2,-1/2}(1)\, Y_{3/2,3/2}(2)$.

Chapter 9

9.3. 56.1 cm. **9.4.** 0.059 barn. **9.5.** $\nabla^2\psi + k^2\psi = 0$. **9.7.** 2.98 MeV.

9.8. $E_R \pm 0.167\Gamma$. **9.9.** $A^*A \dfrac{f^*f}{r^2} v$. **9.10.** $\dfrac{1}{[a^2 + (2\gamma t)^2]^{\frac{1}{2}}} \exp - \dfrac{(x - vt)^2}{a + (2\gamma t)^2/a}$.

9.11. 0.402. **9.12.** $(2l+1) \dfrac{\sin kr}{kr} P_l(\cos\theta)$ when l even; $(2l+1) \dfrac{\cos kr}{ikr} P_l(\cos\theta)$ when l odd.

9.13. $\nabla^2\psi + k^2\psi = 0$. **9.14.** $ka/2,\ \dfrac{1}{2} e^{i(ka - \frac{1}{2}\pi)} + \dfrac{1}{2} i$. **9.15.** 3.14×10^{-13} cm.

9.16. $4\pi\lambda\!\!\!^{-2} (2l + 1) \exp - \dfrac{4(\ln 2)(E - E_R)^2}{2}$.

Chapter 10

10.1. $\dfrac{\hbar^2}{8\mu}$ 1.0132, last factor decreases to 1.0125. **10.2.** $\dfrac{(1+\alpha)^4}{\alpha} = \dfrac{12A\mu a^2}{\hbar^2}$.

10.3. -2.16 MeV. **10.6.** $\dfrac{1}{\sqrt{2}}[1s(1)\,2s(2) \mp 1s(2)\,2s(1)]\dfrac{1}{\sqrt{2}}[\alpha(1)\beta(2) \pm \alpha(2)\beta(1)]$.

10.8. $\lambda^*\lambda\psi^{(S)*}\psi^{(S)} + \mu^*\mu\psi^{(A)*}\psi^{(A)}$. **10.9.** One 0, the other 1.

10.10. $\alpha = \dfrac{\sqrt{f\mu}}{\hbar}, E = \dfrac{1}{2}\hbar\sqrt{\dfrac{f}{\mu}}$. **10.11.** $\alpha = \dfrac{\sqrt{f\mu}}{\hbar}, E = \dfrac{3}{2}\hbar\sqrt{\dfrac{f}{\mu}}$.

10.12. $\alpha = \dfrac{Ze^2\mu}{4\pi\epsilon_0\hbar^2}, E = -\dfrac{Z^2e^4\mu}{32\pi^2\epsilon_0^2\hbar^2}$.

10.13. $u_1(1)\,u_1(2)\,u_1(3)$,

$\dfrac{1}{\sqrt{3}}[u_1(1)\,u_1(2)\,u_2(3) + u_1(3)\,u_1(1)\,u_2(2) + u_1(2)\,u_1(3)\,u_2(1)]$,

$\dfrac{1}{\sqrt{6}}[u_1(1)\,u_2(2)\,u_3(3) + u_1(3)\,u_2(1)\,u_3(2) + u_1(2)\,u_2(3)\,u_3(1) +$

$+ u_1(1)\,u_2(3)\,u_3(2) + u_1(3)\,u_2(2)\,u_3(1) + u_1(2)\,u_2(1)\,u_3(3)]$.

10.14. $\dfrac{1}{\sqrt{2}}[1s(1)\,2\bar{s}(2) \pm 1s(2)\,2\bar{s}(1)], \dfrac{1}{\sqrt{2}}[1\bar{s}(1)\,2s(2) \pm 1\bar{s}(2)\,2s(1)]$. **10.15.** $\dfrac{s+1}{s}$.

10.16. (a) odd J, (b) even J. **10.17.** Forbidden.

Name Index

Subject Index